ADVANCES IN EXPERIMENTAL MEDICINE AND BIOLOGY

Recent Volumes in this Series

FUEL HOMEOSTASIS AND THE NERVOUS SYSTEM

Edited by

Mladen Vranic
University of Toronto
Toronto, Ontario, Canada

Suad Efendic
Karolinska Hospital
Stockholm, Sweden

and

Charles H. Hollenberg
Banting and Best Diabetes Centre
University of Toronto
Toronto, Ontario, Canada

PLENUM PRESS • NEW YORK AND LONDON

Library of Congress Cataloging-in-Publication Data

Toronto-Stockholm Symposium on Perspectives in Diabetes Research (1st
  : 1990 · Toronto, Ont.)
   Fuel homeostasis and the nervous system / edited by Mladen Vranic,
Suad Efendic, and Charles H. Hollenberg.
      p.   cm. -- (Advances in experimental medicine and biology ; v.
291)
   "Proceedings of the First Toronto-Stockholm Symposium on
Perspectives in Diabetes Research, held June 28-29, 1990, in Toronto,
Ontario, Canada."--T.p. verso.
   Includes bibliographical references and index.
   ISBN-13: 978-1-4684-5933-3
   1. Brain--Metabolism--Regulation--Congresses. 2. Energy
metabolism--Regulation--Congresses. 3. Blood sugar--Congresses.
4. Nervous system--Metabolism--Regulation--Congresses.   I. Vranic,
Mladen. II. Efendić, Suad. III. Hollenberg, Charles H., 1930-   .
IV. Title. V. Series.
   [DNLM: 1. Energy Metabolism--congresses. 2. Homeostasis-
-congresses. 3. Nervous System--metabolism--congresses.   W1 AD55S
v. 291 / WL 102 T686f 1990]
   QP376.T64 1990
   612.8'2--dc20
DNLM/DLC
for Library of Congress   91-3652
                                                              CIP

Proceedings of the First Toronto-Stockholm Symposium on
Perspectives in Diabetes Research,
held June 28–29, 1990, in Toronto, Ontario, Canada

ISBN-13: 978-1-4684-5933-3   e-ISBN-13: 978-1-4684-5931-9
DOI: 10.1007/ 978-1-4684-5931-9

© 1991 Plenum Press, New York
Softcover reprint of the hardcover 1st edition 1991
A Division of Plenum Publishing Corporation
233 Spring Street, New York, N.Y. 10013

PREFACE

This book has a dual purpose, to review in depth the control of fuel homeostasis in the brain and the role of the nervous system in the control of fuel deposition in the body. From the methodological point of view the emphasis is on the application of advanced technologies to assess fuel transport and brain metabolism, the role of peptides in the neuroendocrine system and the response of the brain to hypoglycemia. These technologies include positron emmission tomography, nuclear magnetic resonance, immunocytochemistry, molecular biology, autoradiography. To study fuel homeostasis in the body advanced tracer methods that include modelling are set out. From the pathophysiological point of view the emphasis is on abnormalities in stress, brain metabolism in diabetes, eating and degenerative disorders. This book contains contributions from endocrinologists, physiologists, neurologists, psychoneuroendocrinologists, biophysicists, biochemists and experts in nutrition. This authorship represents a unique diversity of researchers who, for the first time, cover comprehensively the interaction between the nervous system and fuel homeostasis, both in health and disease.

We hope this book will be an important source of information for both researchers and practicing clinicians.

Mladen Vranic
Suad Efendic
Charles Hollenberg

ACKNOWLEDGEMENTS

The Symposium from which this volume arose (University of Toronto, June 27-28, 1990) was the first Toronto-Stockholm Symposium on Perspectives in Diabetes Research. These Symposia are organized triennially by the Banting and Best Diabetes Centre, University of Toronto and the Department of Endocrinology, Karolinska Institute, Stockholm. We are very much indebted to our major sponsors the Canadian Diabetes Association, Connaught Novo Ltd., the Juvenile Diabetes Foundation Canada and Novo Nordisk. Thanks are also due to Eli Lilly Canada Ltd. and Miles Laboratories Ltd. for their kind support.

The Editors would like to express their special thanks to Mrs. Norah Rankin and Dr. Patricia Brubaker for their contributions to the organization of the Symposium.

CONTENTS

NEURAL RESPONSES TO ABNORMALITIES IN FUEL HOMEOSTASIS

THE METABOLIC REQUIREMENTS OF FUNCTIONAL ACTIVITY IN THE HUMAN BRAIN:

A POSITRON EMISSION TOMOGRAPHY STUDY

Marcus E. Raichle

Washington University Medical Center
St Louis, Missouri 63110

INTRODUCTION

In the resting state it is well established that the energy demands of the mammalian brain are met by the oxidation of glucose (Siesjo, 1978). Changes in the functional activity of the brain (e.g., motor activity, speech, vision, audition, somesthesis) are accompanied by increases in local blood flow and glucose utilization (for a review of this extensive literature see Raichle, 1987). It had been generally assumed that the energy demands accompanying these phasic changes in neural activity were also supported by glucose oxidation supporting the large energy expenditures required to maintain membrane ionic gradients. The increase in local blood flow had been considered a response to oxygen depletion and metabolite excess in the form of carbon dioxide.

Challenging the conventional formulation that phasic increases in neural activity in the normal brain are accompanied by glucose oxidation, we have reported positron emission tomography (PET) studies in normal human volunteers during visual and somatosensory stimulation that demonstrate large increases in blood flow and glucose utilization but only minimal increases in oxygen consumption (Fox and Raichle, 1986; Fox et al, 1988). In these studies the blood flow so far exceeded the small increase in oxygen consumption that a highly significant decline in the extracted fraction of available oxygen occurred. One must conclude that tissue oxygen tension increased locally during phasic neural activity.

The literature is rich with support for the hypothesis that phasic changes in neural activity are supported by glycolysis. Penfield (1937) noted that local tissue oxygenation (judged by the color of the venous effluent) increased, rather than decreased, during spontaneous focal seizures in human cerebral cortex. Direct measurement of venous oxygen tension during seizures in both animals and humans confirmed this impression (Plum et al, 1968). Cooper and colleagues (966) found a variety of simple motor and sensory tasks produced prompt, dramatic, and highly localized increases in cortical oxygen availability. Rapid eye-movement sleep also causes large increases in blood flow (Reivich et al,1968) and glucose consumption (Heiss et al, 1985) but a decrease in the extraction of oxygen (Santiago et al, 1984). Finally, Veki et al (1988) have recently demonstrated that tissue lactate increases in rat somatosensory cortex during stimulation of the forepaw. This is accompanied by parallel increases in glucose utilization and blood flow in the same area. Thus, available evidence supports the hypothesis that phasic increases in neural activity are supported by "aeorbic" glycolysis.

The fact that phasic neural activity is supported by glycolysis raises a number of interesting questions. For example: What glycolysis? Is glycolysis enough to support the tissue energy demands? What triggers glycolysis? How is the brain organized to support such activity? Do all systems employ glycolysis in this manner? Additionally: Why does blood flow increase? What causes blood flow to increase? In the remainder of this paper I will briefly discuss my thoughts about these questions.

THE LOCAL METABOLIC RESPONSE

Why glycolysis? I can only speculate about the answer to this important question for we have no information. At least one tentative hypothesis is generated when one considers the capillary density of the brain. Capillary density in the brain is less than in the heart. The lesser density of capillaries in the brain aids in maintaining the concept of the blood brain barrier by reducing the available surface area for exchange. Such an advantage is offset by increased distances for diffusion within the so-called tissue cylinders surrounding the capillaries. One might view the situation as a compromise. In such a situation it seems reasonable to suggest that the use of glycolysis to support abrupt changes in local neuronal activity make it unlikely, over short times, that the tissue energy demands will outstrip energy supply. This occurs because the tissue concentrations of glucose and glycogen are sufficient to sustain neural activity over several minutes. This should be contrasted to oxygen which is at near zero concentration in the tissue.

Is glycolysis enough? Despite the large increase in glucose uptake observed during phasic increases in neural activity, the actual acute energy yield must be quite small. Creutzfeldt (1975) reached this conclusion by computing the energy demands of neural activity from heat production and estimating that "only about 0.3 to 3.0%, or even less, of the cortical energy consumption can be accounted for by spike activity of cortical nerve cells". A doubling of neural electrical activity, therefore, should increase oxygen consumption by less than 6% or less, in agreement with the small changes that we have observed (Fox and Raichle, 1986; Fox et al, 1988). Van den Berg (1986) calculated by enzymatic capacity that glucose oxidation is at (or near) maximal capacity at rest and that "increases of glucose oxidation during stimulation, whether natural or experimental, cannot therefore take place, unless by a few percent"

What triggers glycolysis? At the present time we have no direct evidence to implicate any one factor in the stimulation of glycolysis during phasic changes in neural activity. However, dramatic alkalinization of glial intracellular pH has been documented by Chesler and Kraig (1987) during physiological cortical activation and could well serve as a means of triggering glycolysis. An additional implication of such an hypothesis is that a good bit of the glycolysis occurring during phasic changes in neural activity could well be occurring in glia rather than neurons.

How is the brain organized to support such activity? Several factors indicate that the mammalian brain is organized to very efficiently support rapid changes in glycolysis. First, the blood brain barrier has specific carriers that mediate the movement of glucose from the blood to the tissue. Second, the tissue concentration of glucose is not zero. In fact, there is normally sufficient glucose in the tissue to support the energy demands of brief increases in neuronal activity. Finally, the work of Magistretti

(1990) suggests that there exists in the cerebral cortex a neuronally mediated system for the rapid local breakdown of glycogen. This system involves intrinsic cortical neurons which release VIP in cortical columns. The action of these local interneurons is amplified by neurotransmitters such as noradrenaline, histamine and GABA.

THE LOCAL VASCULAR RESPONSE

Why does blood flow increase? In light of these observations, the postulate that blood flow is regulated by and for the sake of the metabolic rate must be reconsidered. The disproportionate increase in blood flow that accompanies physiological neural activation causes tissue oxygen tension and pH to rise and carbon dioxide tension to fall arguing strongly against glucose oxidation as the regulator of blood flow under physiological conditions. It may well be the case that the increase in blood flow is concerned with the regulation of tissue potassium homeostasis as suggested by Newman and Paulson (1987) and the regulation of tissue pH where glycolysis causes a local increase in tissue lactate. In this regard it is interest to note that tissue pH remains constant during phasic changes in neural activity despite an accumulation of lactate in the tissue (Linn and Hossmann, 1988)

What causes blood flow to increase? The literature is full of lengthy reviews on the regulation of the brain blood flow. One must conclude from this voluminous and somewhat unsatisfying literature that it is a multifactorial process. Although the components (i.e., potassium, pH, adenosine, intrinsic and extrinsic neurotransmitters, various endothelial relaxing and constricting factors, etc) are well known, it is entirely unclear just how they act in concert to effect the rapid, beautifully focused local vascular responses to phasic changes in neural activity. This is one of the great challenges in vascular physiology of the brain. A complete understanding will only come from a thoughtful consideration of brain vasculature as one component of a unit consisting of glia, neurons and blood vessels.

REFERENCES

Cooper, R, Crow, HJ, Walter, WG, Winter, AL. Regional control of cerebral vascular reactivity and oxygen supply in man. Brain Res 3:191-174, 1966.

Creutzfeldt, O. Neurophysiological correlates of different functional states of the brain. In BRAIN WORK: THE COUPLING OF FUNCTION, METABOLISM AND BLOOD FLOW IN THE BRAIN. eds DH Ingvar and NA Lassen (Alfred Benzon Symposium VIII, Munksgaard, Copenhagen, 1975), pp 22-47.

Fox, PT, Raichle, ME. Focal physiological uncoupling of cerebral blood flow and oxidative metabolism during somatosensory stimulation in human subjects. Proc Natl Acad Soc (USA) 83:1140-1144, 1986.

Fox, PT, Raichle, ME, Mintun, MA, Dence, C. Nonoxidative glucose consumption during focal physiologic neural activity. Science 241:462-464, 1988.

Heiss, W-D, Pawlik, G, Herholz, K, Wienbard, K. Regional cerebral glucose metabolism in man during wakefulness, sleep and dreaming. Brain Res 327:362-366, 1985.

Magistretti, PJ. VIP neurons in the cerebral cortex. Trends in Pharmacological Sciences 11:250-254, 1990.

Paulson, OB, Newman, EA. Does the release of potassium from astrocyte endfeet regulate cerebral blood flow? Science 237:896-898, 1987.

Penfield, W. The circulation of the epileptic brain. Res Publ Assoc Res Nerv Ment Dis 18:605-637, 1937.

Plum, F, Posner, JB, Troy, B. Cerebral metabolic and circulatory responses to induced convulsions in animals. Arch Neurol 18:1-13, 1968.

Raichle, ME, Circulatory and metabolic correlates of brain function in normal humans. In **HANDBOOK OF PHYSIOLOGY. THE NERVOUS SYSTEM.** ed. VB Mountcastle and F. Plum. (American Physiological Society, Bethesda, 1987), Vol V, pp.643-674.

Reivich M, Isaccs G, Evarts E, Kety S. The effect of slow wave sleep and REM sleep on regional cerebral blood flow in cats. J. Neurochem 15:301-306, 1968.

Santiago TV, Guerra E, Neubauer JA, Eddman NH. Correlation between ventilation and brain blood flow during sleep. J Clin Invest. 73:497-506, 1984.

Siesjo, BK, **BRAIN ENERGY METABOLISM** (Wiley, New York, 1978), pp. 101-110.

Ueki, M, Linn, F, Hossmann, K-A. Functional activation of cerebral blood flow and metabolism before and after global ischemia of rat brain. J Cerebral Blood Flow & Metabol 8:486-494, 1988.

Van den Berg, C. In **ENERGETICS AND HUMAN INFORMATION PROCESSING.** eds GRJ Hockey,, AWK Gaillard, MGH Coles (Nijhoff, Boston, 1986) pp 131-135.

High Resolution NMR Studies of Cerebral Glucose Metabolism in Rats and Humans

R.G. Shulman[§], D.L. Rothman[*], K.L. Behar[‡]
and J.W. Prichard[‡]

Depts. [§]Mol. Biophys. & Biochem.
[*]Medicine and [‡]Neurology
Yale University
New Haven, CT   06510

During the past four years we have been studying human brains by NMR methods that we developed first in animal experiments. These methods allow us to obtain ^1H and ^{13}C NMR spectra from the brain while we also apply ^{31}P NMR methods such as are generally used in the field. The first ^1H NMR spectra of the brain, or in fact of any living animal tissue, were of the rat brain at high magnetic fields of 8.4T (1). To obtain these spectra it was necessary to suppress the intense ^1H NMR peak from H_2O in order to observe the peaks from metabolites. This was done, at first by pre-saturation of the H_2O peak and later by selective pulses that did not excite the H_2O peak (2). Subsequently Rothman introduced editing methods into proton spectroscopy of the brain to simplify the spectrum. First homonuclear editing was used (3) and then heteronuclear in which ^{13}C labeled nuclei were determined from their coupled protons C4.

When we turned to human studies it was necessary to localize the spectrum to a particular volume of interest (VOI). To do this we continued to use surface coils, as in the animal studies, which provide a coarse localization and we refined the localization by the ISIS method (5). This is a differencing method in which peaks in 3 orthogonal planes are inverted so that after subtraction only those nuclei in the intersection of the planes appear in the difference spectrum.

Figure 1 demonstrates the high sensitivity of the surface coil ISIS technique in spectra acquired from 4 cm^3 and 1 cm^3 volumes at 3 cm depths in 5 minutes from the same human volunteer (6). Signals in the 1 cm^3 volume (shown in x4 scale) are reduced by close to 4 fold indicating contributions from other volumes are negligible. The creatine methyl resonance is measured in the 4 cm^3 spectra to have a S/N of ≈25:1. Subsequent spectrometer electronic improvements have increased sensitivity by a factor of 2. With the present sensitivity the

Fuel Homeostasis and the Nervous System, Edited by M. Vranic *et al.*
Plenum Press, New York, 1991

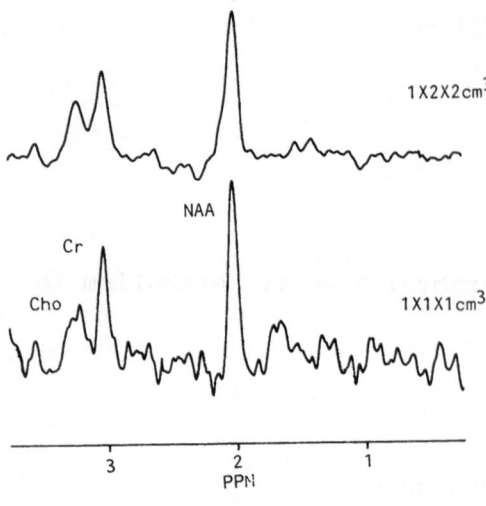

Figure 1

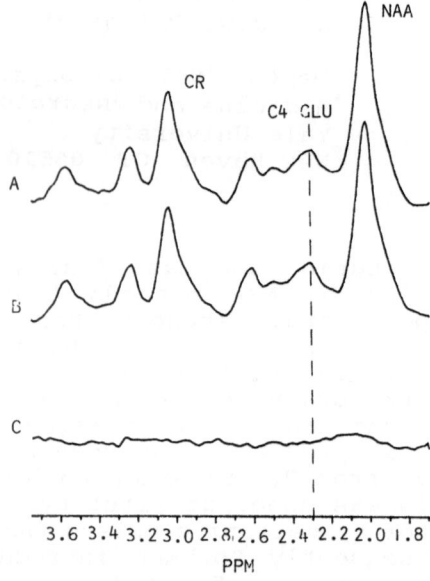

Figure 2

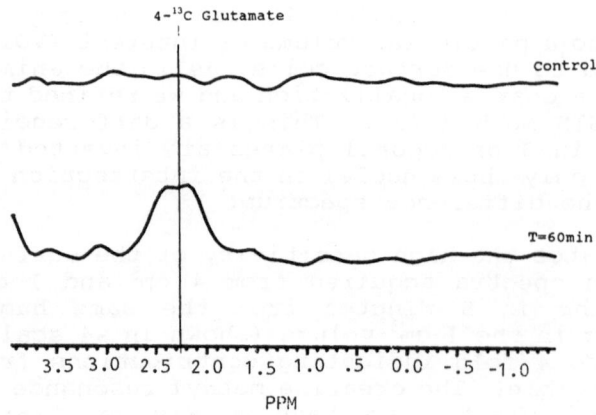

Figure 3

resting lactate level in a 12cc volume in the visual cortex has been measured with good S/N (7). A 2-fold rise in the lactate level during visual stimulation has been measured reproducibly (8).

Recently we have combined the depth pulse/ISIS sequence with ^{1}H observe ^{13}C editing (^{1}H-[^{13}C]NMR) to measure the flow of infused 1-^{13}C glucose into human brain C4 glutamate (9). These experiments are similar to those reported in the rat from which it has been possible, with the help of a model of carbohydrate catabolism, to calculate the glycolytic flux (10). Figure 2 shows the spectra from a control $^{(no\ infusion)1}$H-[^{13}C] ISIS experiment. A 6cm ^{1}H observe surface coil was used along with a 13 cm concentric ^{13}C coil which applied a ^{13}C editing pulse to the C4 ^{13}C resonance of glutamate and a B_2 insensitive ^{13}C decoupling sequence. Excellent subtraction is seen in the difference spectrum containing ^{1}H bound to ^{13}C. Fig 3 shows edited 6 min spectra of 4-^{13}C glutamate obtained before (0 min) and 70 min after the start of a 1-^{13}C glucose infusion (9). The prominent resonance at 2.32 ppm in the 70 min spectrum, which is not present at 0 min, is from 4-^{13}C glutamate. The time resolution (6 min) was sufficient to characterize the kinetics of the glutamate pool turnover which contains information on the rate of pyruvate dehydrogenase. Based on these results the further development of ^{1}H[^{13}C] techniques for measuring ^{13}C turnover and calculating human brain metabolic rates seems very promising.

References

1. K.L. Behar, J.A. den Hollander, M.E. Stromski, T. Ogino, R.G. Shulman, O.A.C. Petroff and J.W. Prichard, "High resolution ^{1}H NMR study of cerebral hypoxia in vivo", Proc. Natl. Acad. Sci. USA, 80:4945 (1983).

2. H.P. Hetherington, M.J. Avison and R.G. Shulman, "^{1}H homonuclear editing of rat brain using semi-selective pulses", Proc. Natl. Acad. Sci. USA, 82:3115 (1985).

3. D.L. Rothman, K.L. Behar, H.P. Hetherington and R.G. Shulman, "Homonuclear ^{1}H double-resonance difference spectroscopy of the rat brain in vivo", Proc. Natl. Acad. Sci. USA, 81:6330 (1984).

4. D.L. Rothman, K.L. Behar, H.P. Hetherington, J.A. den Hollander, M.R. Bendall, O.A.C. Petroff and R.G. Shulman, "^{1}H-observe/and ^{13}C-decouple spectroscopic measurements of lactate and glutamate in the rat brain in vivo", Proc. Natl. Acad. Sci. USA, 82:1633 (1985).

5. R.J. Ordidge, A. Connelly and J.A.B. Lohman, "Image-selected in vivo spectrosocpy (ISIS). A new technique for spatially selective NMR spectroscopy", Jour. Magn. Res., 66:283 (1986).

6. C.C. Hanstock, D.L. Rothman, J.W. Prichard, T. Jue and R.G. Shulman, "Volume-selected proton spectroscopy in the human brain", Jour. Magn. Res., 77:583 (1988).

7. C.C. Hanstock, D.L. Rothman, J.W. Prichard, T. Jue and R.G. Shulman, "Spatially localized ^{1}H NMR spectra of metabolites in the human brain", Proc. Natl. Acad. Sci. USA, 85:1821 (1988).

8. J. Prichard, D. Rothman, E. Novotny, O. Petroff, M. Avison, A. Howseman, C. Hanstock and R. Shulman, "Photic stimulation raises lactate in human visual cortex", Abstract, Soc. Magn. Res. in Med., 8th Annual Mtg., Amsterdam, pg. 1071 (1989)

9. D.L. Rothman, A. Howseman, E.J. Novotny, C.C. Hanstock, G. Lantos, O.A.C. Petroff, J.W. Prichard and R.G. Shulman, "Feasibility of proton-observe carbon-decouple editing of glutamate in the human brain", Abstract, Soc. Magn. Res. in Med., 8th Annual Mtg., Amsterdam, pg. 372 (1989).

10. S.M. Fitzpatrick, H.P. Hetherington, K.L. Behar and R.G. Shulman, "The flux from glucose to glutamate in the rat brain in vivo as determined by ^{1}H-observed, ^{13}C-edited NMR spectroscopy", Jour. Cere. Blood Flow & Metab., 10:170-179 (1990).

USE OF PEPTIDE PROBES TO STUDY

BRAIN REGULATION OF GLUCOSE METABOLISM

Marvin R. Brown

Departments of Medicine and Surgery
University of California, San Diego
San Diego, California

INTRODUCTION

Peptides represent the largest class of biologically active ligand that exist within central nervous system (CNS) neurons and their axonal projections. The physiologic role that these peptides play in the regulation of brain cellular functions, including neurotransmission, has not been determined. Most brain peptides display biological actions when administered into the CNS, thus leading to hypotheses regarding their physiologic roles. In addition to characterization of the physiologic roles of these peptides within the CNS, it is evident that these substances may be utilized as neurochemical probes with unique specificities for select neuronal populations to study both cellular and integrated CNS functions. An area of importance to physiologists has been the use of peptides to modify brain neuroendocrine and autonomic nervous system neuro-humoral effector mechanisms that regulate visceral organ function. This chapter will describe some of the CNS peptides that may be used as probes to study neuroendocrine and autonomic control of glucose metabolism.

BRAIN-DEPENDENT MECHANISMS OF REGULATION

OF GLUCOSE METABOLISM

To study the CNS regulation of metabolism requires knowledge of the afferent and efferent mechanisms involved in this neural-visceral system. Afferent input to the CNS from viscerosensory receptors involved in regulation of glucose metabolism is not well described. Vagal viscerosensory afferents arising from the liver and gastrointestinal tract may relay minute to minute information to the brain regarding blood glucose concentrations.[1,2] Glucose sensitive areas within the hypothalamus and nucleus of the solitary tract and 2-deoxyglucose sensitive areas throughout the brain have been demonstrated.[3] Insulin receptors present in many brain regions, including the circumventricular organs (CVO), may serve

Fuel Homeostasis and the Nervous System, Edited by M. Vranic *et al.*
Plenum Press, New York, 1991

as afferent inputs to brain areas involved in the regulation of glucose metabolism.[4] The best characterized of these afferent sensory systems are the glucose sensitive regions of the hypothalamus. Little is known, however, about the processing of the responses of these glucose sensitive cells into efferent mechanisms that modify blood glucose levels.

Brain efferent mechanisms involved in regulation of metabolism have been more clearly characterized than the afferent pathways. Table 1 outlines the mechanisms by which the brain controls glucose metabolism. Anterior pituitary hormones (growth hormone, ACTH, β-endorphin, and TSH) may modify peripheral glucose metabolism through effects on adipocytes, adrenal cortical secretion of glucocorticoids, thyroid secretion of thyroid hormone and triiodothyronine, and liver production of somatomedins. Posterior pituitary secretion of vasopressin may influence the pancreatic release of insulin and glucagon as a mechanism to regulate glucose metabolism.[5] The possibility that other brain factors may be released into blood has been suggested.[6-8]

The autonomic nervous system is capable of modifying glucose metabolism through changes of parasympathetic, sympathetic, or adrenal medullary function. Pancreatic insulin secretion is under parasympathetic control, acetylcholine stimulates insulin release and sympathetic control, norepinephrine inhibits insulin release.[9] Sympathetic nerves release norepinephrine and neuropeptide Y.[10] Norepinephrine may act at the pancreas, liver, pituitary, or sites of periperhal glucose uptake to modify plasma glucose concentrations. What role neuropeptide Y plays in regulation of glucose metabolism has not been established. The adrenal medulla releases epinephrine and, depending on the species, one of a number of biologically active peptides.[11-13] The role of these peptides in the regulation of glucose metabolism has not been characterized. Epinephrine may exert actions at pancreas, liver, or pituitary to influence glucose metabolism.

An anatomic basis exists for the processing of visceral afferent information and producing coordinated autonomic nervous system and neuroendocrine response.[14]

ORGANIZATION OF NEURAL SYSTEMS INVOLVED IN THE REGULATION

OF GLUCOSE METABOLISM

The first of two mechanisms by which the central nervous system may be involved in regulation of glucose metabolism is that of a simple reflex (see Table 2). Glucose sensors present in specific brain areas, the CVOs, liver, or other sites act via afferent pathways to result in activation of hypothalamic, brain stem, or spinal cord systems that initiate efferent pathways involved in regulation of glucose production. Such a reflex response differs little in organization from that involved in regulation of cardiovascular function through baro-receptor mediated responses. As noted above, a key component of understanding such a reflex system is the identification of the origin and central nervous system site of processing of afferent signals. The efferent limb of this response utilizes those mechanisms described in the section above.

Table 1. Brain Efferent Pathways Involved in the Regulation of Metabolism

Endpoint	Anterior Pituitary	Posterior Pituitary	PSN	SNS	Adrenal Medulla
Neuroendocrine or ANS Chemical Mediator	GH ACTH, β-Endorphin TSH	Vasopressin	ACH	NE NPY	E Peptides
Visceral End-Organ Site of Regulation	Adipocytes Adrenal cortex Thyroid	Pancreas	Pancreas	Pancreas Liver Pituitary sites of glucose uptake	Pancreas Liver Pituitary
Visceral Organ Hormonal Response	↑Glucocorticoids, Thyroid hormone, Somatomedins	↑Glucagon	↑Insulin	↑ACTH/Gluco-corticoids	↑ACTH/Glucagon
Effect on Blood Glucose Levels	↑Glucose	↑Glucose	↓Glucose	↑Glucose	↑Glucose

Abbreviations: ANS, autonomic nervous system; PSN, parasympathetic nervous system; SNS, sympathetic nervous system; GH, growth hormone; ACTH, adrenocorticotropic hormone; TSH, thyroid stimulating hormone; ACH, actylcholine; NE, norepinephrine; NPY, neuropeptide Y; E, epinephrine

11

Table 2.   Neural Mechanisms of Regulation
of Glucose Metabolism I

<u>Reflex Mechanisms</u>

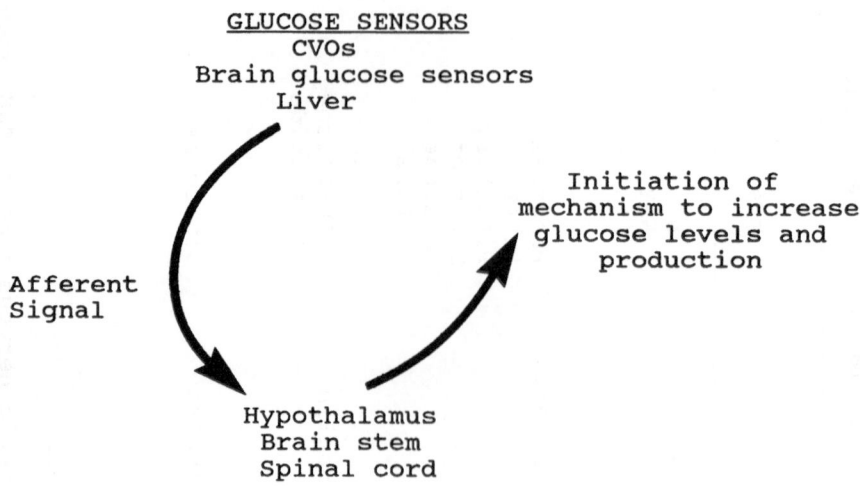

GLUCOSE SENSORS
CVOs
Brain glucose sensors
Liver

Afferent
Signal

Initiation of
mechanism to increase
glucose levels and
production

Hypothalamus
Brain stem
Spinal cord

The second general mechanism by which the central nervous
system participates in regulation of glucose metabolism is
through changes of central command. Central command involves
the insertion of non-reflex signals onto reflex pathways or
directly onto efferent neuroendocrine or autonomic pathways
(Table 3). Central command may utilize sensory or cognitive
inputs to initiate changes in glucose production in the face of
normal glucose levels. Such mechanisms are involved in
stressor induced elevation of plasma glucose concentrations.
The physiologic role of reflex mechanisms in the maintenance of
basal glucose concentrations may be debated since it is likely
that basal glucose concentrations can be maintained in the
absence of central nervous system input. There is no doubt
however that stimuli such as hypoglycemia provoke activation of
reflex neural mechanisms that are capable of elevating blood
glucose concentrations. The role of central command in
regulation of blood glucose levels has certain physiologic
relevance, as demonstrated by changes of plasma glucose
concentrations following exposure of animals or humans to
stressors.

NEUROPEPTIDES THAT INFLUENCE REGULATION

OF GLUCOSE METABOLISM

Table 4 shows some of the peptides that act upon the
anterior pituitary or within the central nervous system to
affect the anterior or posterior pituitary, parasympathetic,
sympathetic, or adrenal medullary function.

Table 3. Neural Mechanisms of Regulation
of Glucose Metabolism II

<u>Central Command</u>

Sensory perception (stressor)

1. Alter reflex regulation
   of glucose production

2. Initiation of mechanism
   to raise glucose levels

Regulation of Pituitary Hormone Secretion

Peptides that may be used to modify anterior pituitary hormone secretion involve those substances that act directly at the level of the pituitary gland in contrast to those agents that act within the central nervous system to modif· pituitary releasing factor secretion. Table 4 illustrates some of the factors that act at the level of the pituitary either to increase or to decrease the secretion of growth hormone, ACTH, or TSH. The use of glucocorticoids, thyroid hormone, or somatostatin may be associated with actions on other unrelated cellular systems that detract from their specific use to inhibit pituitary hormone secretion. Table 5 lists some of the substances that act within the central nervous system to modify the release of growth hormone, ACTH, and TSH.

Administration of somatostatin into the CNS results in increased vasopressin and oxytocin release.[15,16] Substance P administered into the CNS also stimulates pituitary vasopressin release.[17]

Regulation of Parasympathetic Output of the Pancreas

TRF administered into the central nervous system has been reported to increase parasympathetic outflow to the pancreas, resulting in insulin secretion.[18] In rats, TRF administered into the brain produces hyperglycemia through activation of the sympathetic nervous system and adrenal medulla.[19] In mice, TRF lowers plasma glucose concentrations through its effects to stimulate insulin release.[18] Thus, between species TRF may have differential effects on parasympathetic versus sympathoadrenal function. Other peptides, e.g. CRF, may also stimulate changes of parasympathetic innervation of the endocrine pancreas to modify insulin release.

Table 4. Neuropeptides that Act at the Anterior
Pituitary or Within the CNS to Modify
Brain Efferent Pathways Involved in
Regulation of Glucose Metabolism

System	Peptide	Effect
Anterior pituitary	GRF	↑GH
	SS	↓GH
	CRF	↑ACTH
	CRF antagonist	↓ACTH
	TRF	↑TSH
	Thyroid hormone	↓TSH
Posterior pituitary	Somatostatin	↑AVP, oxytocin
	Substance P	↑AVP
Parasympathetic outflow to the pancreas	TRF	↑Insulin
	CRF	↓Insulin
Sympathetic outflow	TRF	↑NE
	CRF	↑NE
	Dynorphin	↓NE
	NPY	↓NE
Adrenal medulla	TRF	↑E
	CRF	↑E
	Bombesin	↑E
	Somatostatin	↓E
	Dynorphin	↓E

Regulation of Sympathetic Outflow

A variety of peptides including TRF, CRF, CCK, substance
P, and bombesin have been reported to increase plasma
concentrations of norepinephrine.[19-23] With the exception of CRF,
little is known about the viscerotropic origin of the
sympathetic terminals that contribute to this increase in
plasma norepinephrine concentration. Both dynorphin and NPY
have been reported to decrease plasma norepinephrine
concentrations.[24,25] Somatostatin has been reported to
decrease sympathetic outflow to thyroid.[26]

Regulation of the Adrenal Medulla

Bombesin may be administered into specific brain
regions of the rat or dog to result in elevation of plasma
epinephrine without significant effects on plasma
norepinephrine concentrations.[23,27] TRF, CRF, and opiate
related peptides have been reported to act within the

Table 5. Substances That Act Within the CNS to Modify Anterior Pituitary Hormone Secretion

Substance	Pituitary Hormone Affected
CRF	↓GH
TRF	↑GH
TRF	↑ACTH
SS	↓ACTH
Dynorphin	↓ACTH
Bombesin	↓TSH

central nervous system to increase adrenal epinephrine secretion.[19,20,28] A physiologic role of CRF in regulating stimulatory pathways to the adrenal medulla has been demonstrated.[29] Both somatostatin and dynorphin administered into the central nervous system suppress adrenal epinephrine secretion.[24,30]

THE USE OF NEUROPEPTIDES TO STUDY CENTRAL NERVOUS SYSTEM

REGULATION OF METABOLISM

To illustrate the use of neuropeptides to study central nervous system mechanisms involved in glucose regulation, two examples will be used. The first of these is the use of an analog of somatostatin to inhibit adrenal epinephrine secretion. The second is the use of CRF to stimulate neural pathways involved in stress-induced changes of neuroendocrine and autonomic function that may impinge upon glucose regulation.

As noted above, peptides related to somatostatin injected into the central nervous system inhibit adrenal epinephrine secretion and lower plasma glucose concentrations.[30-32] Exposure of rats to cold (4°C), ether vapor, or suspending the animals by their tails results in significant elevations of plasma epinephrine and glucose (see Table 6). A somatostatin analog administered icv significantly attenuated plasma concentrations of epinephrine following each of these treatments. However, this somatostatin analog suppressed the elevation of glucose levels following tail suspension, but not following exposure of animals to cold or ether vapor. These results are consistent with the conclusion that the elevation of glucose following tail suspension is dependent on adrenal epinephrine secretion. In contrast, the elevation of glucose concentrations following exposure of animals to cold or ether vapor is not dependent on adrenal epinephrine secretion.

Increasing evidence supports the hypothesis that central nervous system CRF containing pathways are involved in the coordination and activation of neuroendocrine and autonomic responses to stress.[33,34] CRF administered into the central nervous system produces changes of autonomic and visceral organ function that are similar to those following

Table 6. Effects of a Somatostatin Analog on Plasma Concentrations of Glucose and Epinephrine in Animals Exposed to Different Stressors

Treatment	Plasma Concentration of:			
	Glucose		Epinephrine	
	pre	post	pre	post
Cold exposure (4°C)	123±4	147±5**	61±20	206±25**
Cold exposure (4°C) + ODT8-SS (1nmole)	123±3	142±8**	40±18	47±26
Ether vapor exposure	124±5	154±8*	50±10	320±108*
Ether vapor exposure + ODT8-SS (1nmole)	113±2	139±6*	41±7	56±14
Tail suspension	110±8	136±2**	37±10	243±64*
Tail suspension + ODT8-SS (1nmole)	110±3	115±2	71±20	69±27

*$p<0.05$, ** $p<0.01$ compared to pre values

ODT8-SS is desAA^{1,2,4,5,12,13}[D-trp^8]-somatostatin

exposure of animals to a variety of stressors.[33] Based on these findings, CRF has been used to characterize neural efferent pathways involved in regulation of visceral organ physiology following stress. Table 7 demonstrates the use of a CRF receptor antagonist administered into the central nervous system to determine the role of central nervous system CRF pathways in regulating changes of glucose concentrations following exposure of animals to a stressor. As seen in Table 6, exposure of animals to ether vapor resulted in significant elevations of plasma epinephrine and glucose. Central nervous system administration of a CRF receptor antagonist prior to ether exposure prevented the elevation of epinephrine and glucose. These results suggest that a central nervous system CRF pathway is involved in the elevation of blood glucose concentrations following exposure of animals to ether vapor. As noted in Table 6, suppression

Table 7. Effects of a CRF Receptor Antagonist
on Stressor-Induced Change of Plasma
Concentrations of Catecholamines,
ACTH, and Glucose

	EPI	NE	ACTH	GLUCOSE
Ether vapor	↑	↑	↑	↑
Ether vapor + α-hel CRF$^{9-41}$	↓	↑	↓	↓

of adrenal epinephrine secretion using somatostatin
following ether exposure did not result in significant
suppression of blood glucose concentrations. From these
combined pieces of data, it is concluded that ether vapor
induced elevation of blood glucose concentrations, while
dependent on a central nervous system CRF pathway, is not
dependent on adrenal epinephrine secretion.

SUMMARY

Neuropeptides may be used to stimulate or inhibit
neurocircuitry involved in regulation of visceral organ
function, including glucose metabolism. Through the use of
different peptides with different specificities, it may be
possible to characterize the neuroendocrine and autonomic
pathways involved in the physiologic regulation of glucose
homeostasis.

REFERENCES

1. C. L. Lee and R. E. Miller, The hepatic vagus nerve and
    the neural regulation of insulin secretion,
    Endocrinol. 117:307 (1985).
2. A. Niijima, Glucose-sensitive afferent nerve fibers in
    the hepatic branch of the vagus nerve in the guinea
    pig, J. Physiol. 332:315 (1982).
3. O. Yutaka and H. Yoshimatsu, Neural network of glucose
    monitoring system, J. Auton. Nerv. Syst. 10:359
    (1984).
4. B. R. Landau, Y. Takaoka, M. A. Abrams, S. M. Genuth,
    M. Van Houten, B. I. Posner, R. J. White, S. Ohgaku,
    A. Horvat, and E. Hemmelgarn, Binding of insulin by
    monkey and pig hypothalamus, Diabetes 32:284 (1983).
5. B. E. Dunning, J. H. Moltz, and C. P. Fawcett, Actions
    of neurohypophyseal peptides on pancreatic hormone
    release, Amer. J. Physiol. 246:E108 (1984).
6. E. Bobbioni and B. Jeanrenaud, A rat hypothalamic
    extract enhances insulin secretion in vitro,
    Endocrinol. 113:1958 (1983).
7. L. J. Grimes, C. Mok, and J. M. Martin, Effect of a
    bovine hypothalamic extract on glucose utilization
    by rat adipocytes, Amer. J. Physiol. 234:E554
    (1978).

8. L. A. Idahl and J. M. Martin, Stimulation of insulin release by a ventrolateral hypothalamic factor. J. Endocr. 51:601 (1971).

9. G. A. Taborsky and D. Porte, Jr., Stress-induced hyperglycemia, in: "The Neurobiology and Neuroendocrinology of Stress," M. R. Brown, C. Rivier, and G. Koob, eds., Marcel Dekker, Inc., New York (in press).

10. J. Pernow, J. Schwieler, T. Kahan, P. Hjemdahl, J. Oberle, B. G. Wallin, and J. M. Lundberg, Influence of sympathetic discharge pattern on norepinephrine and neuropeptide Y release, Amer. J. Physiol. 257: H866 (1989).

11. J. M. Lundberg, B. Hamberger, M. Schultzberg, T. Hokfelt, P-O Granberg, S. Efendic, L. Terenius, M. Goldstein, and R. Luft, Enkephalin- and somatostatin-like immunoreactivities in human adrenal medulla and pheochromocytoma, Proc. Natl. Acad. Sci. USA 76:4079 (1979).

12. G Terenghi, J. M. Polak, I. M. Varndell, Y. C. Lee, J. Wharton, and S. R. Bloom, Neurotensin-like immunoreactivity in a subpopulation of noradrenaline-containing cells of the cat adrenal gland, Endocrinol. 112:226 (1983).

13. R. Corder, D. F. J. Mason, D. Perrett, P. J. Lowry, V. Clement-Jones, E. A. Linton, G. M. Besser, and L. H. Rees, Simultaneous release of neurotensin, somatostatin, enkephalins and catecholamines from perfused cat adrenal glands, Neuropeptides 3:9 (1982).

14. L. W. Swanson and P. E. Sawchenko, Hypothalamic integration: organization of the paraventricular and supraoptic nuclei, Ann. Rev. Neurosci. 6:269 (1982).

15. M. R. Brown, M. Mortrud, R. Crum, and P. Sawchenko, Role of somatostatin in the regulation of vasopressin secretion, Brain Res. 452:212 (1988).

16. M. R. Brown, R. Crum, and P. Sawchenko, Somatostatin-28 (SS-28) stimulation of vasopressin (AVP) and oxytocin (OT) secretion, Endocrinol. 122(Suppl.): 660 (1988).

17. C. D. Sladek, Regulation of vasopressin release by neurotransmitters, neuropeptides and osmotic stimuli, Prog. Brain Res. 60:71 (1983).

18. S. Amir and P. D. Butler, Thyrotropin-releasing hormone blocks neurally-mediated hyperglycemia through central action, Peptides 9:31 (1988).

19. M. R. Brown, Thyrotropin releasing factor: a putative CNS regulator of autonomic nervous system outflow, Life Sci. 28:1789 (1981).

20. M. R. Brown, L. A. Fisher, J. Spiess, J. Rivier, C. Rivier, and W. Vale, Corticotropin-releasing factor (CRF): actions on the sympathetic nervous system and metabolism, Endocrinol. 111:928 (1982).

21. J. E. Morley and A. S. Levine, Intraventricular cholecystokinin octapeptide produces hyperglycemia in rats, Life Sci. 28:2187 (1981).

22. A. Iguchi, H. Matsunaga, T. Nomura, M. Gotoh, and N. Sakamoto, Glucoregulatory effects of intrahypothalamic injections of bombsin and other peptides, Endocrinol. 114:2242 (1984).

23. M. Brown, Y. Tache, and D. Fisher, Central nervous system action of bombesin: mechanism to induce hyperglycemia, _Endocrinol._ 105:660 (1979).

24. J. M. Overton and L. A. Fisher, Modulations of central nervous system actions of corticotropin-releasing factor by dynorphin-related peptides, _Brain Res._ 488:233 (1989).

25. N. A. Scott, V. Webb, J. H. Boublik, J. Rivier, and M. R. Brown, The cardiovascular actions of centrally administered neuropeptide Y, _Regul. Peptides_ 25:247 (1989).

26. H. Somiya and T. Tonoue, Neuropeptides as central integrators of autonomic nerve activity: effects of TRH, SRIF, VIP and bombesin on gastric and adrenal nerves, _Regul. Peptides_ 9:47 (1984).

27. M. R. Brown, K. Carver, and L.A. Fisher. Bombesin: central nervous system actions to affect the autonomic nervous system, _in_: "Annals of the New York Academy of Sciences, Vol. 547, Bombesin-like Peptides in Health and Disease," Y. Tache, P. Melchiorri, and L. Negri, eds., New York Academy of Sciences, New York (1989).

28. G. R. Van Loon, N. M. Appel, and D. Ho, β-endorphin-induced stimulation of central sympathetic outflow: β-endorphin increases plasma concentrations of epinephrine, norepinephrine, and dopamine in rats, _Endocrinol._ 109:46 (1981).

29. M. R. Brown, L. A. Fisher, V. Webb, W. W. Vale, and J. E. Rivier, Corticotropin-releasing factor: a physiologic regulator of adrenal epinephrine secretion, _Brain Res._ 328:355 (1985).

30. D. A. Fisher and M. Brown, Somatostatin analog: plasma catecholamine suppression mediated by the central nervous system, _Endocrinol._ 107:714 (1980).

31. M. R. Brown and L. A. Fisher, Brain peptide regulation of adrenal epinephrine secretion, _Amer. J. Physiol._ 247:E41 (1984).

32. M. R. Brown, J. Rivier, and W. Vale, Somatostatin: central nervous system actions on glucoregulation, _Endocrinol._ 104:1709 (1979).

33. M. R. Brown and L. A. Fisher, Corticotropin releasing factor: effects on the autonomic nervous system and visceral systems, _Fed. Proc._ 44:243 (1985).

34. C. Rivier and W. Vale, Effects of corticotropin-releasing factor, neurohypophyseal peptides, and catecholamines on pituitary function, _Fed. Proc._ 44:189 (1985).

MEASUREMENT OF LOCAL CEREBRAL GLUCOSE UTILIZATION AND ITS RELATION TO LOCAL FUNCTIONAL ACTIVITY IN THE BRAIN

Louis Sokoloff

National Institute of Mental Health
Building #36, Room 1A-05
Bethesda, Maryland 20892

INTRODUCTION

The average normal, human male brain represents approximately 2% of total body weight but consumes about 20% of the total body basal oxygen consumption (Sokoloff, 1989). The substrate for this high rate of energy metabolism is normally almost exclusively glucose (TABLE 1) (Sokoloff, 1960). In fact, more glucose is consumed than can be oxidized completely to carbon dioxide and water by the oxygen consumption, indicating that the glycolytic rate exceeds the rate of oxidation of the products of glycolysis, normally by about 20% (Table 1). The excess carbon derived from glycolytic utilization of glucose is probably distributed in many, e.g., lactate, pyruvate, and other intermediates of the glycolytic and tricarboxylic acid cycle pathways that leave the brain in amounts too insignificant to be detected in the cerebral blood, and also into several neurotransmitter pools, such as acetylcholine, glutamate, GABA, etc.

Not only is glucose the preferred substrate for the brain's energy metabolism, but it is essential. In its absence, for example in hypoglycemia, brain function, as reflected in level of consciousness, electrical activity, and oxygen metabolism, is impaired (Kety et al., 1948). The loss of consciousness, electroencephalographic slowing, and reduced cerebral O_2 consumption produced by insulin hypoglycemia, for example, are rapidly reversed by glucose administration or prevented by simultaneous administration of glucose along with the insulin. Also, under normal circumstances no substrate other than glucose is taken up from the

Fuel Homeostasis and the Nervous System, Edited by M. Vranic *et al.*
Plenum Press, New York, 1991

blood by the brain in significant amounts, nor is there any potential substrate other than mannose that can reverse the effects of hypoglycemia without first raising the blood glucose level (Sokoloff, 1989). In ketosis accompanying starvation (Owens et al., 1967), fat-feeding (Krebs et al., 1971), diabetes (Gottstein et al., 1972), etc. ketone bodies can substitute for glucose to some extent but cannot completely replace the brain's requirement for glucose.

TABLE 1. *Average oxygen consumption and glucose utilization in the brain as a whole in normal young men*[a]

Cerebral O_2 Consumption	156	μmols/100 g brain/min
Cerebral CO_2 Production	156	μmols/100 g brain/min
Cerebral Respiratory Quotient	0.97	
Cerebral Glucose Utilization	31	μmols/100 g brain/min
O_2/Glucose Ratio	5.5	μmols/μmol
Glucose Utilization Equivalent to Oxygen Consumption	26[b]	μmols/100 g brain/min

[a] Values are medians of numerous values reported in the literature. From Sokoloff (1960).
[b] Calculated on basis of 6 mols of O_2 required for complete oxidation of one mol of glucose to CO_2 and H_2O.

The energy metabolism of the brain as a whole, as determined with the nitrous oxide method of Kety and Schmidt (1948), is relatively constant and changes very little with physiological alterations in cerebral functional activity. Only in conditions that depress the level of consciousness are there any pronounced changes in the oxygen consumption or glucose utilization of the brain as a whole, and then its energy metabolism is decreased (Kety, S.S., 1950; Sokoloff, 1989). During the performance of normal mental functions, such as the mental exercise of solving arithmetical problems (Sokoloff et al., 1955), or during the deranged mental functions in functional psychoses like schizophrenia (Kety et al., 1948; Sokoloff, 1969), the rate of energy metabolism in the brain as a whole is unchanged. In fact, except in convulsions, the whole brain's consumption of oxygen or glucose is almost never increased.

The brain, however, is hardly uniform. In contrast to other organs, like liver, skeletal muscle, and heart, it consists of numerous structural and functional components that subserve different functions and operate more or less independently, sometimes even inversely, of one another. In most tissues rates of energy metabolism of the tissue correlate closely with the work of the tissue. To relate work or functional activity to energy metabolism in brain, it is necessary to examine those regions specifically involved in that particular function and not the brain as a whole. One, for example, would not attempt to relate the work of the heart to its energy metabolism from measurements of the consumption of oxygen by the whole body. A method for measuring local rates of energy metabolism in the brain of conscious, behaving animals or man was needed. This is now possible in animals with the autoradiographic deoxyglucose method (DG) (Sokoloff et al., 1977), and its [18F]fluorodeoxyglucose modification extends the capability to man (Reivich et al., 1979: Phelps et al., 1979).

THEORY OF THE AUTORADIOGRAPHIC DEOXYGLUCOSE METHOD

The method is based on a kinetic analysis of the biochemical behavior of 2-deoxy-D-glucose (DG) and glucose in brain (Fig. 1A) (Sokoloff et al., 1977; Sokoloff, 1982). Both hexoses are transported bi-directionally between blood and brain by a common carrier in the blood-brain barrier. In the tissues hexokinase phosphorylates both to their hexose-6-phosphate derivatives (Sols and Crane, 1954). Glucose and DG are, therefore, competitive substrates for both blood-brain barrier transport and hexokinase-catalyzed phosphorylation. Unlike glucose-6-phosphate (G-6-P), however, which is metabolized further to pyruvate and eventually CO_2 and water and to a lesser extent via the pentosephosphate shunt pathway, deoxyglucose-6-P (DG-6-P) cannot be converted to fructose-6-phosphate and is a poor substrate for G-6-P dehydrogenase (Sols and Crane, 1954). There is very little glucose-6-phosphatase (G-6-Pase) activity in brain (Hers, 1957; Fishman and Karnovsky, 1986) and even less DG-6-phosphatase (DG-6-Pase) activity (Sokoloff et al., 1977). DG-6-P can be converted to DG-1-phosphate, then to uridine-diphospho-deoxyglucose (UDP-DG), and eventually incorporated into glycogen, glycolipids, and glycoproteins, but these reactions are slow. In mammalian brain only a small fraction of the DG-6-P formed proceeds to these products (Nelson et al., 1984). These compounds are relatively stable products of DG-6-P, and all together represent the ultimate products of phosphorylation of DG. DG-6-P and its derivatives, once formed, are, therefore, trapped in the cerebral tissues, not forever but long enough to allow a fairly long period of measurement before significant loss of product occurs. If the interval of time following administration of [14C]DG is kept short enough (e.g., less than 1 hour) to keep the loss of $[^{14}C]DG\text{-}6\text{-}P$ and its secondary products from the tissues negligible, then the quantity of labeled products accumulated in the tissue at any given time following the injection of $[^{14}C]DG$ circulation equals the integral of the rate of $[^{14}C]DG$ phosphorylation by hexokinase in that tissue during the same interval. This integral is in turn related to the amount of glucose that has been phosphorylated over the same interval, depending on the time courses of the $[^{14}C]DG$ and glucose concentrations (e.g., specific activity) in the precursor pools and the Michaelis-Menten kinetic constants for hexokinase with respect to $[^{14}C]DG$ and glucose. With cerebral glucose consumption in a steady state, the amount of glucose phosphorylated during the interval of time equals the steady state flux of glucose through the hexokinase-catalyzed step times the duration of the interval, and the net rate of flux of glucose through this step equals the rate of glucose utilization.

These relationships can be kinetically analyzed and an operational equation derived if the following assumptions are made: 1) constancy of plasma glucose concentration and rate of glucose consumption during the experimental period; 2) homogeneous tissue compartment with respect to blood flow, blood-brain transport, rate of glucose utilization, and $[^{14}C]DG$ and glucose concentrations; and 3) tracer concentrations of $[^{14}C]DG$ and $[^{14}C]DG\text{-}6\text{-}P$ (i.e., molecular concentrations essentially zero). The operational equation defines R_i, the rate of glucose consumption per unit mass in tissue i, in terms of measurable variables (Fig. 1B). It should be noted that the model and the form of the operational equation were specifically designed for application to quantitative autoradiography, which measures only total concentration of isotope in the tissue and not those of individual labeled chemical species.

Rate Constants. The rate constants, K_1^*, k_2^*, and k_3^*, are determined in a separate group of animals by a non-linear, least squares best-fit of an equation, which relates the time course of total tissue ^{14}C concentration, C_i^*, to the time, history of plasma $[^{14}C]DG$ concentration, and the rate constants, to the measured time courses of the concentrations of $[^{14}C]DG$ in arterial plasma and total ^{14}C in the tissues. The procedure for the determination of these constants has been described previously (Sokoloff et al., 1977), and their values in normal rats are presented in TABLE 2.

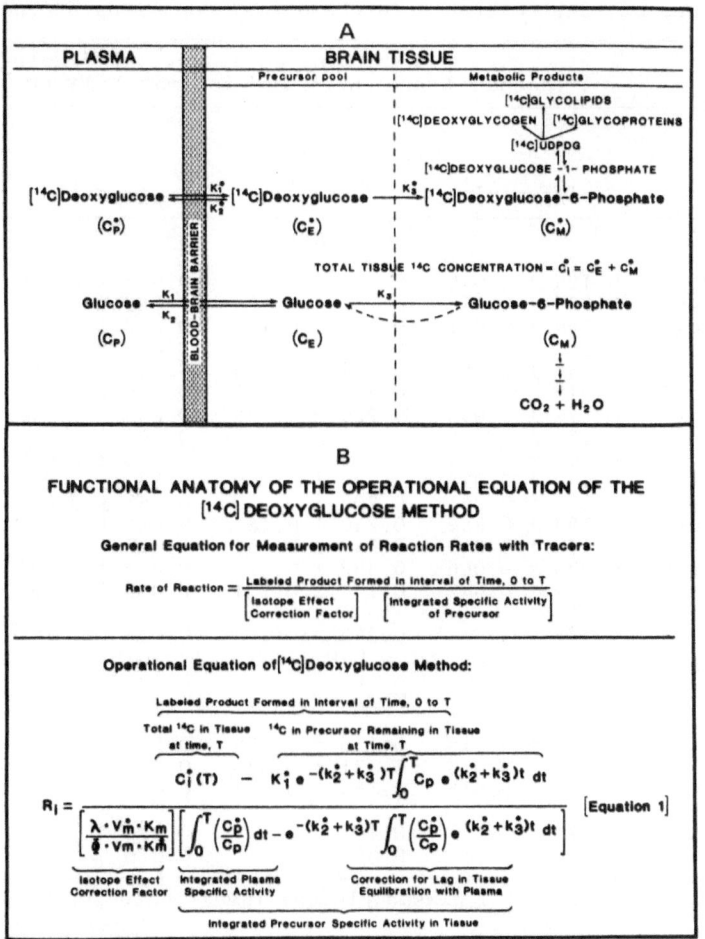

Fig 1. A: Model of DG method. C_i^* = total ^{14}C concentration in tissue; C_P^* and C_P = $[^{14}C]$DG and glucose concentrations in arterial plasma; C_E^* and C_E are their concentrations in tissue pools that serve as substrates for phosphorylation by hexokinase or transport back to plasma; C_M^* = combined concentrations of $[^{14}C]$DG-6-P and its derivatives in the tissue. K_1^*, k_2^*, and k_3^* = rate constants for transport of $[^{14}C]$DG from plasma to tissue and tissue to plasma, and for phosphorylation, respectively; K_1, k_2, and k_3 = equivalent rate constants for glucose. $[^{14}C]$DG and glucose compete for carrier that transports both between plasma and tissue and for hexokinase. Dashed arrow represents possibility of G-6-P hydrolysis by G-6-Pase activity, if any. B: Operational equation of DG method. R_i = rate of glucose utilization; T equals duration of experimental period; λ is the ratio of the distribution space of DG in the tissue to that of glucose; Φ equals fraction of G-6-P that, once formed, continues down glycolytic and pentosephosphate shunt pathways (e.g., with zero G-6-Pase activity Φ = 1); K_m^* and V_m^* are the Michaelis-Menten kinetic constants of hexokinase for DG; and K_m and V_m are the equivalent kinetic constants for glucose.

TABLE 2. *Values of rate constants in the normal conscious albino rat*[a,b]

Structure	Rate Constants (Means ± Stand. Errors of Estimate)			Half-Life of Precursor Pool
	K_1^* (ml/g/min^{-1})	k_2^* (min^{-1})	k_3^* (min^{-1})	$\dfrac{\log_e 2}{(k_2^* + k_3^*)}$ (min)
Gray Matter				
Visual cortex	0.189 ± 0.048	0.279 ± 0.176	0.063 ± 0.040	2.03
Auditory cortex	0.226 ± 0.068	0.241 ± 0.198	0.067 ± 0.057	2.25
Parietal cortex	0.194 ± 0.051	0.257 ± 0.175	0.062 ± 0.045	2.17
Sensorimotor cort.	0.193 ± 0.037	0.208 ± 0.112	0.049 ± 0.035	2.70
Thalamus	0.188 ± 0.045	0.218 ± 0.144	0.053 ± 0.043	2.56
Medial genic.body	0.219 ± 0.055	0.259 ± 0.164	0.055 ± 0.040	2.21
Lat.genic. body	0.172 ± 0.038	0.220 ± 0.134	0.055 ± 0.040	2.52
Hypothalamus	0.158 ± 0.032	0.226 ± 0.119	0.043 ± 0.032	2.58
Hippocampus	0.169 ± 0.043	0.260 ± 0.166	0.056 ± 0.040	2.19
Amygdala	0.149 ± 0.028	0.235 ± 0.109	0.032 ± 0.026	2.60
Caudate-putamen	0.176 ± 0.041	0.200 ± 0.140	0.061 ± 0.050	2.66
Sup. colliculus	0.198 ± 0.054	0.240 ± 0.166	0.046 ± 0.042	2.42
Pontine gray	0.170 ± 0.040	0.246 ± 0.142	0.037 ± 0.033	2.45
Cerebellar cortex	0.225 ± 0.066	0.392 ± 0.229	0.059 ± 0.031	1.54
Cerebellar nucl.	0.207 ± 0.042	0.194 ± 0.111	0.038 ± 0.035	2.99
Mean	0.189	0.245	0.052	2.39
± S.E.M.	± 0.012	± 0.040	± 0.010	± 0.40
White Matter				
Corp. callosum	0.085 ± 0.015	0.135 ± 0.075	0.019 ± 0.033	4.50
Genu corp. callos.	0.075 ± 0.013	0.131 ± 0.075	0.019 ± 0.034	4.62
Internal Capsule	0.077 ± 0.015	0.134 ± 0.085	0.023 ± 0.039	4.41
Mean	0.079	0.133	0.020	4.51
± S.E.M.	± 0.008	± 0.046	± 0.020	± 0.90

[a] Normoglycemic rats; arterial plasma glucose levels between 7 and 12 mM.
[b] From Sokoloff et al. (1977)

Lumped Constant. The λ, Φ, and Michaelis-Menten constants are lumped together to constitute a single constant (Fig. 1B). This lumped constant can be shown to equal the asymptotic value of the product of the ratio of the cerebral extraction ratios of [14C]DG and glucose and the ratio of the arterial blood to plasma specific activities when the arterial plasma [14C]DG concentration is maintained constant (Sokoloff et al., 1977). The lumped constant is also determined in a separate group of animals from arterial and cerebral venous blood samples drawn while the brain and blood are in a steady state with respect to their concentrations of [14C]DG and glucose. The steady state for [14C]DG is achieved by a programmed infusion intravenously that produces and maintains the arterial plasma [14C]DG concentration constant (Sokoloff et al., 1977). The steady state for glucose usually occurs spontaneously in normoglycemic animals but can be achieved in other conditions by a glucose clamp procedure (DeFronzo et al. (1979). Under normal conditions The lumped constant varies with the species of animal but within each species appears to change relatively little over a wide range of physiological conditions (TABLE 3).

TABLE 3. *Values of lumped constant in various species*

Animal	No. of Animals	Mean ± S.D.	S.E.M.
Albino rat:			
Conscious	15	0.464 ± 0.099[a]	± 0.026
Anesthetized	9	0.512 ± 0.118[a]	± 0.039
Conscious (5% CO2)	2	0.463 ± 0.122[a]	± 0.086
Combined	26	0.481 ± 0.119	± 0.023
Monkey (Conscious)	7	0.344 ± 0.095	± 0.036
Cat (Anesthetized	6	0.411 ± 0.013	± 0.005
Dog (beagle puppy)			
(Conscious)	7	0.558 ± 0.082	± 0.031
Sheep:			
Fetus	5	0.416 ± 0.031	± 0.014
Newborn	4	0.382 ± 0.024	± 0.012
Combined	9	0.400 ± 0.033	± 0.011
Humans (Conscious)	6	0.568 ± 0.105	± 0.043

[a] No statistically significant differences between the normal conscious and anesthetized rats ($0.3 < p < 0.4$) and the normal conscious rats and those breathing 5% CO2 ($p > 0.9$).

Note: Values obtained from the literature. For individual references see Sokoloff (1985).

PROCEDURE OF THE DEOXYGLUCOSE METHOD

The operational equation of the method (Fig. 1B) identifies the variables that need to be measured and the parameters that must be known in order to determine R_i, the rate of glucose consumption. The variables measured in each experiment are: 1) history of the arterial plasma [^{14}C]DG concentration, C_P^*, from the time of injection to time T, the time at the end of the experimental period; 2) arterial plasma glucose concentration, C_P; and 3) concentration of ^{14}C in tissue i at time T, $C_i^*(T)$. The rate constants, K_1^*, k_2^*, and k_3^*, and the lumped constant, $\lambda V_m^* K_m / \Phi V_m K_m^*$, are parameters that have been evaluated in other groups of animals as described above and presented in TABLES 2 and 3. It is in the values of these parameters that sensitivities to imperfections in the model reside.

Rate constants. The operational equation is applicable with all types of time courses of arterial plasma [^{14}C]DG concentration, but its form suggests that a declining curve approaching zero by the time of killing is the best choice to minimize potential errors associated with the rate constants. The values of K_1^*, k_2^*, and k_3^* vary from one structure to another, especially between gray and white matter, and from condition to condition, and the values in TABLE 2 for normal conscious rats may not apply to all physiological, pharmacological, and pathological states. K_1^* and k_2^* depend on tissue blood flow, blood-brain barrier transport of [^{14}C]DG, and, because of competition for the transport carrier, also on the glucose concentrations in the plasma and tissue. k_3^* is the rate constant for phosphorylation of [^{14}C]DG and must, therefore, change when glucose concentration or glucose utilization in the tissue is altered. The procedure was, therefore, specifically designed to minimize errors arising from uncertainties or inaccuracies in the values of the rate constants.

Quantitative autoradiography measures only total ^{14}C concentration in tissue and does not distinguish between [^{14}C]DG-6-P and [^{14}C]DG. It is [^{14}C]DG-6-P concentration, however, that must be known to determine glucose consumption (Fig. 1B). Therefore, in the operational equation (Fig. 2B) [^{14}C]DG-6-P concentration is calculated in the numerator by subtracting from the measured total tissue ^{14}C content, $C_i^*(T)$, the residual [^{14}C]DG content in the tissue estimated by the second term in the numerator that contains the exponential factor, $e^{-(k_2^*+k_3^*)T}$ (Fig. 1B). The denominator of the operational equation also contains a term with the same exponential factor; this term corrects for the lag in the equilibration of the tissue specific activity with that of the plasma and makes it possible to calculate the required integrated precursor specific activity in the tissue at the enzyme site from the integrated specific activities measured in the plasma (Fig. 1B). It is only in these two terms that the rate constants appear in the operational equation, and both are exponential terms with the property of approaching zero with increasing time if C_p^* is also allowed to approach zero.

One solution to the problem of uncertainties in the exact values of the rate constants is to redetermine them for each structure in each condition to be studied. An alternative is to administer the [^{14}C]DG as an intravenous pulse at zero time and then allow sufficient time for the free [^{14}C]DG to clear from the plasma and tissues and for the two exponential terms containing the rate constants to decline sufficiently to have little influence on the final calculated value for R_i. Following a pulse these terms would eventually fall to zero; the rate constants would then become negligible and disappear altogether from the equation. To prolong the experimental period until these terms reach zero is, however, not practical because loss of $[^{14}C]DG-6-P$ from the tissues, though negligible at early times, does become significant later after a pulse (Sokoloff, 1982; Nelson et al., 1986, 1987; Dienel et al.,1988). An experimental period of 30-45 minutes has been found to be optimal for normal rats (Mori et al., 1989). Because of the insensitivity to the exact values of the rate constants at this late time, it is acceptable to use the average rate constants for gray matter and white matter in TABLE 2 rather than the individual rate constants for each tissue.

In conditions, like hypoglycemia, in which the rate constants are increased, the normal values for the constants (TABLE 2) can be used with experimental periods of 30-45 minutes (Suda et al., 1990). In conditions, such as hyperglycemia or ischemia, in which the rate constants may be markedly reduced below normal, the rate constants should be redetermined for those special conditions (TABLE 4) (Orzi et al., 1988).

TABLE 4. *Values of rate constants for various levels of hyperglycemia in the rat[a,c]*

Rate Constants	Arterial Plasma Glucose Level (Mean ± S.D.)		
	20 ± 1 mM	25 ± 1 mM	31 mM
Gray Matter:			
K_1^* (ml/g/min^{-1})	0.127 ± 0.006	0.116 ± 0.005	0.085 ± 0.002
k_2^* (min^{-1})	0.235 ± 0.019	0.240 ± 0.022	0.231 ± 0.015
k_3^* (min^{-1})	0.042 ± 0.004	0.026 ± 0.003	0.027 ± 0.002
$Log_e 2/(k_2^*+k_3^*)$ (min)[b]	2.5	2.6	2.7
White Matter:			
K_1^* (ml/g/min^{-1})	0.057 ± 0.005	0.052 ± 0.004	0.041 ± 0.002
k_2^* (min^{-1})	0.168 ± 0.034	0.120 ± 0.018	0.148 ± 0.023
k_3^* (min^{-1})	0.030 ± 0.008	0.004 ± 0.006	0.013 ± 0.006
$Log_e 2/(k_2^*+k_3^*)$ (min)[b]	3.5	5.6	6.3

[a] The values represent the means ± S.E.M. of the values for each rate constant obtained in 16 gray and 2 white structures. The rate constants for each structure were obtained by non-linear least squares fitting to the data obtained from 30 animals. The values for the S.E.M.s were calculated from the standard errors of the estimates of the individual rate constants and their covariances.
[b] $Log_e 2/(k_2^*+k_3^*)$ represents half-life of free $[^{14}C]DG$ in the tissue.
[c] From Orzi et al. (1988)

Lumped Constant. The lumped constant is a combination of 6 separate constants (Fig. 1B). One, Φ, reflects the steady-state hydrolysis of glucose-6-phosphate back to free glucose and phosphate, but because there normal brain tissue contains little such phosphohydrolase activity (Hers, 1957), it is normally equal to approximately 1.0. The other components are arranged in three ratios: 1) λ, the ratio of the distribution spaces in the tissue for free DG and glucose; 2) V_m^*/V_m; and 3) K_m/K_m^*. Although each individual constant may vary from structure to structure and condition to condition, the ratios tend to remain stable under normal physiological conditions. The lumped constant also appears to be normally relatively uniform throughout the brain (Gjedde and Diemer, 1983). Although stable under physiological conditions, the lumped constant does change in some pathophysiological states, e.g., markedly in hypoglycemia (Suda et al., 1990) and moderately in hyperglycemia (Schuier et al., 1990) (Fig. 2).

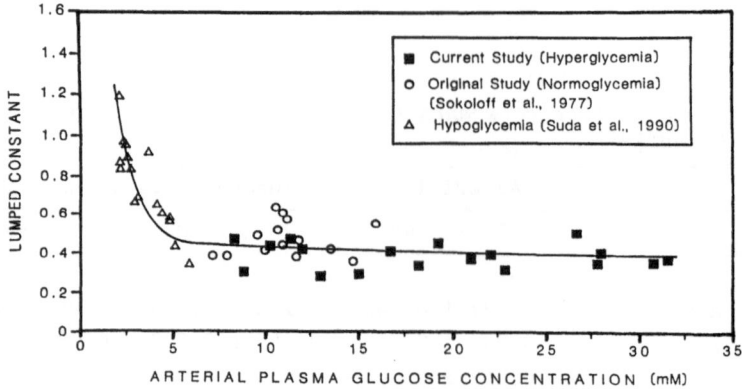

Fig. 2. Values of lumped constant over full range of arterial plasma glucose levels from hypoglycemia through normoglycemia to hyperglycemia From Sokoloff et al. (1977); Suda et al.(1990); and Schuier et al.(1990).

Experimental Protocol. The procedure was designed with consideration of the theoretical and practical issues just discussed. Animals are prepared for the experiment by insertion of polyethylene catheters into any artery and vein under light halothane anesthesia. At least 2 hours are allowed for recovery from surgery and anesthesia before initiation of the experimental procedure. At zero time a pulse of 125 μCi (contained in no more than 2.5 μmoles to adhere to tracer conditions) of [^{14}C]DG per kg of body weight is injected intravenously. Arterial sampling is initiated with the onset of the pulse, and timed samples of arterial blood are collected consecutively as rapidly as possible during the early period so as not to miss the peak of the arterial curve. Arterial sampling is continued at less frequent intervals later in the experimental period but at sufficient frequency to define fully the time course of the arterial curve. The arterial blood samples are immediately centrifuged to separate the plasma, which is stored on ice until assayed for [^{14}C]DG concentrations by liquid scintillation counting and glucose concentrations by standard enzymatic methods. At about 45 minutes preferably or 30 minutes if necessary, the animal is decapitated, and the brain is removed and frozen in isopentane maintained between -50° and -60°C. with liquid N_2. The frozen brain is stored at -70°C until sectioned and autoradiographed.

The ^{14}C concentrations in localized regions of the brain are measured by a quantitative autoradiographic technique (Reivich et al., 1969). The frozen brain is coated with chilled embedding medium, fixed to object-holders appropriate to the microtome to be used, and cut into sections, precisely 20 μm in thickness, in a cryostat maintained at -20°C to -22°C. The brain sections are thaw-mounted on glass cover-slips, dried on a hot plate at 60°C for at least 5 minutes, and placed in an X-ray cassette together with a set of [^{14}C]methylmethacrylate standards, which include a blank and a series of progressively increasing ^{14}C concentrations. These standards must previously have been calibrated for their autoradiographic equivalence to the ^{14}C concentrations in brain sections, 20 μm in thickness, prepared as described above. The method of calibration has been previously described (Reivich et al., 1969).

Autoradiograms are prepared from these sections directly in the X-ray cassette with medical X-ray film (e.g., Kodak Type SB-5 or OM-1). The exposure time varies with the film used, but is usually 5 days to 3 weeks with the doses described above. The autoradiograms provide a pictorial representation of the relative ^{14}C concentrations in the various cerebral structures and the plastic standards (Fig. 3). A calibration curve of the relationship between optical density and tissue ^{14}C concentration for each film is obtained by densitometric measurements of the portions of the film representing the various standards. The local tissue concentrations are then determined from the calibration curve and the optical densities of the film in the regions representing the cerebral structures of interest.

Local cerebral glucose utilization is calculated from the local tissue concentrations of ^{14}C and the plasma [^{14}C]DG and glucose concentrations according to the operational equation (Fig. 1B). Image-processing systems can be used to carry out the quantitative densitometry automatically and to convert the autoradiograms into quantitative color-coded maps of the distribution of local rates of glucose utilization in the brain (Goochee et al., 1980); these are convenient but not essential as quantitative densitometry can be readily carried out with manual densitometers.

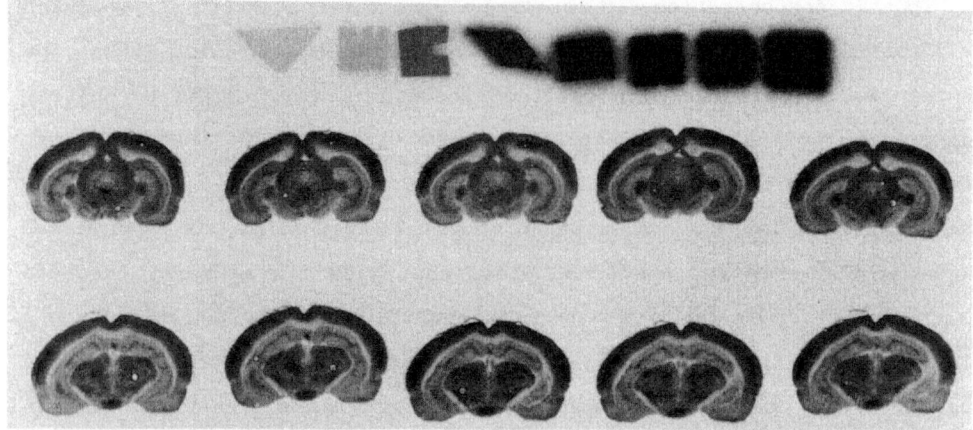

Fig.3. [^{14}C]Deoxyglucose autoradiograms of coronal sections of conscious rat brain and of [^{14}C]methylmethacrylate standards used to quantify ^{14}C concentrations in tissues by quantitative densitometry.

NORMAL RATES OF LOCAL CEREBRAL GLUCOSE UTILIZATION

Local rates of cerebral glucose utilization in normal conscious adult rats vary widely throughout the brain (Sokoloff et al., 1977) (TABLE 5). Values in white matter are fairly uniform and considerably below those of gray structures. Values in gray matter are more diverse and approximately 2 to 5 times those in white matter. Individual values in gray matter vary from about 50 to 200 μmoles/100 g/min. The highest values are in structures involved in auditory functions with the inferior colliculus clearly the most metabolically active structure in the brain. Barbiturate anesthesia markedly reduces rates of glucose utilization in all structures of the brain, e.g., by about 15-30% in white matter and 25-50% in gray matter, particularly in components of sensory pathways (Sokoloff et al., 1977).

Rates of local cerebral glucose utilization in the conscious monkey show similar heterogeneity but are about one-third to one-half of those in corresponding structures of the rat brain (Kennedy et al., 1978) (TABLE 5). The differences in rates in the rat and monkey are consistent with the different cellular packing densities in the brains of these species.

In both the rat and monkey, the average rates of glucose utilization in the brain as a whole are in good agreement with those obtained by measurements of average cerebral energy metabolism in these species with the Kety-Schmidt method (TABLE 5).

METABOLIC MAPPING OF LOCAL FUNCTIONAL ACTIVITY

Numerous studies with the DG method involving experimentally induced alterations of local functional activity in a wide variety of neural systems have demonstrated that in the nervous system, as in other tissues, there is a close relationship between local functional activity and energy metabolism in the nervous system (Kennedy et al., 1975; Sokoloff, 1977, 1981, 1982). Increased functional activity is associated with an increased rate of glucose utilization; decreased functional activity is accompanied by decreased glucose utilization. The effects are often so pronounced that they can be visualized directly in the autoradiograms, which are really pictorial representations of the relative rates of local glucose utilization throughout the brain. The effects are even more easily imaged with computerized image-processing and quantitative color-coding of the autoradiograms (Goochee et al., 1980). This technique of autoradiographic and tomographic imaging of altered local metabolism offers a powerful tool with which to map normal and abnormal functional neural pathways in the nervous system. An early demonstration of its potential was its use to image the full extent and distribution of the ocular dominance columns in the striate cortex of primates with binocular visual systems (Kennedy et al., 1976) (Fig. 4).

TABLE 5. *Representative values for local cerebral glucose utilization in the normal, conscious albino rat and monkey*

Structure	Albino Rat (10)[a]	Monkey (7)[b]
Gray Matter		
Visual Cortex	107 ± 6	59 ± 3
Auditory Cortex	162 ± 5	79 ± 4
Parietal Cortex	112 ± 5	48 ± 4
Sensory-Motor Cortex	120 ± 5	44 ± 3
Thalamus: Lateral Nucleus	116 ± 5	54 ± 2
Thalamus: Ventral Nucleus	109 ± 5	43 ± 2
Medial Geniculate Body	131 ± 5	65 ± 3
Lateral Geniculate Body	96 ± 5	39 ± 1
Hypothalamus	54 ± 2	25 ± 1
Mammillary Body	121 ± 5	57 ± 3
Hippocampus	79 ± 3	39 ± 2
Amygdala	52 ± 2	25 ± 2
Caudate-Putamen	110 ± 4	52 ± 3
Nucleus Accumbens	82 ± 3	36 ± 2
Globus Pallidus	58 ± 2	26 ± 2
Substantia Nigra	58 ± 3	29 ± 2
Vestibular Nucleus	128 ± 5	66 ± 3
Cochlear Nucleus	113 ± 7	51 ± 3
Superior Olivary Nucleus	133 ± 7	63 ± 4
Inferior Colliculus	197 ± 10	103 ± 6
Superior Colliculus	95 ± 5	55 ± 4
Pontine Gray Matter	62 ± 3	28 ± 1
Cerebellar Cortex	57 ± 2	31 ± 2
Cerebellar Nucleus	100 ± 4	45 ± 2
White Matter		
Corpus Callosum	40 ± 2	11 ± 1
Internal Capsule	33 ± 2	13 ± 1
Cerebellar White Matter	37 ± 2	12 ± 1
Weighted Average for Whole Brain		
	68 ± 3	36 ± 1

Note: The values are Means ± SEM. (in μmoles/100 g/min) of values determined in number of animals indicated in parentheses.
[a] From Sokoloff et al. (1977); [b] From Kennedy et al. (1978)

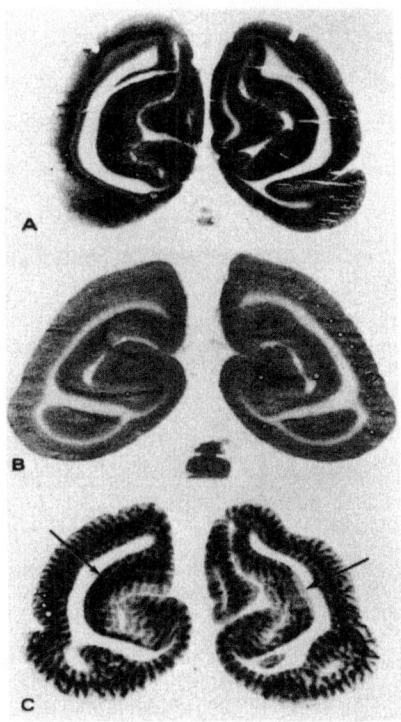

Fig. 4. Autoradiograms of coronal brain sections from Rhesus monkeys at level of striate cortex. **A:** Animal with normal binocular vision. Note laminar distribution of the density; the dark band corresponds to Layer IV. **B:** Animal with bilateral visual occlusion. Note the almost uniform and reduced relative density, especially the virtual disappearance of the dark band corresponding to Layer IV. **C:** Animal with right eye occluded. The half-brain on the left side of the photograph represents the left hemisphere contralateral to the occluded eye. Note alternate dark and light striations, each approximately 0.3-0.4 mm in width that represent the ocular dominance columns. These columns are most apparent in the dark band (Layer IV) but extend through the entire thickness of the cortex. The arrows point to regions of bilateral asymmetry where the ocular dominance columns are absent. These are presumably areas with normally only monocular input. The one on the left, contralateral to occluded eye, has a continuous dark lamina corresponding to Layer IV which is completely absent on the side ipsilateral to the occluded eye. These regions are believed to be the loci of the cortical representations of the blind spots of the visual fields. From Kennedy et al. (1976).

EFFECTS OF INSULIN

Effects on transport of hexoses between blood and brain. Insulin is known to enhance hexose transport from blood into a variety of insulin-sensitive tissues, and the insulin-sensitivity of brain has long been a matter of controversy. Therefore, before the 2-DG method could be applied to studies of the effects of insulin on local cerebral glucose utilization, it was necessary to determine its effects on the blood-brain transport of glucose and 2-DG, specifically the rate constants for their bi-directional transport, K^1 and k_2. 3-O-Methyl-D-glucose is a non-metabolizable analogue of glucose that shares the same blood-brain carrier for transport between blood and brain with glucose and 2-DG and is, therefore, useful for examining transport phenomena uncontaminated by metabolic factors. Namba et al. (1987) have determined the effects of insulin under normoglycemic conditions on the rate constants for inward and outward transport of [^{14}C]methylglucose, K_1 and k_2, respectively (TABLE 6). They found that, contrary to expectations, insulin decreased both rate constants but k_2 more than K_1 so that the distribution space was increased by about 10-11%. Insulin, of course, is known to increase the distribution spaces for hexoses in a number of tissues. The effects in brain were, however, so small that it is doubtful that effects of insulin on hexose transport from blood to the tissue is of any significant influence on the metabolism of glucose in the brain.

TABLE 6. *Values of rate constants for 3-O-methylglucose transport across blood-brain barrier in the rat.*[a,c]

Rate Constants	Control[b]	Insulin[b]	% Change
Average Gray Matter			
K_1 ($ml \cdot g^{-1} \cdot min^{-1}$)	0.222 ± 0.066	0.165 ± 0.039	-24
k_2 (min^{-1})	0.462 ± 0.159	0.307 ± 0.084	-31
Distribution Ratio			
(K_1/k_2) (ml/g)	0.49 ± 0.03	0.54 ± 0.03	+11
Average White Matter			
K_1 ($ml \cdot g^{-1} \cdot min^{-1}$)	0.070 ± 0.009	0.065 ± 0.012	- 7
k_2 (min^{-1})	0.155 ± 0.009	0.131 ± 0.019	-16
Distribution Ratio			
(K_1/k_2) (ml/g)	0.45 ± 0.04	0.50 ± 0.02	+10

[a] The values are means ± SD of rate constants determined in 13 gray and 2 white structures of 12 rats in each group.
[b] Arterial plasma insulin levels (Means ± SEM of 12 rats): Controls, 14 ± 2 $\mu U/ml$; insulin-treated, 229 ± 24 $\mu U/ml$.
[c] From Namba et al. (1987).

Effects of insulin on cerebral glucose utilization under normoglycemic conditions. Lucignani et al. (1987) applied the $[^{14}C]DG$ method to rats under normoglycemic conditions maintained by the glucose clamp technique DeFronzo et al. (1979). They found that even with arterial plasma insulin concentrations 15 times the normal level, insulin had no significant effect on average glucose utilization in the brain as a whole (TABLE 7). At the local level there were no significant effects of insulin, except for a small stimulation of glucose utilization in the anterior hypothalamus (+19%) and a very small decrease in the lateral thalamic nucleus (−11%).

Effects of insulin hypoglycemia. Suda et al. (1990) used the $[^{14}C]DG$ method to study the effects of moderate insulin-induced hypoglycemia on local cerebral glucose utilization in the rat. They determined and used the appropriate lumped constants for the degree of hypoglycemia (Fig. 2) in their studies. The arterial plasma glucose level was reduced by insulin administration from a control level of 7.1 ± 1.2 (Mean ± SEM) to 2.4 ± 0.3, and at this relatively moderate level of hypoglycemia average glucose utilization in the brain as a whole was reduced by about 14% (TABLE 7). Reductions were observed in almost all structures, gray and white, but the effects were statistically significant only in several structures in the brain stem that normally have high rates of glucose utilization. There were no regions in which glucose utilization was stimulated. It would appear from these results that the cerebral structures with normally the greatest demand for glucose are those that are first affected by reduced glucose supply, and no specific effects attributable to the insulin *per se* were observed.

EFFECTS OF HYPERGLYCEMIA

The rate constants (Orzi et al., 1988) and the lumped constants (Schuier et al., 1990) have been determined for a wide range of levels of hyperglycemia in the rat (TABLE 4, Fig. 2), and these have made possible the examination of the effects of hyperglycemia on local cerebral glucose utilization (Orzi et al., 1988). There were no statistically significant effects of hyperglycemia on average glucose utilization in the brain as whole (TABLE 7), but at arterial plasma glucose levels of about 20 mM and above significant increases in glucose utilization were observed in the anterior hypothalamic nucleus (approximately +33%), ventromedial hypothalamic nucleus (approximately 50%), central amygdala (+24% to +46%), and the globus pallidus (about +22% to +44%). The significance of these local changes is unclear.

TABLE 7. *Effects of insulin in normoglycemia, insulin-induced hypoglycemia, and hyperglycemia on local cerebral glucose utilization in the rat[a]*

Condition	Arterial Plasma Insulin Level (μU/ml)	Arterial Plasma Glucose Concentration (mM)	Average Glucose Utilization in Brain as a Whole (μmols/100 g/min)
Hyperinsulinemia & Normoglycemia[b]:			
Controls (6)	19 ± 4	6.9 ± 0.4	69 ± 2
Hyperinsulinemia (4)	287 ± 42**	7.5 ± 0.6	71 ± 4
Moderate Hypoglycemia[c]:			
Controls (6)	—	7.1 ± 1.2	71 ± 5
Hypoglycemia (9)	—	2.4 ± 0.3***	61 ± 9*
Hyperglycemia[d]:			
Controls (4)	16 ± 2	7.4 ± 1.2	69 ± 2
Mild hyperglycemia (4)	105 ± 14***	19.7 ± 0.6***	74 ± 6
Medium hyperglycemia (4)	131 ± 11***	25.3 ± 1.9***	70 ± 6
Severe Hyperglycemia (4)	132 ± 13***	31.1 ± 0.9***	71 ± 5

[a] Values are means ± SEM of number of animals indicated in parentheses.
[b] From Lucignani et al. (1987).
[c] From Suda et al. (1990).
[d] From Orzi et al. (1988).
Statistically significant differences from control values:
  * $p < 0.05$; ** $p < 0.01$; *** $p < 0.001$.

THE [18F]FLUORODEOXYGLUCOSE TECHNIQUE

Inasmuch as the [14C]DG method requires the measurement of local concentrations of radioactivity in brain, it could not be used in man as originally designed because autoradiography could not be applied to man and the ß-radiation of ^{14}C could not be detected outside the body. Computerized emission tomography, however, made it possible to measure local tissue concentrations of γ-emitting isotopes in vivo in man. A γ-emitting analogue of deoxyglucose, 2-[18F]fluoro-2-deoxy-D-glucose, was, therefore, synthesized and found to retain the necessary biochemical properties of 2-DG. It was first used to measure local cerebral glucose utilization in man with a single photon scanner (Reivich et al., 1979). Because ^{18}F is a positron-emitter, the method was adapted further for use with positron emission tomography (PET), which provides better spatial resolution and more accurate estimates of local concentrations of isotope in the tissues (Phelps et al., 1979). Recently, 2-$[^{11}C]$DG, a positron-emitting version of 2-DG, has been synthesized, and its lumped and rate constants for the human brain have been determined (Reivich et al., 1985). The 20-minute physical half-life of $[^{11}C]$DG offers considerable advantages because it allows repeated studies in the same subject within a relatively short time-span.

REFERENCES

DeFronzo, R.A., Tobin, J.D., Andres, R. (1979) Glucose clamp technique: a method for quantifying insulin secretion and resistance. *Am. J. Physiol.* **237**: E214-E223.

Dienel, G.A., Nelson, T., Cruz, N.F., Jay, T., Crane, A.M., and Sokoloff, L. (1988) Over-estimation of glucose-6-phosphatase activity in brain in vivo: Apparent difference in rates of $[2-^{3}H]$glucose and $[U-^{14}C]$glucose utilization is due to contamination of precursor pool with ^{14}C-labeled products and incomplete recovery of ^{14}C-labeled metabolites. *J. Biol. Chem.* **263**: 19697-19708.

Fishman, R.S.and Karnovsky, M.L. (1986) Apparent absence of a translocase in the cerebral glucose-6-phosphatase system. *J. Neurochem.* **46**: 371-378.

Gjedde, A. and Diemer, N.H. (1983) Autoradiographic determination of regional brain glucose content. *J. Cereb. Blood Flow Metab.* **4**: 303- 310.

Goochee, C., Rasband, W.,and Sokoloff, L. (1980) Computerized densitometry and color-coding of $[^{14}C]$deoxyglucose autoradiographs. *Ann. Neurol.* **7**: 359-370.

Gottstein, U.U., Held, K., Müller, W., and Berghoff, W. (1972) Utilization of ketone bodies by the human brain. In: *Research on the Cerebral Circulation. Fifth International Conference, 1970,* J.S. Meyer, M. Reivich, H. Lechner, and O. Eichhorn, eds. Charles C. Thomas, Springfield, IL, pp. 137-145.

Hers, H.G. (1957) *Le Metabolisme du Fructose.* Editions Arscia, Bruxelles, p. 102.

Kennedy, C., Des Rosiers, M.H., Reivich, M., Sharp, F., Jehle, J.W., and Sokoloff, L. (1975) *Science,* **187**: 850-853.

Kennedy, C., Des Rosiers, M.H., Sakurada, O., Shinohara, M., Reivich, M., Jehle, J.W., and Sokoloff, L (1976) Metabolic mapping of the primary visual system of the monkey by means of the autoradiographic $[^{14}C]$deoxyglucose technique. *Proc. Natl. Acad. Sci., USA,* **73**: 4230-4234.

Kennedy, C., Sakurada, O., Shinohara, M., Jehle, J., and Sokoloff, L. (1978) Local cerebral glucose utilization in the normal conscious Macaque monkey. *Ann. Neurol.* 4: 293-301.

Kety, S.S. (1950) Circulation and metabolism of the human brain in health and disease. *Am. J. Med.* 8: 205-217.

Kety. S.S., Polis, B.D., Nadler, C.S., and Schmidt, C.F. (1948) Blood flow and oxygen consumption of the human brain in diabetic acidosis. *J. Clin. Invest.* 27: 500-510.

Kety, S.S. and Schmidt, C.F. (1948) The nitrous oxide method for the quantitative determination of cerebral blood flow in man: theory, procedure, and normal values. *J. Clin. Invest.* 27: 476-483.

Kety, S.S., Woodford, R.B., Harmel, M.H., Freyhan, F.A., Appel, K.E., and Schmidt, C.F. (1948) Cerebral blood flow and metabolism in schizophrenia. The effects of barbiturate semi-narcosis, insulin coma, and electroshock. *AM. J. Psychiat.* 104: 765-770.

Krebs, H.A., Williamson, D.H., Bates, M.W., Page, M.A., and Hawkins, R.A. (1971) The role of ketone bodies in caloric homeostasis. *Adv. Enzyme Regul.* 9: 387-409.

Lucignani, G., Namba, H., Nehlig, A., Porrino, L. J., Kennedy, C., and Sokoloff, L. (1987) Effects of insulin on local cerebral glucose utilization in the rat. *J. Cereb. Blood Flow Metab.* 7: 309-314.

Mori, K., Schmidt, K., Jay, T., Palombo, E., Nelson, T., Lucignani, G., Pettigrew, K., Kennedy, C., and Sokoloff, L. (1990) Optimal duration of experimental period in measurement of local cerebral glucose utilization with the deoxyglucose method. *J. Neurochem.* 54: 307-319.

Namba, H., Lucignani, G., Nehlig, A., Patlak, C., Pettigrew, K., Kennedy, C., and Sokoloff, L. (1987) Effects of insulin on hexose transport across blood-brain barrier in normoglycemia. *Am. J. Physiol.* 252 (Endocrinol. Metab. 15): E299-303.

Nelson T, Kaufman, E.E., and Sokoloff, L. (1984) 2-Deoxyglucose incorporation into rat brain glycogen during measurement of local cerebral glucose utilization by the 2-deoxyglucose method. *J. Neurochem.* 43: 949-956.

Nelson, T., Lucignani, G., Goochee, J., Crane, A. M., and Sokoloff, L. (1986) Invalidity of criticisms of the deoxyglucose method based on alleged glucose-6-phosphatase activity in brain. *J. Neurochem.* 46: 905-919.

Nelson, T., Dienel, G.A., Mori, K., Cruz, N.F., and Sokoloff, L. (1987) Deoxyglucose-6-phosphate stability in vivo and the deoxyglucose method: Response to comments of Hawkins and Miller. *J. Neurochem.* 49: 1949-1960.

Orzi, F., Lucignani, G., Dow-Edwards, D., Namba, H., Nehlig, A., Patlak, C. S., Pettigrew, K., Schuier, F., and Sokoloff, L. (1988) Local cerebral glucose utilization in controlled graded levels of hyperglycemia in the conscious rat. *J. Cereb. Blood Flow Metab.* 8: 346-356.

Owen, O.E., Morgan, A.P., Kemp, H.G., Sullivan, J.M., Herrera, M.G., and Cahill, G.F. (1967) Brain metabolism during fasting. *J. Clin. Invest.* **46**: 1589-1595.

Phelps, M. E., Huang, S. C., Hoffman, E. J., Selin, C., Sokoloff, L., and Kuhl, D. E. (1979) Tomographic measurement of local cerebral glucose metabolic rate in humans with (F-18)2-fluoro-2-deoxy-D-glucose: validation of method. *Ann. Neurol.* **6**: 371-388.

Reivich, M., Jehle, J., Sokoloff, L., Kety, S.S. (1969) Measurement of regional cerebral blood flow with antipyrine-^{14}C in awake cats. *J. Appl. Physiol.* **27**: 296-300.

Reivich, M., Kuhl, D., Wolf, A., Greenberg, J., Phelps, M., Ido, T., Cassella, V., Fowler, J., Hoffman, E., Alavi, A., Som, P., and Sokoloff, L. (1979) The [^{18}F]fluoro-deoxyglucose method for the measurement of local cerebral glucose utilization in man. *Circulation Res.* **44**: 127-137.

Reivich, M., Alavi, A., Wolf, A., Fowler, J., Russell, J., Arnett, C., MacGregor, R.R., Shiue, C.Y., Atkins, H., Anand, A., Dann, R., and Greenberg, J.H. (1985) Glucose metabolic rate kinetic model parameter determination in humans: the lumped constants and rate constants for [^{18}F]fluorodeoxyglucose and [^{11}C]deoxyglucose. *J. Cereb. Blood Flow Metab.* **5**: 179-192.

Schuier, F., Orzi, F., Suda, S., Lucignani, G., Kennedy, C., and Sokoloff, L. (1990) Influence of plasma glucose concentration on lumped constant of the deoxyglucose method: Effects of hyperglycemia in the rat. *J. Cereb. Blood Flow Metab.* In press.

Sokoloff, L. (1960) Metabolism of the central nervous system in vivo. In: *Handbook of Physiology - Neurophysiology, Vol. III*, J. Field, H.W. Magoun, and V.E. Hall, eds. American Physiological Society, Washington, D. C., pp. 1843-1864.

Sokoloff, L. (1969) Cerebral circulation and behavior in man: strategy and findings. In: *Psychochemical Research in Man*, A.J.Mandell and M.P. Mandell, eds. Academic Press, New York City, pp. 237-252.

Sokoloff, L. (1981) Localization of functional activity in the central nervous system by measurement of glucose utilization with radioactive deoxyglucose. *J. Cereb. Blood Flow Metab.* **1**: 7-36.

Sokoloff, L. (1982) The radioactive deoxyglucose method: theory, procedure, and applications for the measurement of local glucose utilization in the central nervous system. In: *Advances in Neurochemistry, Vol. 4*, B.W. Agranoff and M.H. Aprison, eds. Plenum Publishing Corp., New York, pp. 1-82.

Sokoloff, L. (1985) Basic principles in imaging of regional cerebral metabolic rates. In: *Brain Imaging and Brain Function*, Research Publication of Association for Research in Nervous and Mental Disease, Vol. 63, L. Sokoloff, ed. Raven Press, New York, pp. 21-49.

Sokoloff, L. (1989) Circulation and energy metabolism of the brain. In: *Basic Neurochemistry, Fourth Edition*, G. Siegel, B. Agranoff, R.W. Albers, and P. Molinoff, eds. Raven Press, New York, pp. 565-590.

Sokoloff, L., Mangold, R., Wechsler, R. L., Kennedy, C., and Kety, S.S. (1955) The effect of mental arithmetic on cerebral circulation and metabolism. *J. Clin. Invest*. **34**: 1101-1108.

Sokoloff, L., Reivich, M., Kennedy, C., Des Rosiers, M.H., Patlak, C.S., Pettigrew,K.D., Sakurada, O., and Shinohara, M. (1977) The [^{14}C]deoxyglucose method for the measurement of local cerebral glucose utilization: theory, procedure, and normal values in the conscious and anesthetized albino rat. *J. Neurochem*. **28**: 897-916.

Sols, A. and Crane, R.K. (1954) Substrate specificity of brain hexokinase. *J. Biol. Chem*. **210**: 581-595.

Suda, S., Shinohara, M., Miyaoka, M., Lucignani, G., Kennedy, C., and Sokoloff, L. (1990) The lumped constant of the deoxyglucose method in hypoglycemia: Effects of moderate hypoglycemia on local cerebral glucose utilization in the rat. *J. Cereb. Blood Flow Metab*. In press.

BLOOD-BRAIN BARRIER TRANSPORT OF GLUCOSE, FREE FATTY ACIDS, AND KETONE BODIES

William M. Pardridge

Department of Medicine and Brain Research Institute
UCLA School of Medicine
Los Angeles, California 90024

INTRODUCTION

Fuel homeostasis is important to the regulation of central nervous system (CNS) function because cerebral pathways of metabolism of glucose, ketone bodies, and amino acids are dependent upon plasma substrate availability. Therefore, brain function and brain metabolism are under nutritional regulation and the interface between diet or fuel homeostasis and brain function is the limiting transport barrier between blood and brain, which is the brain capillary endothelial wall, i.e., the blood-brain barrier (BBB) (Figure 1).

Studies with positron emission tomography (PET) have demonstrated that human brain function, both motor and cognitive, is controlled by cerebral metabolic utilization of glucose,[1] which constitutes essentially the only carbon fuel that the brain combusts under normal conditions.[2] However, under conditions of hyperketonemia, such as fasting, a high fat diet, the neonatal period, or diabetic ketoacidosis, a substantial portion of brain oxygen consumption is derived from ketone body combustion.[3] Depending on the physiologic condition, either brain glucose or ketone body utilization may be rate limited by an enzyme that is not saturated in vivo by precursor substrates (i.e., glucose or ketone bodies). Therefore under these conditions, brain metabolism is a function of precursor nutrient availability.[3] The latter is determined by (a) BBB nutrient transport and (b) plasma nutrient concentrations. Plasma nutrient concentrations are in turn regulated by diet and interorgan metabolism (Figure 1).

Ketone body oxidation in brain is normally rate limited by BBB transport.[5,6] Under normal conditions, cerebral glucose utilization is rate limited by hexokinase phosphorylation, as intracellular glucose is readily measurable in brain in vivo.[7] However, when rates of BBB glucose transport fall below ambient rates of glucose utilization, as in hypoglycemia, then cerebral glucose utilization is rate limited by BBB transport of glucose.[8] Conversely, plasma glucose may be normal or even elevated, but if brain glucose utilization rates are increased above the ambient rate of glucose influx into brain from blood, such as might occur in seizures or salicylate intoxication, then brain glucose utilization may be rate limited by BBB transport of glucose.[3]

Since brain glucose or ketone body utilization is, under many conditions, rate limited by BBB transport of the precursor nutrient, then it is important to understand the mechanisms regulating nutrient flux across the BBB membrane. Indeed, since this is either a rate limiting or a rate affecting step in the cerebral metabolic pathways for a number of nutrients, then it is to be expected that modulations of nutrient transport into brain will occur at the BBB membrane. The lumenal and antilumenal sides of the brain capillary endothelium are embedded with specific carrier systems that transport individual classes of nutrients.

Fuel Homeostasis and the Nervous System, Edited by M. Vranic *et al.*
Plenum Press, New York, 1991

These individual nutrient transport systems and a representative substrate are shown in Table 1. The BBB transport of neutral or basic amino acids, choline, purine bases, or nucleosides, has been shown in other studies to be important rate affecting steps in the overall cerebral utilization of these circulating nutrients, as has been reviewed elsewhere.[3,9] This review will discuss recent information on two of these transporters: the glucose carrier and the monocarboxylic acid carrier, which transports ketone bodies.

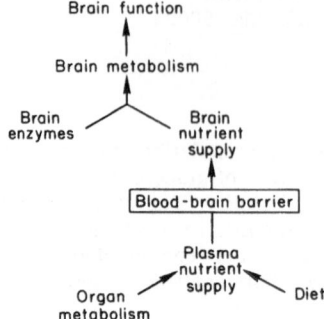

DIET, BLOOD-BRAIN BARRIER,
BRAIN METABOLISM

Figure 1. Scheme relating behavior and diet views the blood-brain barrier as both the interface between blood and brain and between nutritional inputs and brain output, e.g., behavior.
From [4] with permission.

BLOOD-BRAIN BARRIER GLUCOSE TRANSPORT

The transport of glucose from blood into brain involves the carrier-mediated movement of this molecule across two membranes in series, i.e., the endothelial or BBB membrane, and the brain cell (neuronal or glial) membrane (Figure 2). Since the surface area of the neuronal and glial membranes is log orders greater than the surface area of the endothelial membrane, nutrient transport across the BBB is the rate limiting step in the overall movement of nutrient flux between blood and brain intracellular spaces. The BBB glucose transporter belongs to a supergene family of sodium independent glucose transporters (Table 2). The GLUT-1[13] or brain-type isoform was cloned from a brain cDNA library prepared by polysome precipitation using an antiserum specific to the human erythrocyte glucose transporter.[15] Although it is generally recognized that the GLUT-1 isoform is enriched in microvascular endothelium,[16] it is also generally assumed that the GLUT-1 isoform mediates glucose transport into neurons and glial cells.[15,16] However, recent studies from this laboratory using Northern blots of RNA obtained from brain following capillary depletion of bovine brain homogenate have shown that

Table 1. Blood-Brain Barrier Nutrient and Thyroid Hormone Carriers[a]

Carrier	Representative Substrate	K_m (μM)	V_{max} (nmol min^{-1}g^{-1})
Hexose	Glucose	11,000 $\pm$ 1,400	1,420 $\pm$ 140
Monocarboxylic acid	Lactic acid	1,800 $\pm$ 600	91 $\pm$ 35
Neutral amino acid	Phenylalanine	26 $\pm$ 6	22 $\pm$ 4
Amine	Choline	340 $\pm$ 70	11 $\pm$ 1
Basic amino acid	Arginine	40 $\pm$ 24	5 $\pm$ 3
Nucleoside	Adenosine	25 $\pm$ 3	0.75 $\pm$ 0.08
Purine base	Adenine	11 $\pm$ 3	0.50 $\pm$ 0.09
Thyroid hormone	T_3[b]	1.7 $\pm$ 0.7	0.19 $\pm$ 0.08

[a]From[10].
[b]T_3 = triiodothyronine.

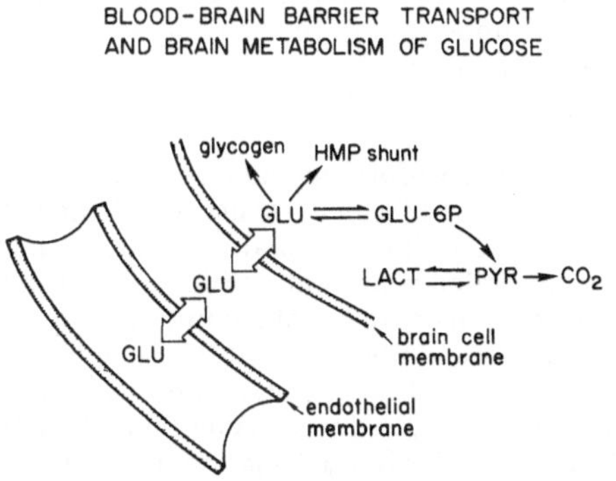

BLOOD-BRAIN BARRIER TRANSPORT
AND BRAIN METABOLISM OF GLUCOSE

Figure 2. Scheme for brain transport and metabolism of circulating glucose (GLU). Glucose carriers (arrows) are situated at both the endothelial membrane and the brain cell membrane, although the glucose transporters at these two discrete membranes are two different glucose transporter proteins.[17] Once in brain intracellular space, glucose may be converted to glycogen, to CO_2 via the hexose monophosphate (HMP) shunt, or to glucose-6-phosphate (GLU-6P), which is subsequently converted to pyruvate (PYR) or lactate acid (LACT).

Table 2. Glucose Transporter (GLUT) Isoforms

Class	Primary Tissue Origin	Human Chromosome	Homology to GLUT-1
GLUT-1	Brain/red cell	1p35	100%
GLUT-2	Liver/β-cell	3q26	68%
GLUT-3	Fetal muscle	12p13	74%
GLUT-4	Muscle/fat	17p13	76%
GLUT-5	Small intestines	1p31	58%

From[11-14].

no detectable GLUT-1 transcript is found in neuronal/glial RNA preparations (Figure 3).[17] These studies led to the hypothesis that the GLUT-1 isoform is selectively expressed at the BBB in brain and, as yet unidentified, glucose transporter isoforms function on neuronal and glial membranes. This hypothesis has recently been confirmed using both in situ hybridization and quantitative Western blotting studies.[18] With the latter methodology, the availability of purified human erythrocyte glucose transporter as an assay standard allowed for quantitation of the concentration of GLUT-1 isoform in bovine brain capillary plasma membranes. These studies showed that 10.8 ± 0.9 pmol/mg protein of GLUT-1 isoform was present in bovine BBB membranes, and this value was not statistically different from the number of D-glucose displaceable cytochalasin B binding sites in the same membrane preparation, i.e., 11.7 ± 3.5 pmol/mg protein. Therefore, essentially 100% of glucose transport across the BBB is mediated by the GLUT-1 isoform.[18] The finding that all of glucose transport across the BBB occurs via the GLUT-1 isoform is important to future studies investigating the molecular regulation of BBB glucose transport. This is because the various antisera or cDNA reagents used in molecular studies of glucose transporters are isoform-specific.[11-14]

Modulations of the BBB GLUT-1 isoform would be expected to have profound effects on brain glucose utilization. One paradigm of down-regulation of the BBB glucose transporter is chronic hyperglycemia and diabetes mellitus. Diabetic patients with longstanding poor glycemic control who are brought under rapid reinstitution of insulin therapy and improvement in glycemic control often complain of symptoms of hypoglycemia, despite relatively normal concentrations of blood sugar.[19,20] It has been hypothesized that these symptoms of hypoglycemia, despite blood sugar concentrations in the 100-150 mg% range, are bona fide and represent glucopenia owing to down-regulation of the BBB glucose transporter. That is, the concentration of glucose in brain is a function, not only of the concentration of sugar in blood, but also of the BBB glucose transporter activity. Several years ago two groups provided physiologic evidence that rats with streptozotocin-induced experimental diabetes mellitus demonstrated a decrease in the activity of the BBB glucose transporter.[21,22] This paradigm has recently been challenged by two other groups who find no decrease in the rate of glucose influx across the BBB in rats with experimental diabetes as compared to rats with acute hyperglycemia.[23,24] However, these two most recent measurements of BBB permeability to glucose were made at plasma glucose concentrations of over 500 mg%, or about 30 mM. Since this concentration is several-fold greater than the Km of the BBB glucose transporter (Table 1), these recent measurements of transporter activity in experimental diabetes were performed at maximal saturation of the transporter. Under these conditions, it is difficult to assess differences in BBB permeability to glucose since influx rates are primarily a function of the nonsaturable component of sugar transport.

In order to define the changes in BBB glucose transport in experimental diabetes, recent studies have been performed in this laboratory aimed at measuring BBB glucose transporter activity in vivo in rats with experimental diabetes under physiologic concentrations of glucose, e.g., 10 mM, using the in situ internal carotid artery perfusion technique.[25] This technique is advantageous in that either cerebral blood flow or cerebral blood volume may be measured

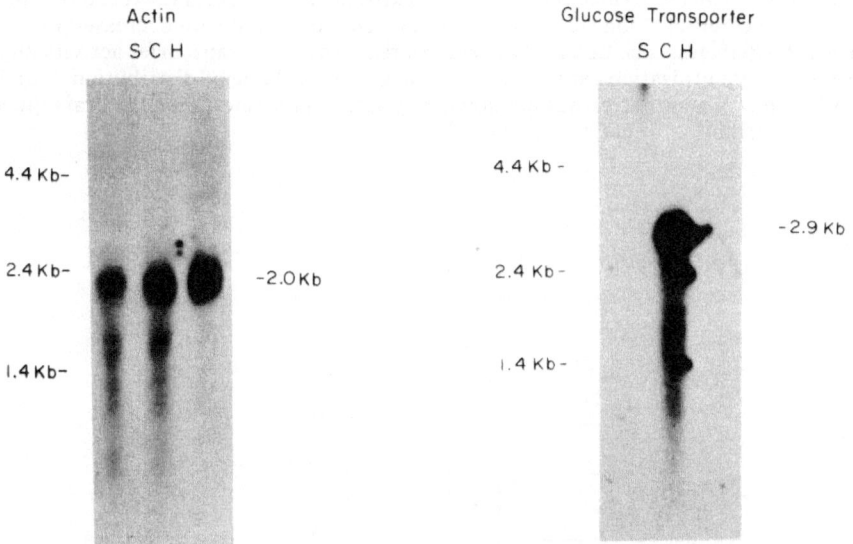

Figure 3. Northern blots of RNA obtained from bovine brain homogenate (H), bovine brain capillaries (C), or capillary-depleted bovine brain supernatant (S). Following RNA blotting, the filters were probed with [^{32}P]-labeled full length cDNA to the rat brain glucose transporter, or probed with a cDNA to mouse actin. Although the glucose transporter cDNA readily hybridizes to a 2.9 Kb transcript in the capillary fraction, there is no detectable glucose transporter transcript in the supernatant fraction containing neuronal or glial RNA that is more than 98% depleted of capillary derived RNA. Therefore, the faint detection of the 2.9 Kb glucose transporter transcript in the homogenate fraction represents dilution of microvascular derived glucose transporter transcript. The failure to find glucose transporter mRNA in the neuronal/glial RNA preparations in the supernatant fraction is not due to degradation of the RNA, since this RNA preparation readily hybridizes with the actin cDNA, revealing the 2.0 cytoplasmic actin transcript present in this fraction. From [17].

simultaneously with rates of glucose clearance. This is of importance since previous studies have reported a decrease in cerebral blood flow associated with experimental diabetes.[26] As shown in Figure 4, the serum glucose concentration was increased more than four-fold in the diabetic animals as compared to control animals. Cerebral blood flow is measured in the control and diabetic animals by monitoring the brain clearance of [^{3}H]-diazepam, a fluid microsphere. Blood flow was found to be decreased 44% in the rats with experimental diabetes, whereas cerebral blood volume, measured by determining the brain clearance of [^{3}H]-sucrose, was unchanged in rats with experimental diabetes, e.g., 7.8 ± 0.8 $\mu L/g$, as compared to the brain volume, 8.0 ± 0.2 $\mu L/g$, in control rats. Therefore, the decrease in brain blood flow in diabetes was caused by an increase in the brain capillary transit time, from 0.20 ± 0.01 to 0.35 ± 0.06 seconds (Figure 4). In parallel to the decrease in brain blood flow, there was a 44% decrease in BBB permeability⋅surface area (PS) product for glucose in experimental diabetes (Figure 4). Thus, these studies show that the BBB glucose transport activity is, in fact, down-regulated in experimental diabetes, and confirm the initial results of

Gjedde and Crone[21] and McCall et al.[22] Given the general linkage between BBB glucose transporter activity, cerebral blood flow, and brain glucose utilization,[27,28] the finding of a decrease in BBB glucose transporter activity in experimental diabetes is expected. For example, both cerebral blood flow and cerebral glucose utilization are decreased in experimental diabetes. The linkage between decreased glucose transporter activity and decreased glucose utilization explains the normal glucose volume of distribution in diabetes.[29] If the BBB glucose transporter was not down-regulated in diabetes, then the brain glucose volume of distribution would be increased.[3]

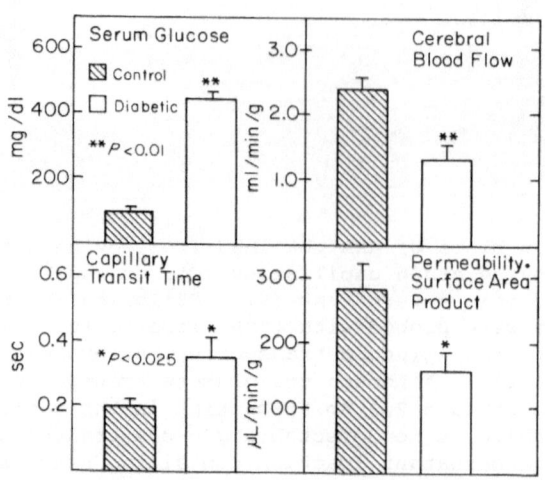

Figure 4. Serum glucose is shown to be increased approximately four-fold in the control and diabetic rats, whereas cerebral blood flow is decreased 44% in the diabetic animals in association with a 44% decrease in BBB permeability·surface area product for glucose. Cerebral blood volume was unchanged and the decrease in cerebral blood flow was due to a prolongation of the capillary transit time in diabetes. The normal blood volume in diabetes indicates the decrease in permeability·surface area (PS) product is due to a decrease in permeability (P) and not due to a decrease in surface area (S). Moreover, Western blotting studies showed that the concentration of BBB glucose transporter is decreased in the diabetic state. From [25].

The decreased BBB PS product to glucose in experimental diabetes could be due to either decreased capillary surface area or to a decrease in BBB permeability to glucose. However, the brain blood volume is unchanged in diabetes (Figure 4), and this indicates the overall capillary surface area is also unchanged. Therefore, the decreased PS product is due to decreased BBB permeability to glucose, and this may represent either decreased carrier mobility within the membrane or diminished concentration of GLUT-1 isoform at the plasma membrane of brain capillary endothelial cells in diabetes. The latter was observed using quantitative Western blotting of control and diabetic brain capillaries.[25] For these studies, a synthetic peptide corresponding to the thirteen amino acids at the C-terminus of the rat brain glucose transporter[15] was synthesized and an antiserum to this synthetic peptide was prepared in rabbits. This antiserum selectively illuminated the microvasculature in brain using immunocytochemistry and also specifically reacted with the human erythrocyte glucose transporter in Western blotting with a broad band at 52,000 Daltons. Microvessels were isolated from control and diabetic rat brain as described previously in studies showing down-regulation of the insulin receptor in capillaries isolated from rats with experimental diabetes.[30] The antiserum prepared against the C-terminus of the GLUT-1 isoform specifically reacted with the 52 KDa protein in rat brain capillaries. Quantitation of the Western blot with a scanning laser densitometer revealed a $77 \pm 9\%$ reduction in immunoreactive glucose transporters in the diabetic brain capillaries. Therefore, these Western blotting studies corroborate the physiologic results shown in Figure 4 and indicate the decrease in BBB transport of glucose in diabetes is due to decreased production of the GLUT-1 isoform at the BBB in experimental hyperglycemia.[25]

In order to determine whether the down-regulation of the BBB glucose transporter in diabetes occurs at the transcriptional level, we also isolated RNA from microvessels obtained from either control rats or animals with experimental diabetes.[31] Blots were probed with [^{32}P]-labeled cDNA corresponding to the rat brain glucose transporter generously provided by Ora M. Rosen, M.D., of the Memorial Sloan-Kettering Cancer Center of New York, as well as a cDNA to actin, a housekeeping gene kindly provided by Michael J. Getz, Ph.D. of the Mayo Foundation in Rochester, Minnesota. These results showed surprisingly that the GLUT-1 transcript in rat brain microvessels in diabetes was increased 152% following normalization for actin transcript content. Whether the increased BBB glucose transporter mRNA levels in experimental diabetes reflect increased transcription rates or decreased RNA degradation rates cannot be determined at the present time. Nuclear run-on assays would discriminate between these two processes, and these studies are ongoing in our laboratory. However, the finding of increased concentrations of the 2.9 Kb GLUT-1 transcript in diabetic brain capillaries in conjunction with decreased amounts of GLUT-1 protein indicate there is a prominent block in translation of the GLUT-1 transcript in diabetic brain capillary cytosol.[25,31] Presumably, cytosolic proteins that bind GLUT-1 mRNA and sterically hinder translation are increased in diabetic brain capillary endothelia. A summary of the changes with diabetes in the BBB glucose transporter at the various cellular levels is given in Table 3.

BLOOD-BRAIN BARRIER TRANSPORT OF FREE FATTY ACIDS AND KETONE BODIES

Although free fatty acids are an important carbon source for cellular combustion in tissues such as skeletal muscle, fat, or liver in the postabsorptive state, brain does not significantly combust circulating free fatty acid, even after several weeks of prolonged starvation.[32] This failure to oxidize circulating free fatty acids is not due to a deficiency of the relevant free fatty acid oxidizing enzymes in brain since labeled free fatty acids are readily

Table 3. Summary of Changes in BBB Transport of Glucose in Experimental Diabetes

Parameter	Change in Diabetes
Cerebral blood flow	Decreased 44%
Cerebral blood volume	Unchanged
Cerebral capillary transit time	Increased 75%
BBB PS product for glucose	Decreased 44%
BBB GLUT-1 concentration	Decreased 77%
BBB (GLUT-1 mRNA/actin mRNA)	Increased 152%

From[25,31].

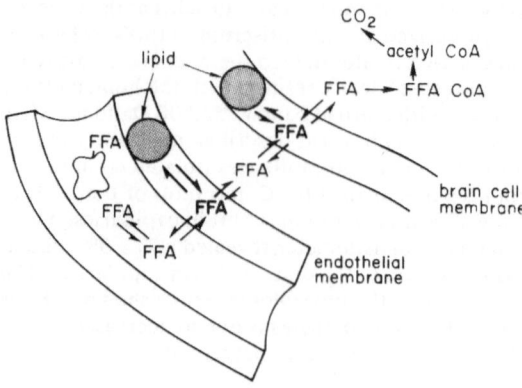

Figure 5. Scheme of transport and utilization of circulating free
fatty acids (FFA). Subsequent to enhanced dissociation
from circulating albumin binding sites, the free fatty
acid may be transported across either the endothelial
membrane or the brain cell membrane and intracellular
free fatty acid may be converted into free fatty acyl-
coA, acetyl-coA, and CO_2, were it not for rapid
membrane-bound esterification systems that provide an effective enzymatic barrier to circulating free fatty acid.
Owing to this effective enzymatic barrier, circulating
free fatty acid is not combusted in brain even under
states of prolonged starvation.[32]

converted to CO_2 following the intracerebral administration of $[^{14}C]$-labeled free fatty acid,[33] and small amounts of circulating free fatty acids are converted to Krebs cycle intermediates[34]. Rather, the failure of brain to utilize circulating free fatty acids as an important source of combustible carbon is due, in part, to a slow transport through the BBB. In the absence of plasma proteins, both medium chain and long chain free fatty acids are rapidly transported through the BBB.[35] However, free fatty acids are more than 99% bound by high affinity binding sites on circulating albumin, and only approximately 5% of plasma free fatty acid is unidirectionally extracted by brain on a single pass through the cerebral microcirculation.[36] Moreover, there is a prominent enzymatic barrier to the utilization of the circulating free fatty acids,[3] as depicted in Figure 5. There is rapid esterification into membrane-bound triglyceride of circulating free fatty acid at either the endothelial membrane or the brain cell membrane. Thus, in the steady state, an equal amount of free fatty acid taken up by brain and esterified in the endothelial or brain cell membranes is released to blood via hydrolysis of membrane-bound triglyceride via brain microvascular lipoprotein lipase.[3] This enzymatic barrier protecting brain intracellular space from circulating free fatty acids is very well developed and breaks down only under pathologic conditions in brain.

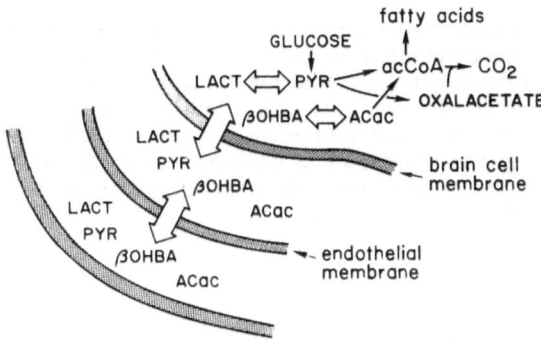

Figure 6. Transport of circulating monocarboxylic acids such as lactate (LACT), pyruvate (PYR), β-hydroxybutyrate (β-OHBA), or acetoacetate (ACac) involve transport across the endothelial and brain cell membranes. Once in brain, ketone bodies are converted to CO_2 in the presence of sufficient oxalacetate 4-carbon pool reserves. An important source of the oxalacetate is carboxylation of glucose- or lactate-derived pyruvate.[39] Alternatively, the acetoacetate may be converted into fatty acids in brain owing to shuttling into fatty acid synthetic pathways.

Despite this enzymatic barrier preventing brain from oxidizing circulating free fatty acid, brain is still endowed with the metabolic machinery for deriving energy from carbon stored in fat depot sites. Following conversion of peripheral fat into ketone bodies in liver, the ketone bodies (acetoacetate or β-hydroxybutyrate) are released to blood and are transported across the BBB by the monocarboxylic acid carrier (Figure 6). This carrier transports lactate, pyruvate, β-hydroxybutyrate, and acetoacetate across the BBB, as well as the α-ketoacids of a number of neutral amino acids.[37,38] Once in brain, the ketone bodies are transported across the brain cell membrane and rapidly deliver two carbon units to the Kreb's cycle. The combustion of the two carbon units derived from ketone bodies by the Kreb's cycle, however, is a function of the pool size of the four carbon skeletons such as oxalacetate in brain mitochondria. Since the Kreb's cycle only allows for the combustion of two carbon units and does not allow for the net synthesis of the four carbon skeletons of the cycle,[3] the oxalacetate pool size, which can be limiting for ketone body utilization, is maintained through pyruvate carboxylase conversion of glucose- or lactate-derived pyruvate into oxalacetate.[39] Therefore, glucose has a permissive effect on ketone body utilization in brain and conditions associated with decreased production of oxalacetate in brain are characterized by the development of coma and seizures with fasting despite high serum ketone body concentrations, such as in the case of ketotic hypoglycemia or pyruvate carboxylase deficiency.[39] The fact that the brain fails to completely shut off glucose utilization, even in the face of high rates of ketone body utilization, was originally observed by Cahill and co-workers,[32] who found that 30% of oxygen consumption in prolonged fasting in brain is still attributed to glucose utilization, whereas the oxygen consumption, owing to ketone body combustion, does not exceed 60%.

Apart from providing three carbon precursors for oxalacetate, glucose conversion to pyruvate or lactate in fasting brain may play a second permissive role in ketone body utilization by providing anion in the form of lactate to exchange with blood borne ketone body anion.[40] The precise mechanisms by which electrical neutrality is maintained at the monocarboxylic acid carrier is not known at present. However, it is likely that this carrier transports negatively charged anion and, for each ketone body extracted by brain from blood, one glucose-derived pyruvate or lactate molecule must be exported from brain to blood to maintain electrical neutrality.

ACKNOWLEDGMENTS

Dawn Brown and Michele Smotony skillfully prepared the manuscript. The author is indebed to Dr. Ruben J. Boado for many valuable discussions.

REFERENCES

1. M.E. Phelps, D.E. Kuhl, and J.C. Mazziotta, Metabolic mapping of the brain's response to visual stimulation: Studies in humans, Science 211:1445 (1981).
2. L. Sokoloff, Localization of functional activity in the central nervous system by measurement of glucose utilization with radioactive deoxyglucose, J. Cereb. Blood Flow Metab. 1:7 (1981).
3. W.M. Pardridge, Brain metabolism: A perspective from the blood-brain barrier, Physiol. Rev. 63:1481 (1983).
4. W.M. Pardridge, Blood-brain barrier transport of nutrients, Nutr. Rev. 44:15 (1986).
5. N.B. Ruderman, P.S. Ross, M. Berger, and M.N. Goodman, Regulation of glucose and ketone-body metabolism in brain of anaesthetized rats, Biochem. J. 138:1 (1974).
6. R.A. Hawkins, A.M. Mans, and D.W. Davis, Regional ketone body utilization by rat brain in starvation and diabetes, Am. J. Physiol. 250:E169 (1986).
7. W.M. Pardridge, P.D. Crane, L.J. Mietus, and W.H. Oldendorf, Kinetics of regional blood-brain barrier transport and brain phosphorylation of glucose and 2-deoxyglucose in the barbiturate-anesthetized rat, J. Neurochem. 38:560 (1982).
8. J.B.G. Ghajar, F. Plum, and T.E. Duffy, Cerebral oxidative metabolism and blood flow during acute hypoglycemia and recovery in unanesthetized rats, J. Neurochem. 38:397 (1982).
9. W.M. Pardridge and W.H. Oldendorf, Transport of metabolic substrates through the blood-brain barrier, J. Neurochem. 28:5 (1977).
10. W.M. Pardridge, Recent advances in blood-brain barrier transport, Ann. Rev. Pharmacol. Toxicol. 28:25 (1988).
11. G.I. Bell, T. Kayano, J.B. Buse, C.F. Burant, J. Takeda, D. Lin, H. Fukumoto, and S. Seino, Molecular biology of mammalian glucose transporters, Diabetes Care 13:198 (1990).
12. M.A. Kasanicki and P.F. Pilch, Regulation of glucose-transporter function, Diabetes Care 13:219 (1990).
13. B. Thorens, M.J. Charron, and H.F. Lodish, Molecular physiology of glucose transporters, Diabetes Care 13:209 (1990).
14. A.G. de Herreros and M.J. Birnbaum, The acquisition of increased insulin-responsive hexose transport in 3T3-L1 adipocytes correlates with expression of a novel transporter gene, J. Biol. Chem. 264:19994 (1989).
15. J.J. Birnbaum, H.C. Haspel, and O.M. Rosen, Cloning and characterization of a cDNA encoding the rat brain glucose-transporter protein, Proc. Natl. Acad. Sci. USA 83:5784 (1986).
16. J.S. Flier, M. Mueckler, A.L. McCall, and H.F. Lodish, Distribution of glucose transporter messenger RNA transcripts in tissues of rat and man, J. Clin. Invest. 79:657 (1987).
17. R.J. Boado and W.M. Pardridge, The brain-type glucose transporter mRNA is specifically expressed at the blood-brain barrier, Biochem. Biophys. Res. Comm. 166:174 (1990).

18. Pardridge, W.M., Boado, R.J., and Farrell, C.R., Brain-type glucose transporter is selectively localized to the blood-brain barrier. Studies with quantitative Western blotting and in situ hybridization. J. Biol. Chem., in press.

19. R.A. DeFronzo, R. Hendler, and N. Christensen, Stimulation of counterregulatory hormonal responses in diabetic man by a fall in glucose concentration, Diabetes 289:125 (1980).

20. P.J. Boyle, N.S. Schwartz, S.D. Shah, W.E. Clutter, and P.E. Cryer, Plasma glucose concentrations at the onset of hypoglycemic symptoms in patients with poorly controlled diabetes and in nondiabetes, N. Engl. J. Med. 218, 1487 (1988).

21. A. Gjedde and C. Crone, Blood-brain glucose transfer: Repression in chronic hyperglycemia, Science 214:456 (1981).

22. A.L. McCall, W.R. Millington, and R.J. Wurtman, Metabolic fuel and amino acid transport into the brain in experimental diabetes mellitus, Proc. Natl. Acad. Sci. USA 79:5406 (1982).

23. S.I. Harik and J.C. LaManna, Vascular perfusion and blood-brain glucose transport in acute and chronic hyperglycemia, J. Neurochem. 51:1924 (1988).

24. R.B. Duckrow, Glucose transfer into rat brain during acute and chronic hyperglycemia, Metabol. Brain Dis. 3:201 (1988).

25. W.M. Pardridge, D. Triguero, and C.R. Farrell, Down-regulation of blood-brain barrier glucose transporter in experimental diabetes, Diabetes, in press (1990).

26. R.B. Duckrow, D.C. Beard, and R.W. Brennan, Regional cerebral blood flow decreases during chronic and acute hyperglycemia, Stroke 18:52 (1987).

27. J.E. Cremer, D.E. Ray, G.S. Sarna, and V.J. Cunningham, A study of the kinetic behaviour of glucose based on simultaneous estimates of influx and phosphorylation in brain regions of rats in different physiological states, Brain Res. 221:331 (1981).

28. R.A. Hawkins, A.W. Mans, D.W. Davis, L.S. Hibbard, and D.M. Lu, Glucose availability to individual cerebral structures is correlated to glucose metabolism, J. Neurochem. 40:1013 (1983).

29. R.B. Duckrow and R.M. Bryan, Jr., Regional cerebral glucose utilization during hyperglycemia, J. Neurochem. 48:989 (1987).

30. H.J.L. Frank, W.M. Pardridge, T. Jankovic-Vokes, H.V. Vinters, and W.L. Morris, Insulin binding to the blood-brain barrier in the streptozotocin diabetic rat, J. Neurochem. 47:405 (1986).

31. T.B. Choi, R.J. Boado, and W.M. Pardridge, Blood-brain barrier glucose transporter mRNA is increased in experimental diabetes mellitus, Biochem. Biophys. Res. Comm. 164:375 (1989).

32. G.F. Cahill, Jr., M.G. Herrera, A.P. Morgan, J. Soeldner, J. Steinke, P.L. Levy, G.A. Richard, Jr., and D.M. Kipnis, Hormone-fuel interrelationships during fasting, J. Clin. Invest. 45:1751 (1966).

33. E.J. Fritschka, L. Ferguson, and J.J. Spitzer, Increased free fatty acid turnover in CSF during hypotension in dogs, Am. J. Physiol. 236:H802 (1979).

34. J.C. Miller, J.M. Gnaedinger, and S.I. Rapoport, Utilization of plasma fatty acid in rat brain: Distribution of [^{14}C]palmitate between oxidative and synthetic pathways, J. Neurochem. 49:1507 (1987).

35. R. Spector, Fatty acid transport through the blood-brain barrier, J. Neurochem. 50:639 (1988).

36. W.M. Pardridge and L.J. Mietus, Palmitate and cholesterol transport through the rat blood-brain barrier, J. Neurochem. 34:463 (1980).

37. W.H. Oldendorf, Carrier-mediated blood-brain barrier transport of short-chain monocarboxylic organic acids, Am. J. Physiol. 224:1450 (1973).

38. A.R. Conn, D.I. Fell, and R.D. Steele, Characterization of α-keto acid transport across blood-brain barrier in rats, Am. J. Physiol. 245:E253 (1983).

39. D.C. De Vivo, The effects of ketone bodies on glucose utilization, in: "Cerebral Metabolism and Neural Function," R.A. Hawkins, W.D. Lust, and F.A. Welsh, eds., Williams and Wilkins, Baltimore (1980).

40. R.G. Kammula, Metabolism of ketone bodies by ovine brain in vivo, Am. J. Physiol. 231:1490 (1976).

THE BLOOD-BRAIN BARRIER AND THE REGULATION OF AMINO ACID

UPTAKE AND AVAILABILITY TO BRAIN

Quentin R. Smith

Laboratory of Neurosciences
National Institute on Aging
National Institutes of Health
Bethesda, Maryland, USA

INTRODUCTION

The brain requires, in addition to glucose, a continuous and balanced supply of a number of essential nutrients to sustain normal cerebral development and function. Principal among these are the amino acids which serve diverse metabolic and structural roles throughout the central nervous system.

Though not used primarily for cerebral energy metabolism, the amino acids are required by brain as the building blocks of proteins and peptides and as substrates for for a number of key brain metabolic pathways (1). Several, such as glutamate, aspartate, glycine and GABA, have roles as neurotransmitters or neuromodulators within the central nervous system, whereas others (tryptophan, tyrosine, histidine) serve as precursors to neurotransmitters, including serotonin, dopamine, norepinephrine and histamine (2). Through transamination and decarboxylation, the amino acids are closely linked to brain intermediary metabolism and the Krebs cycle. And lastly, the amino acids have important roles in the removal of brain ammonia and in the regulation of cerebral osmotic and anionic balance.

While the brain requires over 20 separate amino acids for normal development and function, only approximately half can be synthesized within the central nervous system. Those that can be formed within brain or other tissues of the body at rates sufficient to meet metabolic demand are termed nutritionally "dispensable" or "nonessential" and include proline, glycine, alanine, serine, asparagine, glutamine, glutamate, aspartate, cysteine and tyrosine (3). The remaining amino acids - phenylalanine, tryptophan, leucine, isoleucine, valine, methionine, histidine, threonine, lysine and arginine - cannot be formed at adequate rates and therefore must be supplied from the diet. This involves consumption, gastrointestinal absorption, peripheral distribution and metabolism, and transport across the brain capillary, neuronal and glial cell membranes (4). Transport into brain is restricted and mediated by a small number of saturable carriers that are exquisitely sensitive to competition effects. For the essential amino acids, brain concentrations are determined in large part by plasma concentrations and transport across the blood-brain barrier (5,6).

Interest in the factors that regulate brain amino acid concentrations has been stimulated in recent years by the finding that a number of key brain amino acid metabolic pathways are critically sensitive to changes in precursor supply. This has been best established for the monoamine neurotransmitters - serotonin, norepinephrine, and histamine - the rates of synthesis and brain concentrations of which have been clearly shown to be influenced by brain levels of respective precursor amino acids, tryptophan, tyrosine and histidine (7-9). Though less well documented, several studies have suggested that brain rates of synthesis of S-adenosyl methionine from methionine (10), of phenethylamine and tyramine from phenylalanine and tyrosine (11), and of the kynurenines - quinolinic acid and kynurenic acid - from tryptophan (12,13), are also critically influenced by precursor availability. This dependence arises because the key rate-limiting enzymes in the metabolic pathways are not saturated with amino acid precursor at normal brain concentrations. Thus, factors that change brain amino acid supply and concentration are directly transformed into alterations in rates of synthesis. Imbalances in brain amino acid concentrations are known to occur in a number of conditions, including liver disease, diabetes, uremia, severe dietary disturbances and the aminoacidurias (14-17).

This chapter will review the factors that regulate the uptake and distribution of amino acids within the central nervous system. Special emphasis will be placed on the blood-brain barrier and the specific carriers that mediate transport into brain. Finally, some questions will be raised concerning problems and issues that need to be addressed in the future.

BRAIN AMINO ACID TRANSPORT

Amino acid uptake into brain depends critically on the type of amino acid, essential vs nonessential, and on the presence of the blood-brain barrier. Nonessential amino acids are derived primarily from within the nervous system by intracerebral synthesis and show minimal uptake from the circulation. Synthesis is from glucose and Krebs cycle intermediates, and thus is influenced, at least in part, by cerebral energy metabolism. Essential amino acids, in contrast, are derived exclusively from the circulation (except for those released from protein breakdown; 18) and cross the blood-brain barrier readily by carrier-mediated transport. Flux rates for essential amino acids are appreciable so that half times for equilibration of free pools are on the order of minutes (19-20). Transfer rates for both groups are restricted and regulated by the blood-brain barrier.

BLOOD-BRAIN BARRIER

The blood-brain barrier is series of tissue sites that together limit and modulate the exchange of hydrophilic solutes between plasma and the central nervous system (21). The barrier is formed primarily by the cerebrovascular endothelium which comprises the largest single surface for solute uptake into brain (surface area = 120 cm^2/g; 22). Additional sites include the arachnoid membrane, which surrounds and envelopes the brain, and the choroid plexus epithelium, which secretes cerebrospinal fluid. At each of these sites there is at least one layer of cells that are joined by bands of tight junctions (23). These junctions essentially "fuse" the plasma membranes of adjacent barrier cells thereby blocking intercellular diffusion. As there are no alternative paracellular routes, the barrier displays many of the properties of a continuous "tight" epithelium, including high electrical resistance ($\sim$2000 Ohm.cm^2)(24), limited passive ion permeation (25), and selective permeability for "lipid-soluble" as compared to "water-soluble" compounds (26). The barrier apparently serves to protect the brain from circulating toxins and to help maintain a more stable ionic and nutrient internal environment for optimal brain function.

Critical hydrophilic nutrients that are required for brain metabolism, such as glucose, ketone bodies and essential amino acids, are transported into brain by saturable carriers at the blood-brain barrier (27). Oldendorf, in a classic series of studies in the 1970's, identified three carriers for amino acids at the blood-brain barrier; one for large neutral amino acids (PHE, TRP, LEU, ISL, VAL and TYR), one for basic amino acids (LYS, ARG and ORN), and one for acidic amino acids (GLU and ASP)(28,29). Transfer activities were greatest for the large neutral and basic amino acid carriers consistent with the fact that these carriers mediate the uptake of amino acids that are nutritionally essential (27).

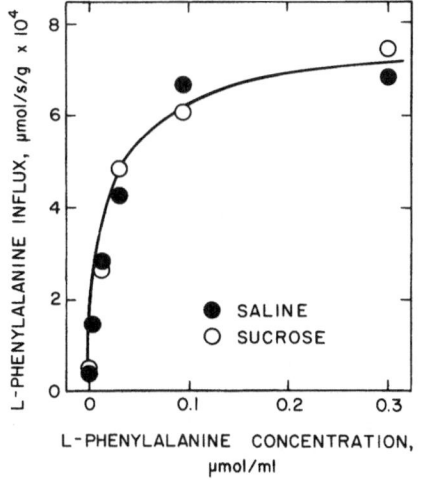

Fig. 1. Saturable L-phenylalanine transport across the rat blood-brain barrier. Influx was measured during brain perfusion with physiological saline or isotonic sucrose solution. Each point represents the mean for 3 or more animals. Recalculated from (39).

During the past ten years, considerable progress has been made in characterizing and defining blood-brain barrier amino acid transporters using a variety of refined in vivo and in vitro procedures. Quantitative autoradiographic techniques have been developed that allow accurate estimates of regional amino acid influx rates into brain in experimental animals under a number of different physiologic and pathologic conditions (30). Further, extended brain perfusion procedures have been created that allow absolute control of perfusate composition and flow for detailed studies of transport kinetics and characteristics under conditions that are not easily obtainable in vivo, i.e., for studying transport in the

absence of sodium or in the absence of competing amino acids) (31-33).
Similarly, in vitro procedures using isolated brain microvessels and
cultured brain endothelial monolayers have been explored for identifying
transport carriers and examining the polarity of the blood brain barrier
(i.e., transport differences between the luminal and abluminal endothelial
cell membranes) (34-36). These latter procedures will be ideal in the
future for eventually isolating and cloning barrier amino acid transport
proteins.

MECHANISMS OF TRANSPORT

Recent studies have greatly extended our knowledge of barrier amino
acid transporters and factors that regulate their activity under differing
pathological and physiological conditions. In addition, new information
has been gained on previously unidentified carriers at brain capillaries
(Table 1) and on the transport characteristics of other barrier sites. A
brief summary of transport mechanisms is give below.

Large Neutral Amino Acid Carrier - System L1

Of the blood-brain barrier amino acid transporters, the carrier for
large neutral amino acids shows the greatest transport capacity and
mediates the uptake of the largest number of amino acids into the central
nervous system. In vivo studies have shown that this carrier has
measurable affinity for at least 14 of the 16 plasma neutral amino acids
and is the primary mechanism of brain uptake for all, or most all,
essential neutral amino acids (28,29,37,38).

Table 1. Brain capillary amino acid transport systems

Transport System	Activity In Vivo	Activity In Vitro	Na Dependence	Representative Substrates
Neutral Amino Acid				
L1	++++	++	−	Phe, Trp, Leu, Met, Ile, Tyr, Val, BCH
ASC	+	++	+	Thr, Ala, Ser, Cys
A	−	++	+	Pro, Ala, Ser, MeAIB
Basic Amino Acid				
y+	+++	++	−	Lys, Arg, Orn, GPA
Acidic Amino Acid				
X−	+	++	(?)	Glu, Asp

Refer to text for appropriate references. Scales are relative and
meant only to give a general idea of transport activity.

Table 2. Kinetic parameters for neutral amino acid transport into brain

Amino Acid	Plasma Concentration (uM)	Km (uM)	Vmax (nmol/min/g)	Km(app) (uM)	Influx (nomol/min/g)
Phenylalanine	81	11	41	170	13.2
Tryptophan	82	15	55	330	8.2a
Leucine	175	29	59	500	14.5
Methionine	64	40	25	860	1.7
Isoleucine	87	56	60	1210	4.0
Tyrosine	63	64	96	1420	4.1
Histidine	95	100	61	2220	2.5
Valine	181	210	49	4690	1.8
Threonine	237	220	17	4860	0.8
Glutamine	485	880	43	19900	1.0
		Average Vmax	51 $\pm$ 7	Total Influx	51.8

Recalculated from Smith et al. (1987). aEstimated assuming ~70% of albumin bound tryptophan contributes to uptake.

The presence of the large neutral amino acid transporter at the blood-brain barrier has been demonstrated both in vivo and in vitro, and it is believed that the transporter is located on both the capillary luminal and abluminal cell membranes (Table 1)(34). Transport by this mechanism is saturable, stereospecific and sodium/energy independent (28,33,37,40), consistent with facilitated diffusion (Fig. 1). The carrier appears to express many of the same characteristics of the classic L system of Oxender and Christensen (41,42) including primary affinity for amino acids with large, bulky side chains, such as PHE and LEU, trans-stimulation, and inhibition by the L system-specific analogue - 2-aminobicyclo[2.2.1]heptane-2-carboxylic acid (BCH) (37,43-45). When amino acids are ordered in terms of affinity for the cerebrovascular large neutral amino acid carrier and ranked on a scale of 0-4 (Table 2), values agree perfectly for 10 of the 14 amino acids with values reported by Christensen for the L system (37,46). Only one amino acid, tryptophan, differs by two units or more. The L system carrier is thought to be present in most all tissues of the body and appears to mediate bidirectional, equilibrative transport. It's activity in peripheral tissues is notably unresponsive to hormonal regulation or starvation (41).

Although the cerebrovascular large neutral amino acid carrier expresses many characteristics of the classical L system, correspondence is not absolute. The affinity of the cerebrovascular neutral amino acid transporter exceeds that reported for the L system in most tissues by as much as 100-1000 fold. For example, the Km for L-phenylalanine uptake into rat brain is ~10 uM, whereas the value in most other organs is 1-10 mM or greater (40,47). Recent studies by Hargreaves and Pardridge (48) using isolated brain microvessels suggest that in human capillaries the Km may be even lower (0.3 uM).

High affinity L transport systems have been reported for some blood cells, tumor cells and cultured tissue cells (49,50), and Weissbach et al. (51) have even suggested the designations L1 and L2 to distinguish the high affinity and low affinity forms. These two carriers may represent different transport proteins as expression of the L1 form by cultured hepatocytes can be blocked by inhibitors of RNA and protein synthesis (51). Interestingly, the Ll system of the cultured hepatocyte shows excellent affinity for tryptophan, similar to the cerebrovascular large neutral amino acid carrier and quite unlike the classical, low-affinity L transport mechanism (Table 2). These facts may suggest that the cerebrovascular large neutral amino acid carrier corresponds to the high affinity Ll mechanism, though a more complete identification awaits isolation and cloning of the transporter. The fact that the cerebrovascular endothelium expresses special glucose and neutral amino acid transport proteins, distinct from those expressed in most other tissues of the body (52), highlights the unique role of the blood-brain barrier in regulating and maintaining the brain internal environment.

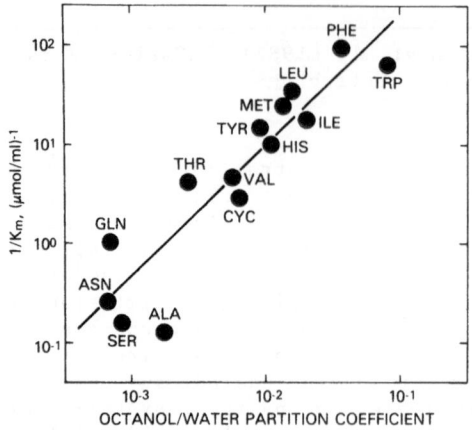

Fig. 2. Relation of apparent affinity (1/Km) of the cerebrovascular Ll carrier to octanol/water partition coefficient for 14 neutral amino acids. The octanol/water partition coefficient is one measure of side chain hydrophobicity. From Smith et al. (37).

Table 2 summarizes values of Vmax and Km for saturable amino acid transport across the blood-brain barrier as measured using the in situ brain perfusion technique (37). These values are likely more accurate than previous estimates obtained with the brain uptake index technique because of reduced errors from mixing and efflux of "cold" amino acid as the perfusate passes through the brain capillaries (32,40,53). Vmax values differ slightly among neutral amino acids but average ~50 nmol/min/g (8 x 10^{-4} umol/s/g). In contrast, Km values vary to a greater extent, differing by >100 fold.

Affinity for the large neutral amino acid transporter appears to depend critically on amino acid side chain hydrophobicity, as shown in Fig. 2. The relation between apparent affinity (1/Km) and amino acid octanol/water partition coefficient, a measure of side chain hydrophobicity (54), is linear over 2-3 orders of magnitude (37). Other factors influence affinity as well. The transporter is stereoselective and prefers L, as compared to D, amino acids (27,40). In addition, substrates for affinity must have an free carboxyl group with an unsubstituted amino group on the alpha carbon.

Small Neutral Amino Acid Carriers - Systems A and ASC

While for many years it was thought that the large neutral amino acid carrier was the only transporter for neutral amino acids at the blood-brain barrier, recent studies employing both in vivo and in vitro techniques have forced a re-evaluation of that view. Betz and Goldstein (34) demonstrated in isolated brain microvessels the presence of amino acid transport system A which mediates the sodium-dependent uptake of small neutral amino acids, such as alanine, serine and proline, and is inhibited by the A system specific analogue, methylaminoisobutyric acid (MeAIB). In 1987, Tayarani et al. (55) presented in vitro evidence for a third carrier, system ASC, which is also sodium dependent, transports primarily small neutral amino acids (alanine, serine, cysteine, and threonine) and is inhibited by neither the system L nor system A specific analogues (BCH or MeAIB).

Based on the known low rates of uptake of small neutral amino acids into brain and minimal evidence for blood-brain barrier system A or ASC activity in vivo (28,29,56,57), both Betz and Goldstein (34) and Tayarani et al. (55) suggested that the system A and ASC transporters may be located selectively on the abluminal capillary membrane and may function primarily to transport amino acids out of brain. Some transport activity, however, may be located at the luminal membrane as well, as Sershen and Lajtha (58) and more recently Tovar et al. (38) found evidence for small components of brain uptake of alanine, serine and threonine that display the characteristics of system ASC transport. From the study of Tovar et al., it would appear that in the absence of competing amino acids and at tracer L-threonine concentrations, system ASC mediates ~25% of L-threonine uptake with ~40% by the Ll system and the remainder by a noninhibitable (presumably nonsaturable) mechanism. This may not , however, reflect the actual contributions for transport from plasma as the Ll carrier is highly saturated with amino acids at normal plasma concentrations and thus transports individual amino acids from plasma at rates 1/10-1/20 of those predicted in the absence of competing amino acids (37,40).

Wade and Brady (59) previously found no evidence for a significant contribution of system ASC in the transport of L-cysteine into brain. Further, most all studies agree that large neutral amino acids, such as phenylalanine and leucine, are taken up into brain almost exclusively by the Ll carrier mechanism (28,38,40,45; Q. Smith, unpublished observation). Tayarani et al. (55) estimated a Km of 1.3 mM for ASC system transport of L-alanine in isolated brain microvessels.

The blood-brain barrier permeability to [14C]MeAIB, the A system specific analogue, is quite low in adult animals ($\sim 3 \times 10^{-5}$ ml/s/g), lower than most all other plasma amino acids and consistent with the hypothesis that system A contributes minimally to amino acid uptake into brain (60). In infants animals, however, there is evidence for significant component of A system transport into brain that is sodium dependent, saturable and inhibitable by MeAIB (61).

Systems A and ASC, like system Ll, show rather weak selectivity and have measurable affinity for a number of amino acids (42). Transport activities for both the A and ASC systems are strongly influenced by hormones and diet in peripheral tissues.

Basic Amino Acid Carrier - System y$^+$

Next to the cerebrovascular L1 carrier, the basic amino acid carrier (also termed system y$^+$) exhibits the greatest transport capacity for amino acids at the blood-brain barrier. Preliminary estimates with the brain perfusion technique indicate that the Vmax is ~25 nmol/min/g with a Km of 65 uM for L-lysine (Q. Smith, unpublished observations). Like the L1 carrier, system y$^+$ transport is stereospecific, sodium independent, and thought to be mediated by facilitated exchange (28,62,63). Affinity is greatest for alpha amino acids with a cationic side chains, including lysine, arginine, ornithine and histidine, though in some tissues short chain neutral amino acids also interact with moderate affinity (64).

4-Amino-1-guanylpiperidine-4-carboxylic acid (GPA) has been developed as a specific model substrate for this system (64). Histidine, which is approximately 2-3% charged at physiological pH (7.4), apparently binds but is not transported by the blood-brain barrier y$^+$ carrier (65). Instead, histidine is taken up into brain as the neutral zwitterion by the cerebrovascular large neutral amino acid carrier.

System y$^+$ transport activity has been demonstrated by the brain capillaries both in vivo and in vitro (62,63,66), and it is thought that the carrier is present at both the capillary luminal and abluminal membranes. The y$^+$ carrier is widely distributed throughout the body and is considered relatively unresponsive to hormonal and dietary regulation (64).

Acidic Amino Acid Carrier - System X$^-$

The acidic amino acids, glutamate and aspartate, are transported into brain by the blood-brain barrier acidic amino acid carrier, as originally described by Oldendorf and colleagues in the 1970's (29). Transport capacity for this carrier is markedly less than either the L1 or y$^+$ systems as might be expected since both glutamate and aspartate are excitatory neurotransmitters and can be synthesized within the central nervous system.

The transporter is apparently saturated at normal plasma glutamate and aspartate concentrations so that the influx rates are essentially independent of plasma levels under physiological conditions (67). Little is known about the exact characteristics of this carrier as no detailed kinetic studies have been performed.

The acidic amino acid carrier has been suggested to function primarily to transport amino acids out of the central nervous system (68), based on the reported net efflux of GLU from brain as measured with the isolated, perfused dog brain technique (69). However, as there was no evidence for a significant net brain efflux of aspartate in the same study, and as significant net brain glutamate effluxes have not been observed in some other brain arterio-venous difference studies (70), this hypothesis may require further verification.

Brain acidic amino acid concentrations have been noted to be essentially independent of plasma concentrations, except in nonbarrier regions, under a number of physiological and pathological conditions (71). This homeostasis may reflect saturated influx, regulated efflux or regulated intracerebral synthesis and catabolism. Evaluation of the homeostatic mechanisms for acidic amino acids at the blood brain barrier is further complicated by the fact that brain glutamate and aspartate concentrations exceed plasma concentrations by some 20-100 fold (71). This large gradient may mask small changes in brain interstitial fluid concentrations as produced by altered plasma concentrations and transport across the blood-brain barrier.

Nonsaturable Uptake

Most studies of the concentration dependence of amino acid uptake into brain have found evidence for a "nonsaturable" component of uptake that varies in magnitude depending on the amino acid and the method employed to measure blood-brain barrier transport (1,32,33,40,72). In some reports, the nonsaturable component comprises up to 50% of total amino acid uptake from plasma (73), whereas in others, the contribution is much less significant (5-10%)(40,72).

A clear explanation of this phenomena has not been put forward. The nonsaturable component, of course, may represent simple passive diffusion of amino acid across the blood-brain barrier or transport by a specific, low affinity carrier. Amino acids are known to cross black lipid membranes by passive diffusion (74), and cerebrovascular permeability coefficients reported for the nonsaturable component at the blood-brain barrier with the brain perfusion technique match well those predicted by amino acid lipid solubilities (as measured by the octanol/water partition coefficient) and the empirical relation between lipid solubility and passive blood-brain barrier permeation (40,53).

Still, Christensen (75) has challenged the possible role of passive diffusion and it may be that alternative explanations are necessary. At the blood-brain barrier, the process cannot be explained as an artifact of tracer metabolism as nonsaturable components have been reported for both metabolizable and nonmetabolizable tracers.

We recently suggested that the component may arise as an error due to failure to correct for residual intraendothelial tracer in uptake experiments (45). Such an error, it was proposed, would lead to overestimation of tracer influx especially at high perfusate concentrations where the fraction of tracer that had actually crossed the blood-brain barrier would be the lowest. This explanation, however, cannot account for the discrepancy as computer modeling of the endothelium suggests that the endothelial cell tracer content will decrease with increasing perfusate unlabeled amino acid concentration so as to keep the total fractional content approximately equal in all experiments (Q. Smith, unpublished observations). Still, the component may arise as an artifact of failure to adequately correct for brain intravascular (blood) tracer or from use of tracers that are not sufficiently pure. A careful re-examination of the nonsaturable component of D-glucose uptake into brain revealed that the nonsaturable component was not as high as previously reported and in fact was on the same order of magnitude as that expected for passive diffusion as measured using L-glucose or sucrose (76).

TRANSPORT CARRIERS AT OTHER BARRIER SITES

Although the brain capillaries are the primary interface for solute exchange between plasma and brain interstitial fluid, other barrier sites also contribute and have important roles in the production of cerebrospinal fluid and in the removal of toxins from the nervous system.

Preston et al., (77) recently demonstrated the presence of systems L1 and ASC for transport of amino acids from blood into the choroid plexus epithelium. Similarly, earlier studies by Wright suggested that leucine and other amino acids may be transported out of cerebrospinal fluid by saturable processes at the arachnoid membrane (78,79).

Facilitated transport systems for essential large neutral amino acids have also been demonstrated at the blood-nerve and blood-retinal barriers (80). In addition, Abbott et al. (81) showed that there is an L system for facilitated uptake of large neutral amino acids at the glial blood-brain barrier of the dogfish. As at the cerebral capillary barrier, facilitated transport carriers at these other barrier sites are likely required to overcome the limitations of passive permeability and to supply neural tissue with sufficient essential amino acids to meet the needs of metabolism and protein synthesis.

FACTORS THAT INFLUENCE AMINO ACID TRANSPORT INTO BRAIN

Brain concentrations of essential amino acids are determined by plasma concentrations and transport across the blood-brain barrier. Unidirectional influx rates for amino acid uptake from plasma can be estimated using the Michaelis-Menten equation with a correction for multiple substrate competition (1,37), as

$$Influx = V_{max}C/[K_m(1 + \Sigma (C_i/K_i)) + C]$$

where C = plasma concentration of the amino acid of interest, V_{max} and K_m are the transport constants of the amino acid, and C_i and K_i are the plasma concentration and corresponding half-saturation constant of each competing amino acid.

Correction for competition is critical because the blood-brain barrier large-neutral, basic and acidic amino acid carriers each transport multiple amino acids at high affinity and are each essentially saturated with amino acids at normal plasma concentrations. For example, the brain capillary L1 transporter expresses affinity for at least 14 of the 16 plasma neutral amino acids, with nine having plasma concentrations equal to or greater than K_m (Table 2). As a consequence, $\Sigma (C_i/K_i)$ is ≥ 9, and thus predicts a degree of saturation of $\geq 90\%$ (37). Similar calculations can be performed for the blood-brain barrier basic and acidic amino acid carriers from reported K_m values (1,67).

Competition has two important consequences. First, it makes the total influx for amino acids in a group essentially independent of concentration because of transport saturation. And secondly, it makes the influx of each individual amino acid dependent not only on its own concentration but on the concentrations of all competitors. Thus, a large increase in the concentration of one amino acid will decrease the flux rates of all competitors by transport inhibition. This latter relation has been demonstrated extensively both in vivo and in vitro following dietary or pharmacologic manipulations of plasma concentrations (16,82). Because of competition, the effective in vivo K_m for an amino acid, termed the "apparent" K_m and defined as $K_m[1 + \Sigma(C_i/K_i)]$, exceeds the true K_m by 10-20 fold (Table 2).

From inspection of Eq. 1 it can be seen that the influx of an amino acid into brain depends on four factors; the plasma concentration, the concentrations of competitors, the transport V_{max} and the transport K_m's. At normal plasma concentrations, influx will be essentially proportional to plasma concentration and inversely related to the summed concentration of competitors. This latter relation was first recognized by Fernstrom and Wurtman in 1972 (83) and termed the "plasma ratio", defined as $C/\Sigma C_i$.

The plasma ratio has since proved a valuable predictor of brain amino acid concentrations.

Plasma amino acid concentrations are influenced by a number of factors and can become imbalanced in several disease states, including liver disease, uremia and diabetes. Large changes in the plasma amino acid ratio can lead to significant alterations in brain amino acid concentrations and disruptions in brain amino acid metabolism. Diet-induced changes in the plasma ratio are, for the most part, rather small as most foods contain balanced levels of proteins and amino acids. Consumption of a high carbohydrate, low protein meal has been noted to increase brain tryptophan and the plasma tryptophan ratio by up to 50%, whereas consumption of a high-protein meal produces the opposite effects (7).

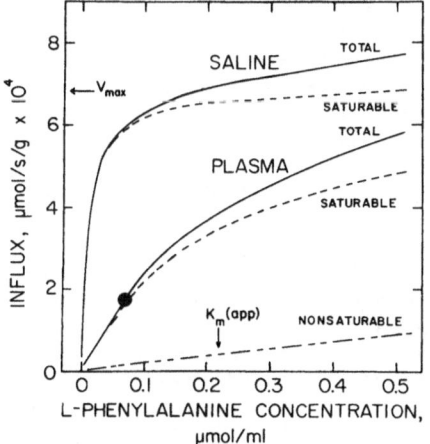

Fig. 3. Concentration dependence of L-phenylalanine influx into rat cerebral cortex from saline and plasma. The curves represent total, saturable and nonsaturable influxes.

Influx is also influenced by the transport Vmax and Km. Vmax has been reported to vary with anesthesia and the level of brain capillary perfusion (capillary "recruitment")(73,84). Further, the degree of expression of the transporter has been noted to change with disease and development. Liver disease apparently increases the capacity of the blood-brain barrier L1 carrier and decreases that of the y⁺ carrier(14,66,85). The change in transport capacity for the L1 carrier in liver disease has been related to trans-stimulation from increased brain glutamine concentration (44,86,87) and to altered blood ammonia (88).

Infant rats express greater blood-brain barrier transport capacities for each of the L1, ASC, A and y^+ carriers (61). Chronic hyperglycemia, which has been reported to reduce blood-brain barrier glucose transport capacity, apparently does not alter the transport capacity or affinity of the blood-brain barrier amino acid carriers (89,90).

Protein binding also has been noted to influence the brain uptake of tryptophan, the one plasma amino acid that binds reversibly and significantly to serum albumin (~70-90% bound at normal concentrations) (7). However, the reported influence of binding is less than that predicted by the free fraction (1,91) and apparently varies with blood flow (92). At normal flow rates, the effective in vivo free fraction exceeds the measured in vitro free fraction by two-to-three fold (50-70% vs 18-25%), whereas at high flow rates, such as obtained with the brain perfusion technique, the two are essentially equal (92).

SUMMARY AND DIRECTIONS FOR FUTURE RESEARCH

Although considerable progress has been made during the past 20 years in the study of amino acid uptake and metabolism in brain, large areas remain essentially unexplored. While transport systems have been defined and characterized in vivo, the transport proteins have not been identified and cloned. Further, the implications of the polarity of blood-brain barrier amino acid transport have not been addressed and an effective model of transport across the brain capillary endothelium has not been presented. Cerebrospinal fluid and brain interstitial fluid concentrations of amino acids have long been known to be a fraction of respective plasma concentrations (93,94). Yet, the mechanism that maintains these gradients has not been elucidated. Davson et al. (95) in a recent study showed that the gradients are not an artifact of cerebral metabolism as similar low cerebrospinal fluid concentrations are maintained for the nonmetabolizable amino acids, aminoisobutyric acid and cycloleucine. This suggests that other mechanisms must be operative, such as active efflux across the blood-brain barrier. Finally, while considerable effort has been extended towards characterizing the kinetics of uptake, little effort has been placed on defining the degree to which transport is modulated or controlled in vivo. Preliminary reports suggest that there may be some hormonal and neural regulation (96-98). Even the mechanism that mediates the enhanced transport capacity of the L1 system in hepatic encephalopathy has not been fully elucidated.

The fact that the high affinity L1 carrier of the blood-brain barrier likely differs from that expressed in most tissues of the body suggests a unique role of this protein in regulating brain amino acid availability. Newly accumulating information will permit the investigation of possible dysfunctions of barrier transport systems under pathological conditions and the potential roles of these carriers in central nervous system diseases.

REFERENCES

1. W. M. Pardridge, Brain metabolism: a perspective from the blood-brain barrier, Physiol. Rev. 63:1481 (1983).
2. C. J. Pycock and P. V. Taberner, "Central Neurotransmitter Turnover," University Park Press, Baltimore (1981).
3. W. C. Rose, Amino acid requirements of man, Fed. Proc. 8:546 (1949).
4. A. E. Harper and J. K. Tews, Nutritional and metabolic control of

brain amino acid concentrations, in: "Amino Acid Availability and Brain Function in Health and Disease," G. Huether, ed., Springer-Verlag, Berlin, pp. 3-12 (1988).

5. J. D. Fernstrom and D. V. Faller, Neutral amino acids in the brain: changes in response to food ingestion, J. Neurochem. 30:1531 (1978).

6. J. K. Tews, J. Greenwood, O. E. Pratt, and A. E. Harper, Valine entry into rat brain after diet-induced changes in plasma amino acids, Am. J. Physiol. 252:R78 (1987).

7. J. D. Fernstrom, Role of precursor availability in control of mono-amine biosynthesis in brain, Physiol. Rev. 63:484 (1983).

8. I. N. Acworth, M. J. During, and R. J. Wurtman, Processes that couple amino acid availability to neurotransmitter synthesis and release, in: "Amino Acid Availability and Brain Function in Health and Disease," G. Huether, ed., Springer-Verlag, Berlin, pp. 117-136 (1988).

9. J. C. Scwartz, C. Lampart, and C. Rose, Histamine formation in rat brain in vivo: effects of histidine loads, J. Neurochem. 19:801 (1972).

10. R. A. Rubin, L. A. Ordonez, and R. J. Wurtman, Physiological dependence of brain methionine and S-adenosylmethionine concentrations on serum amino acid patterns, J. Neurochem. 23:227 (1974).

11. S. N. Young, The significance of tryptophan, phenylalanine, tyrosine and their metabolites in the nervous system, in: "Handbook of Neurochemistry, Vol. 3, Metabolism in the Nervous System," A. Lajtha, ed., Second Edition, Plenum, New York, pp. 559-581 (1983).

12. M. J. During, M. P. Heyes, A. Freese, S, P. Markey, J. B. Martin, and H. Roth, Quinolinic acid concentrations in striatal extracellular fluid reach potentially neurotoxic levels following systemic L-tryptophan loading, Brain Res. 476:384 (1989).

13. F. Moroni, P. Russi, G. Lombardi, M. Beni, and V. Carla, Presence of kynurenic acid in the mammalian brain, J. Neurochem. 51: 177 (1988).

14. J. H. James, J. Escourrou, J. E. Fischer, Blood-brain neutral amino acid transport activity is increased after portacaval anastomosis, Science 200:1395 (1978).

15. A. M. Mans, M. R. DeJoseph, D. W. Davis, and R. A. Hawkins, Regional amino acid transport into brain during diabetes: effect of plasma amino acids, Am. J. Physiol. 253:E575 (1987).

16. C. M. McKean, D. E. Boggs, and N. A. Peterson, The influence of high phenylalanine and tyrosine on the concentrations of essential amino acids in brain, J. Neurochem. 15:235 (1968).

17. F. A. Hommes, Aminoacidemias and brain dysfunction, in: "Handbook of Neurochemistry, Vol. 10, Pathological Neurochemistry," A. Lajtha, ed., Second Edition, Plenum, New York, pp. 15-41 (1983).

18. C. B. Smith, G. E. Deibler, N. Eng, K. Schmidt, and L. Sokoloff, Measurement of local cerebral protein synthesis in vivo: influence of recycling of amino acids derived from protein, Proc. Natl. Acad. Sci. USA 85:9341 (1988).

19. M. Kirikae, M. Diksic, and Y. L. Yamamoto, The transfer coefficients for L-valine and the rate of incorporation of L-[1-14C]valine into proteins in normal adult rat brain, J. Cereb. Blood Flow Metab. 8:598 (1988).

20. R. A. Hawkins, S. C. Huang, J. R. Barrio, R. E. Keen, D. Feng, J. C. Mazziotta, and M. E. Phelps, Estimation of local cerebral protein synthesis rates with L-[1-14C]leucine and PET: methods, model, and results in animals and humans, J. Cereb. Blood Flow Metab. 9: 446 (1989).

21. J. D. Fenstermacher and S. I. Rapoport. Blood-brain barrier, in: "Handbook of Physiology: The Cardiovascular System IV," S. R.

Greger, ed., American Physiological Society, Bethesda, pp. 969–1000 (1984).

22. P. M. Gross, N. M. Sposio, S. E. Pettersen, and J. D. Fenstermacher, Differences in function and structure of the capillary endothelium in gray matter, white matter and a circumventricular organ of rat brain, Blood Vessels 23:261 (1986).

23. M. M. Brightman and T. S. Reese, Junctions between intimately apposed cell membranes in the vertebrate brain, J. Cell Biol. 40:648 (1969).

24. C. Crone and S. P. Olesen, Electrical resistance of brain microvascular endothelium, Brain Res. 241:49 (1982).

25. Q. R. Smith and S. I. Rapoport, Cerebrovascular permeability coefficients to sodium, potassium and chloride, J. Neurochem. 46:1732 (1986).

26. K. Ohno, K. D. Pettigrew, and S. I. Rapoport, Lower limits of cerebrovascular permeability to nonelectrolytes in the conscious rat, Am. J. Physiol. 235:H299 (1978).

27. W. M. Pardridge and W. H. Oldendorf, Transport of metabolic substrates through the blood-brain barrier, J. Neurochem. 28:5 (1977).

28. W. H. Oldendorf, Brain uptake of radiolabeled amino acids, amines, and hexoses after arterial injection, Am. J. Physiol. 221:1629 (1971).

29. W. H. Oldendorf and J. Szabo, Amino acid assignment to one of three blood-brain barrier amino acid carriers, Am. J. Physiol. 230:94 (1976).

30. R. A. Hawkins, A. M. Mans, and J. F. Biebuyck, Amino acid supply to individual cerebral structures in awake and anesthetized rats, Am. J. Physiol. 242:E1 (1982).

31. Y. Takasato, S. I. Rapoport, and Q. R. Smith, An in situ brain perfusion technique to study cerebrovascular transport in the rat, Am. J. Physiol. 247:H484 (1984).

32. Q. R. Smith, Y. Takasato, and S. I. Rapoport, Kinetic analysis of L-leucine transport across the blood-brain barrier, Brain Res. 311:167 (1984).

33. J. Greenwood, A.S. Hazell, and O. E. Pratt, The transport of leucine and aminocyclopentanecarboxylate across the intact, energy-depleted rat blood-brain barrier, J. Cereb. Blood Flow Metab. 9:226 (1989).

34. A. L. Betz and G. W. Goldstein, Polarity of the blood-brain barrier: neutral amino acid transport into isolated brain capillaries, Science 202:225 (1978).

35. P. A. Cancilla and L. E. DeBault, Neutral amino acid transport properties of cerebral endothelial cells in vitro, J. Neuropath. and Exp. Neurol. 42:191 (1983).

36. K. L. Audus and R. T. Borchardt, Characteristics of the large neutral amino acid transport system of bovine brain microvessel endothelial cell layers, J. Neurochem. 47:484 (1986).

37. Q. R. Smith, S. Momma, M. Aoyagi, and S. I. Rapoport, Kinetics of neutral amino acid transport across the blood-brain barrier, J. Neurochem. 49:1651 (1987).

38. A. Tovar, J. K. Tews, N. Torres, and A. E. Harper, Some characteristics of threonine transport across the blood-brain barrier of the rat, J. Neurochem. 51:1285 (1988).

39. S. Momma, M. Aoyagi, S. I. Rapoport, and Q. R. Smith. Phenylalanine transport across the blood-brain barrier as studied with the in situ brain perfusion technique, J. Neurochem. 48:1291 (1987).

40. D. L. Oxender and H. N. Christensen, Distinct mediating systems for the transport of neutral amino acids by the Ehrlich cell, J. Biol. Chem. 238:3686 (1963).

41. M. A. Shotwell, M. S. Kilberg, and D. L. Oxender, The regulation of neutral amino acid transport in mammalian cells, Biochim. Biophys. Acta 737:267 (1983).

42. H. N. Christensen, Role of amino acid transport and countertransport in nutrition and metabolism, Physiol. Rev. 70:43 (1990).

43. L. A. Wade and R. Katzman, Synthetic amino acids and the nature of L-dopa transport at the blood-brain barrier, J. Neurochem. 25:837 (1975).

44. C. Cangiano, P. Cardelli-Cangiano, J. H. James, F. Rossi-Fanelli, M. A. Patriz, K. A. Brackett, R. Strom, and J. E. Fischer, Brain microvessels take up large neutral amino acids in exchange for glutamine, J. Biol. Chem. 258:8949 (1983).

45. M. Aoyagi, B. W. Agranoff, L. C. Washburn, and Q. R. Smith, Blood-brain barrier transport of l-aminocyclcohexanecarboxylic acid, a nonmetabolizable amino acid for in vivo studies of brain transport, J. Neurochem. 50:1220 (1988).

46. H. N. Christensen, On the development of amino acid transport systems, Fed. Proc. 32:19 (1973).

47. J. Lerner and D. L. Larimore, Comparative aspects of the apparent Michaelis constant for neutral amino acid transport in several animal tissues, Comp. Biochem. Physiol. 84B:235 (1986).

48. K. M. Hargreaves and W. M. Pardridge, Neutral amino acid transport at the human blood-brain barrier, J. Biol. Chem. 263:19392 (1988).

49. M. A. Shotwell, P. M. Mattes, D. W. Jayme, and D. L. Oxender, Regulation of amino acid transport system L in Chinese hamster ovary cells, J. Biol. Chem. 257:2974 (1982).

50. D. T. Vistica, Cellular pharmacokinetics of the phenylalanine mustards, Pharmac. Ther. 22:379 (1983).

51. L. Weissbach, M. E. Handlogten, H. N.Christensen, and M.S. Kilberg, Evidence for two Na-independent neutral amino acid transport systems in primary cultures of rat hepatocytes, J. Biol. Chem. 257: 12006 (1982).

52. H. Weiler-Guttler, H. Zinke, B. Mockel, A. Frey, and H. G. Gassen, cDNA cloning and sequence analysis of the glucose transporter from porcine blood-brain barrier, Biol. Chem. Hoppe-Seyler 370:467 (1989).

53. Q. R. Smith, Y. Takasato, D. J. Sweeney, and S. I. Rapoport, Regional cerebrovascular transport of leucine as measured by the in situ brain perfusion technique, J. Cereb. Blood Flow Metab. 5:300 (1985).

54. L. B. Yunger and R. D. Cramer, Measurement and correlation of partition coefficients of polar amino acids, Mol. Pharmacol. 20:602 (1981).

55. I. Tayarani, J. M. Lefauconnier, F. Roux, and J. M. Bourre, Evidence for an alanine, serine, and cysteine system of transport in isolated brain capillaries, J. Cereb. Blood Flow Metab. 7:585 (1987).

56. G. Banos, P. M. Daniel, S. R. Moorhouse, and O. E. Pratt, The influx of amino acids into the brain of the rat in vivo: the essential compared with some nonessential amino acids, Proc. R. Soc. Lond. B. 183:59 (1973).

57. G. Banos, P. M. Daniel, S. R. Moorhouse, and O. E. Pratt, The requireof the brain for some amino acids, J. Physiol. (Lond.) 246:539 (1975).

58. H. Sershen and A. Lajtha, Inhibition patterns by analogs indicates the presence of ten or more transport systems for amino acids in brain cells, J. Neurochem. 32:719 (1979).

59. L. A. Wade and H. M. Brady, Cysteine and cystine transport at the blood-brain barrier, J. Neurochem. 37:730 (1981).

60. R. G. Blasberg, J. D. Fenstermacher, and C. S. Patlak, Transport of alpha-aminoisobutyric acid across brain capillary and cellular membranes, J. Cereb. Blood Flow Metab. 3:8 (1983).

61. T. Nagashima, J. M. Lefauconnier, and Q. R. Smith, Developmental changes in neutral amino acid transport across the blood-brain barrier, J. Cereb. Blood Flow Metab. 7:S501 (1987).

62. C. Cangiano, P. Cardelli-Cangiano, F. Ceci, A. Fiori, M. Mulieri, M. Muscaritoli, C. Barberini, R. Strom, and R. Fanelli, Uptake of amino acids by brain microvessels isolated from rats with experimental chronic renal failure, J. Neurochem. 51:1675 (1988).

63. G. Banos, P. M. Daniel, and O. E. Pratt, Saturation of a shared mechanism which transports L-arginine and L-lysine into the brain of the living rat. J. Physiol. (Lond.) 236:29 (1974).

64. M. F. White, The transport of cationic amino acids across the plasma membrane of mammalian cells, Biochim. Biophys. Acta 822:355 (1985).

65. W. H. Oldendorf, P. D. Crane, L. D. Braun, E. A. Gosschalk, J. D. Diamond, and M. A. Hill, pH dependence of histidine affinity for blood-brain barrier carrier transport systems for neutral and cationic amino acids, J. Neurochem. 50:857 (1988).

66. A. M. Mans, J. F. Biebuyck, K. Shelly, and R. A. Hawkins, Regional blood-brain barrier permeability to amino acids after portacaval anastomosis, J. Neurochem. 38:705 (1982).

67. H. Davson, M. N. Lipovac, J. B. Mackic, J. E. Preston, M. B. Segal, G. Tang, and B. V. Zlokovic, Kinetics of L-glutamic acid uptake by the luminal side of the blood-brain barrier studied using an in situ perfused brain of the anesthetized guinea pig, J. Physiol. (Lond.) 423:36P (1990).

68. W. M. Pardridge, Regulation of amino acid availability to brain: Selective control mechanisms for glutamate, in: "Glutamic Acid: Advances in Biochemistry and Physiology," L. J. Filer, Jr., ed., Raven Press, New York, pp. 125-137 (1979).

69. L. R. Drewes, W. P. Conway, and D. D. Gilboe, Net amino acid transport between plasma and erythrocytes and perfused dog brain, Am. J. Physiol. 233:E320 (1977).

70. J. M. Pell and E. N. Bergman, Cerebral metabolism of amino acids and glucose in fed and fasted sheep, Am. J. Physiol. 244:E282 (1983).

71. M. T. Price, M. E. Pusateri, S. E. Crow, S. Buchsbaum, J. W. Olney, and O. H. Lowry, Uptake of exogenous aspartate into circumventricular organs but not other regions of adult mouse brain, J. Neurochem. 42:740 (1984).

72. P. M. Daniel, O. E. Pratt, and P. A. Wilson, The transport of leucine into the brain of the rat in vivo: saturable and nonsaturable components of influx, Proc. Roy. Soc. (Lond.) B. 196:333 (1977).

73. L. P. Miller, W. M. Pardridge, L. D. Braun, and W. H. Oldendorf, Kinetic constants for blood-brain barrier amino acid transport in conscious rats, J. Neurochem. 45:1427 (1985).

74. R. A. Klein, M. J. Moore, and M. W. Smith, Selective diffusion of neutral amino acids across lipid bilayers, Biochim. Biophys. Acta 233:420 (1971).

75. H. N. Christensen, Distinguishing amino acid transport systems of a given cell or tissue, Methods of Enzymology 173:576 (1989).

76. M. Aoyagi, Y. Takada, M. Matocha, S. I. Rapoport, and Q. R. Smith, Glucose transport across the blood-brain barrier: a kinetic analysis using the in situ brain perfusion technique. J. Neurochem. Submitted (1990).

77. J. E. Preston, M. B. Segal, G. J. Walley, and B. V. Zlokovic, Neutral amino acid uptake by the isolated perfused sheep choroid plexus, J. Physiol. (Lond.) 408:31 (1989).

78. E. M. Wright, Active transport of glycine glycine across the frog arachnoid membrane, Brain Res. 76:354 (1974).

78. E. M. Wright, G. J. Nogueira, and E. Levin, Role of the pia mater in the transfer of substances in and out of the cerebrospinal fluid, Exp. Brain Res. 13:294 (1971).

80. K. C. Wadhwani, Q. R. Smith, and S. I. Rapoport, Facilitated transport of L-phenylalanine across blood-nerve barrier of rat peripheral nerve, Am. J. Physiol. 258:R1436 (1990).

81. N. J. Abbott, J. Hart, L. Rogac, M. Taylor, and B. V. Zlokovic, Amino acid transport by the glial blood-brain barrier of the anesthetized dogfish, J. Physiol. (Lond.) 407:25P (1988).

82. O. E. Pratt, Transport inhibition in the pathology of phenylketonuria and other inherited metabolic diseases, J. Inherited Metabol. Dis. 5:75 (1982).

83. J. D. Fernstrom and R. J. Wurtman, Brain serotonin content: physiological regulation by plasma neutral amino acids, Science 178: 414 (1972).

84. J. C. LaManna and S. I. Harik, Regional studies of blood-brain barrier transport of glucose and leucine in awake and anesthetized rats, J. Cereb. Blood Flow Metab. 6:717 (1986).

85. P. Cardelli-Cangiano, C. Cangiano, J. H. James, B. Jeppsson, W. Brenner, and J. E. Fischer, Uptake of amino acids by brain microvessels isolated from rats after portacaval anastomosis, J. Neurochem. 36:627 (1981).

86. T. Jonung, P. Rigotti, J. H. James, K. Brackett, and J. E. Fischer, Effect of hyperammonemia and methionine sulfoxime on the kinetic parameters of blood-brain transport of leucine and phenylalanine, J. Neurochem. 45:308 (1985).

87. P. Rigotti, T. Jonung, J. C. Peters, J. H. James, and J. E. Fischer, Methionine sulfoxime prevents the accumulation of large neutral amino acids in brain of portacaval-shunted rats, J. Neurochem. 44:929 (1985).

88. A. M. Mans, J. F. Biebuyck, and R. A. Hawkins, Ammonia selectively stimulates neutral amino acid transport across blood-brain barrier, Am. J. Physiol. 245:C74 (1983).

89. A. L. McCall, W. R. Millington, and R. J. Wurtman, Metabolic fuel and amino acid transport into the brain in experimental diabetes mellitus, Proc. Natl. Acad. Sci. USA 79:5406 (1982).

90. J. T. Brosnan, R. G. P. Forsey, M. E. Brosnan, Uptake of tyrosine and leucine in vivo by brain of diabetic and control rats, Am. J. Physiol. 247:C450 (1984).

91. C. A. Fenerty and W. E. Lindup, Brain uptake of L-tryptophan and diazepam: the role of plasma protein binding, J. Neurochem. 53:416 (1989).

92. Q. R. Smith, S. Fukui, P. J. Robinson, and S. I. Rapoport, Influence of cerebral blood flow on tryptophan uptake into brain, in: "Amino Acids: Chemistry, Biology and Medicine," G. Lubec and G. A. Rosenthal, ed., ESCOM, Amsterdam, pp. 364-369 (1990).

93. T. L. Perry, S. Hansen, J. Kennedy, CSF amino acids and plasma-CSF amino acid ratios in adults, J. Neurochem. 24:587 (1975).

94. P. H. Hutson, G. S. Sarna, B. D. Kantamaneni, and G. Curzon, Monitoring the effect of a tryptophan load on brain indole metabolism in freely moving rats by simultaneous cerebrospinal fluid sampling and brain dialysis, J. Neurochem. 44:1266 (1985).

95. H. Davson, D. J. Begley, D. G. Chain, F. O. Briggs, and M. T. Shepherd, Steady-state distribution of cycloleucine and alpha-aminoisobutyric acid between plasma and cerebrospinal fluid, Exp. Neurol. 91:163 (1986).

96. P. Brust, Changes in regional blood-brain transfer of L-leucine elicited by arginine-vasopressin, J. Neurochem. 46:534 (1986).

97. P. Brust and J. Zicha, Kinetics of regional blood-brain barrier transport of L-leucine in Brattleboro rats, Biomed. Biochim. Acta 12:1013 (1988).

98. T. Eriksson and A. Carlsson, Beta-adrenergic control of amino acid uptake of large neutral amino acids, Life Sci. 42:1583 (1988).

ENERGY AND MACRONUTRIENT INTAKE REGULATION: INDEPENDENT OR INTERRELATED MECHANISMS?

G. Harvey Anderson and Richard M. Black

Department of Nutritional Sciences
Faculty of Medicine
University of Toronto, Toronto, Canada

INTRODUCTION

The initiation and termination of feeding are complex processes involving many signals and their integration in the central nervous system. Fluxes of metabolic fuels should logically be determinants of both the initiation and termination of feeding to meet energy needs of the organism, but the relationships of these fluxes to the regulatory centres in the central nervous system are presently undefined. Furthermore, it has become clear that the ingestion, digestion and metabolism of carbohydrate, fat and proteins provides a wide array of signals to the central nervous system and that these signals influence not only the quantity of food consumed but also the composition of the chosen foods. The purpose of this presentation is to highlight these influences and their role in food intake regulation and to provide examples of linkages between body fuel composition, central nervous system neurotransmitters and feeding behaviour.

OVERVIEW OF FOOD TO BRAIN SIGNALS

The brain serves as the prime organizer and integrator of biologically relevant signals arising from many sources, both from within and outside of the body, to achieve regulation of body energy intake, output and storage (Anderson, 1988). In humans, cultural and social conventions are significant modifiers of the impact of the metabolic and physiologic cues that arise from the perception, ingestion, digestion, absorption and metabolism of food and nutrients. However, powerful determinants of feeding behaviour also arise from the absence of food, which results in the initiation of feeding, or from the assimilation of food eaten, which results in the termination of feeding.

Uncertain at the present time is the relative importance of the many signals provided to the brain, both during and after food ingestion. Hypotheses have, at various times, stressed the importance of signals arising from the ingestion,

Fuel Homeostasis and the Nervous System, Edited by M. Vranıc *et al.*
Plenum Press, New York, 1991

digestion, absorption, or the metabolism of ingested food. Yet the pre-eminence of any one signal has not emerged. Evolution seems to have provided a great deal of redundancy in feeding control mechanisms making it difficult to sort out primary and secondary signals.

METABOLIC FUEL SIGNALS TO THE CENTRAL NERVOUS SYSTEM

For metabolic events to affect feeding behaviour they must be detected by the nervous system. Logically, one might expect the brain to be very sensitive to the availability of energy producing fuel because the brain has a very high metabolic rate relative to most other body organs. In the human, the brain constitutes only 2% of adult body weight, but receives 15% of cardiac output and accounts for 20-30% of the whole body resting metabolic rate. Because the respiratory quotient of the brain is 0.97, and since its glycogen stores are only 0.1% of its weight, the brain must depend on a continuous supply of oxygen and glucose (Sokoloff et al., 1977). Glucose needs are met by the facilitated transport of glucose across the blood-brain barrier by the hexose specific carrier. Under normal circumstances fatty acids supply little of the energy needed by the brain, because their only entry is through passive diffusion. In contrast to fatty acids, amino acids play several key roles in the brain, including the provision of substrate for protein and neurotransmitter synthesis. They are taken up from the blood and cross the blood-brain barrier by rate-limited carrier mediated transport mechanisms (Pardridge, 1984; Pardridge and Oldendorf, 1975). However, the oxidation of amino acids accounts for less than 10% of the total whole-brain energy utilization (Sokoloff et al., 1977). Nonetheless, despite the brain's high demand for energy it is not clear that the availability of energy substrate per se plays a primary role in the regulation of feeding behaviour.

Specific brain areas have been implicated in the control of food intake. For example, the hypothalamus has been identified as playing a central integrative role in the regulation of feeding behaviour. Originally, on the basis of lesion studies, the ventromedial hypothalamus (VMH) was identified as the satiety centre because its destruction led to overeating and obesity. The lateral hypothalamus (LH) was named the feeding centre because destruction of this brain region resulted in aphagia (Carlisle and Stellar, 1969). Other brain regions are also involved however, e.g. the paraventricular nucleus (Leibowitz and Shor-Posner, 1986), suggesting that it is more appropriate to recognize the hypothalamus as an integrative unit functioning in the brain's reception and organization of the many signals arising from food ingestion.

The signals arising from the ingestion and metabolism of foods may be pre-absorptive as well as post-absorptive, and can be unrelated to the energy value of the food. Ingestion and digestion of food has the potential to stimulate glucose specific (Pardridge and Oldendorf, 1977) and amino acid specific (Jeanningros, 1982) chemoreceptors of the vagus nerve. Glucose, amino acids and fat also cause the release of gastrointestinal hormones, many of which, including chole-cystokinin, inhibit feeding (Smith and Gibbs, 1981). After entry into the portal vein and passage to the liver, these nutrients may also provide, via vagal innervation, information to various brain areas associated with feeding (Anil and Forbes, 1987).

The brain can monitor the body's milieu interior through plasma fluctuations in nutrients, which in turn create changes in the concentrations of these nutrients within the brain. Neurons sense these changes in nutrient availability by a variety of mechanisms, including direct interaction through receptors for glucose (Oomura, 1976) and amino acids (Wayner et al., 1975), monitoring of change in rate of glucose utilization (Mayer, 1980), and fluctuation in neurotransmitter synthesis and release (Anderson et al., 1984). The information collected by the neurons eventually becomes integrated and feeding behaviour is regulated to maintain the nutritional homeostasis of the organism (Morley and Levine, 1983).

Clearly, each of the macronutrients, in addition to providing energy substrates for cells, has properties which could provide signals to the central nervous system and thus impact directly or indirectly on feeding behaviour. However, it was perhaps the recognition of cellular needs for energy substrates which in turn led to the emergence of a major hypothesis relating glucose, fat and amino acid metabolism to the regulation of food intake and energy balance.

REGULATION OF FOOD INTAKE AND ENERGY BALANCE

The focus of research on the regulation of energy intake arises from observation on the constancy of adult body weight (Durnin, 1961) and concerns with deviations from this as occur in obesity, anorexia nervosa and bulimia nervosa in humans. Furthermore, it is clear that food intake is rapidly and appropriately adjusted to meet energy requirements under a variety of conditions. For example, food intake is increased if rats are exposed to low ambient temperatures (Sellers et al., 1954; Musten et al., 1974) or to exercise (Collier et al., 1969) such that body weight, or growth, remain normal. Evidence such as this has led to the view that maintenance of energy balance is the primary purpose of food intake regulatory mechanisms. It is logical, therefore, that attempts to define control mechanisms have concentrated on both the maintenance of energy balance as an easily defined motivation for feeding, and the components of food which give rise to energy.

Carbohydrate, fat and protein each produce metabolic energy and each has received extensive investigation as regulators of feeding behaviour. Furthermore, the metabolism of each of these macronutrients provides important information to the central nervous system (CNS).

GLUCOSE

The glucostatic theory of feeding (Mayer, 1953) is based on the view that glucose availability to, and utilization by the brain is a primary determinant of food intake. Consistent with this hypothesis is the observation that decreases in blood glucose concentration occur prior to the initiation of meals by rats, presumably creating a deficit which is read out in some way by the brain (Louis-Sylvestre and Le Magnen, 1980). In later studies, the continuous and simultaneous monitoring of blood glucose and feeding has enabled the measurement of a close temporal association between a drop in blood glucose and the onset of feeding in the rat (Campfield et al., 1985).

However, because hypoglycemia does not always elicit eating, it can be argued that it is not glucose per se, but rather it is the energy derived from it that is the signal relevant to feeding behaviour. For example, if rats are treated with insulin to bring about hypoglycemia, feeding will not occur if fructose is infused into the portal vein (Stricker et al., 1977). This failure to eat in the face of hypoglycemia may be related to changes in peripheral fuel flux, resulting in increased energy production in peripheral tissues. This view is supported by studies of feeding responses in rats after energy producing pathways of the liver are blocked. Blockage of energy production from substrates entering the Krebs cycle results in increased food intake, suggesting a link between peripheral energy utilization and central nervous system regulation of food intake (Friedman et al., 1986). The precise nature of this link has not yet been identified, but one hypothesis proposes an information pathway to the brain utilizing the vagus nerve from the liver (Anil and Forbes, 1987). Brain cellular glucose utilization also appears to be closely related to feeding: a high rate of utilization corresponds with a state of satiation in animals (Glick and Mayer, 1968) and the converse with eating (Muller et al., 1973).

The role of blood glucose concentration per se as a prime determinant of feeding behaviour is also challenged by the fact that hyperglycemia does not necessarily result in failure to eat. Rats receiving a continuous infusion of glucose, sufficient to raise blood glucose by a constant 20-30 mg/dl, did not alter either frequency of eating or the amount consumed (Rezek et al., 1975). In addition, despite the three to four fold elevation in blood glucose concentrations in diabetic rats, these animals display a mild hyperphagia rather than a hypophagia (Shimomura et al., 1990). Therefore, while it is obvious that alterations in blood glucose can affect feeding behaviour, the mechanism(s) underlying this effect have not been elucidated.

LIPID

Because transfer of fatty acids across the blood-brain barrier is relatively slow, the oxidation of this fuel in the brain is unlikely to be a significant factor in food intake regulation. However, an indirect role for lipid in influencing some of the brain areas associated with feeding is evident. For instance, blockage of fatty acid utilization for energy production in the liver results in increased feeding, perhaps due to information on the energy state of the liver being relayed to the brain via the vagus (Friedman et al., 1986). Also, a link between the VMH and lipolysis has been found. Stimulation of the VMH leads to enhanced lipolysis (Shimazu, 1981) and lesions of the VMH (or the ventromedial nucleus of the VMH) lead to an enhanced lipogenesis even in the absence of an elevated intake (Weingarten et al., 1985; Parkinson and Weingarten, in press). Finally, electrical activity of the VMH, but not of the lateral hypothalamus (the "feeding" centre), is high during the light period when the animals are not eating and are drawing on body stores for energy, but is suppressed during the dark period when they are eating (Schmitt, 1973). Thus, while fatty acid metabolism may have little effect on any individual meal, the neural control of fatty acid metabolism may be part of the intrinsic mechanism which governs the circadian feeding behaviour of rats.

Although the mechanism remains unknown, it is possible that enhanced lipolysis provides an energy substrate for the liver which in turn provides information to the brain either directly via the vagus (Friedman et al., 1986), or indirectly through a delay in the fall of blood glucose (LeMagnen, 1983). In either case, the result might be a reduction in the frequency of eating during the day.

There is a crucial role, however, for insulin in the integration of fatty acid mobilization and utilization with feeding behaviour. In insulin deficiency, lipolysis and fatty acid breakdown in the liver occurs at a much higher rate than that occurring in normal rats feeding infrequently during the light hours of the day. Logically, this should lead to an increase in metabolic fuels available to the liver and a subsequent decrease in feeding. Yet both food intake relative to body weight, and daytime feeding, are increased in insulin deficiency (Reuterving and Hagg, 1987). This suggests that in the absence of insulin, signals arising from lipolysis and fatty acid breakdown in the liver affect feeding in a manner fundamentally different from that when insulin levels are normal.

Although the importance of body lipid stores and plasma lipids as a source of energy is unquestioned, it is clear that neither independently controls feeding behaviour. Similarly, no readily apparent breakdown product of lipid metabolism explains feeding behaviour. Plasma levels of free fatty acids, glycerol, triglycerides, and ketone bodies fluctuate with fasting and feeding but there is no evidence that they provide signals to the central nervous system which are instrumental in the regulation of feeding behaviour.

AMINO ACIDS

Interest in the role of amino acids in food intake regulatory mechanisms was stimulated by the observations of Mellinkoff et al (1956). They reported an inverse relationship between serum amino acid concentration after intravenous infusions of an amino acid hydrolysate and appetite in humans. This led to the hypothesis that the effects of dietary protein and amino acids on food intake were mediated by signals to the central nervous system arising from plasma amino acids. In support of this hypothesis are the observations that high protein diets, as well as amino acid deficient or imbalanced diets, cause sharp reductions in the food intake of experimental animals and of humans (for a review, see Anderson, 1979).

Brain regions sensitive to changes in plasma amino acid concentrations lie outside the hypothalamus, one of the centres traditionally recognized in the regulation of food intake. The identification of these regions provides evidence that feeding responses are guided by some aspect of central nervous system response to dietary amino acids. In rats, lesions of the VMH have little effect on the depression of food intake which follows the consumption of a high protein diet (Rogers and Leung, 1973). However, lesions of the prepyriform cortex or the amygdala prevent the food intake suppression normally observed when rats are fed amino acid imbalanced diets.

Precisely why changes in plasma amino acid concentrations are of significance to the regulation of food intake by healthy animals remains undefined. As pointed out earlier, amino acids are not a major energy source for the brain. It is possible that elevations in amino acid concentrations in the brain provide a non-specific indicator of excess production of ammonia or urea resulting from protein intake. But while high concentrations of both ammonia and urea have been shown to decrease food intake, concentrations required are higher than those normally found even after consumption of a high protein diet. Furthermore, studies on the effects of the addition of ammonia or urea to diets do not explain the feeding responses of rats consuming high protein diets. Finally, while changes in plasma amino acid concentration are also reflected in the central nervous system, the relationship between plasma concentrations and brain concentrations is determined by blood-brain barrier transport mechanisms.

How amino acids provide signals to the different brain areas involved in the control of feeding is unknown. In the brain, amino acids not only serve as neurotransmitters, but also as precursors for neurotransmitters, several of which are known to be involved in food intake regulation (Anderson and Johnston, 1983; Anderson et al., 1984). The latter include the monoamines such as norepinephrine, dopamine and serotonin, histamine, aminobutyric acid and an expanding list of neuropeptides such as endorphin, the opiates, cholecystokinin, bombesin, neurotensin, calcitonin, corticotrophin releasing factor and thyrotropin-releasing hormone (Morley and Levine, 1983). Amino acids may also provide direct signals to neurons because receptors for specific amino acids are present on brain neurons.

More recent research would seem to indicate that amino acid concentrations are not signals per se. Rats made insulin deficient through administration of streptozotocin are hyperphagic despite plasma and brain amino acid concentrations which are similar to those occurring after consumption of high protein diets and generally associated with food intake suppression (Glanville and Anderson, 1985). In these rats, plasma gluconeogenic amino acids are decreased due to enhanced hepatic extraction for glucose production, whereas the branched chain amino acids and methionine are increased (Glanville and Anderson, 1985). This elevation of branched chain amino acids in plasma is directly related to their decreased uptake by muscle, which depends upon an insulin facilitated transport combined with decreased protein synthesis. This suggests the possibility that it is the gluconeogenic amino acids which are important in food intake regulation, though this remains to be seen. (It should be noted that the high plasma concentrations of methionine may be related to a specific toxicity effect of streptozotocin on the liver (Glanville and Anderson, 1984a), or some modification of other metabolic responses to diet (Dyer et al., 1988) and indicate a complication of using this drug to create an animal model of diabetes.)

In brains of diabetic rats, methionine and valine concentrations are increased, but threonine, tyrosine and tryptophan concentrations are decreased (Glanville and Anderson, 1985). A good association is found between the calculated brain influx rate and the actual brain concentrations of threonine, methionine, tyrosine and tryptophan. There is no correlation, however, between brain influx rate and brain branched chain amino acid levels, suggesting that the brain tissue is able to either catabolize the increased influx of these amino acids or to increase afflux rates. It is difficult at this time to ascertain the precise role of these amino acids in food intake regulation.

REGULATION OF FOOD SELECTION AND MACRONUTRIENT INTAKE

Animals survive and reproduce in a variety of nutritional environments. To do so they need to obtain an adequate supply not only of energy, but essential nutrients as well. Because adequate food selection is a necessary component of good nutritional health, it has been suggested that mechanisms have evolved which link internal metabolic processes to nutrient need and the consequences of eating.

The concept of nutrient specific appetites was initially suggested by Osborne and Mendel (1918) who observed that "the desire of a young animal for food is something more than a satisfaction of its caloric needs". Nutrient specific appetites were extensively investigated by Richter et al. (1938) and Lat (1967). They suggested that rats were able to maintain relatively constant daily intakes of most nutrients by adjusting their food choices according to nutrient requirements and nutrient availability.

The presence of nutrient specific appetites suggests that mechanisms have evolved to regulate these appetites. If so, the apparent redundancy of control mechanisms leading to the maintenance of energy balance may be a consequence of the experimental approaches used by investigators. That is, most investigators studying food intake regulatory mechanisms feed their test animals diets of fixed composition, and so their conclusions about the regulatory mechanisms under study can relate only to feeding responses to that particular diet, and to the overall energy balance of the animal. It may be, though, that some control mechanisms play a stronger guiding role in food selection than in determining the total amount of food intake.

It is only relatively recently, however, that experimental approaches have been developed to examine the role of putative feeding mechanisms in food selection and macronutrient intake in addition to total food energy intake. Interest in mechanisms was brought about by research in the 1970's which provided additional evidence that rats given appropriate dietary choices have a capacity to regulate not only their energy balance, but also their intakes of the macronutrients protein, carbohydrate and fat (Anderson et al., 1984). Furthermore, it was shown that these nutrient specific appetites differed from the norm in animal models of obesity (Anderson et al., 1979) and were adjusted appropriately during physiologic demands of exercise (Collier et al., 1969) and cold exposure (Musten et al., 1974).

To study food and macronutrient selection, the experimental animal must be given a choice of diets which allows it to regulate both quality and quantity of food. An example of such a study is the systematic investigation of protein specific appetites of young rats reported by Musten et al. (1974). In these studies rats were provided with two diets one high (e.g. 50%) and one low (e.g. 5%) in protein content. Both diets contained 10% by weight of fat and identical concentrations of essential vitamins and minerals. Carbohydrate varied inversely with protein, so that the high protein diet was relatively low in carbohydrate and the low protein diet was high in carbohydrate. Through an equal exchange of protein for carbohydrate, the energy density was identical for the two diet choices. Over a four week period, these young rats maintained their protein intake at approximately 33% of total dietary energy.

Research also indicates that protein intake regulation appears to be separate from energy (or total food) intake regulation. Physiologic situations which increased energy requirements, such as cold exposure (Musten et al., 1974) or increased activity (Collier et al., 1969) resulted in the animal increasing food intake, but by selecting diets which gave it increased carbohydrate energy. Protein consumption in these animals was unchanged in relation to body weight, but was decreased as a proportion of the total dietary energy consumed. Similarly, when given choices of high protein and high carbohydrate diets, rat models of overeating (including hypothalamic (VMH) hyperphagia and the genetically determined hyperphagia of the Zucker fatty rat) exhibited a selective drive to overeat carbohydrate and fat energy, but maintained relative control over protein intake (Anderson et al., 1979).

Carbohydrate may also be a separately regulated entity from protein. Dietary manipulations have led to the demonstration of a constancy of carbohydrate intake or carbohydrate selective appetites under certain conditions (Anderson, 1988). It is also clear that rats will shift food choices depending upon the composition of a recently consumed meal. A high protein meal leads to a carbohydrate preference in the next meal (Li and Anderson, 1982) and vice versa (Wurtman et al., 1983). Thus rats can recognize diet composition, and may adjust their intake as a result of the metabolic consequences arising from food ingestion.

Further evidence for macronutrient specific appetites comes from studies of streptozotocin-induced diabetic rats. These animals selected diets similar in protein concentration to normal rats given a choice of 10 and 60% protein diets (Woodger et al., 1979). However, total food intake in relation to body weight was doubled. The failure of diabetic rats to switch to a preference for the protein diet is surprising, since not only would this have reduced carbohydrate intake, but high protein diets have been shown to reduce blood and urinary glucose in diabetic rats and to result in improved growth and survival. Therefore, while protein intake was regulated in some manner in diabetic rats, this regulation was not necessarily to the rats advantage.

There may be many different brain mechanisms affecting food choice and macronutrient regulation which operate in concert or parallel to each other, but at present very few have received detailed investigation. Systems involving the monoamines, 5 hydroxytryptamine (serotonin, 5HT) and the catecholamines, norepinephrine (NE) and dopamine (DA), have received the most study, perhaps for two reasons. First, they are well-known to play a role in control mechanisms for feeding and second, their synthesis and release in neurons is influenced by brain precursor concentrations which in turn are influenced by diet composition. From these observations it has been argued that a direct link between food composition and a signal to the brain may exist. However, proving that these neurotransmitter systems are important regulators under physiologic conditions of meal to meal feeding and food selection has been difficult and polarized view points have developed (Fernstrom, 1987). On the other hand there is convincing evidence that pharmacologic or other non-physiologic manipulations of these neurotransmitter systems can alter the rat's food selection, providing support for the existence of regulatory mechanisms for macronutrient specific appetites (Luo and Li, 1990; Luo et al., 1990).

Although the relationship between serotonin and physiologic responses to diet is uncertain at the present time, the relationship of serotonin to food intake and selection after pharmacologic manipulation is clear. Agreement on the role of serotonin as a neurotransmitter involved in feeding control mechanisms has developed over the past twenty years. This has been based primarily on studies of rats fed single diets while serotonergic systems have been manipulated pharmacologically or through destruction of serotonin neurons (Leibowitz and Shor-Posner, 1986). Based on such studies, serotonin has come to be known as a neurotransmitter which inhibits feeding. A decrease in the food intake of a rat is found after administration of serotonin agonists such as dexfenfluramine, fluoxetine and RU24969 (Luo and Li, 1990) whereas increased food intake is seen after administration of the antagonists 8-hydroxy-s-(di-n-propylamino)-tetraline (8-OH-DPAT) or buspirone (Luo et al, 1990). Consistent with the notion that serotonin is under precursor control and that this may be physiologically relevant, are reports that tryptophan (a serotonin precursor) given in pills to humans (Hrboticky et al., 1985) or injected intraperitoneally in rats (Morris et al., 1986) suppresses food intake by 20-40% in meals consumed within the following 60 to 90 minutes. Moreover, the dose required is pharmacologic in that it approximates the amount of tryptophan normally consumed in one day.

A role for serotonergic neurons in food selection has also emerged from pharmacologic or destructive manipulations of these neurons. Inhibition of serotonin synthesis with para-chlorophenylalanine or destruction of serotonin neurons by 5,7-dihydroxytryptamine infusion or through mid-brain raphe lesions by a thermal probe, led to a shift in diet choice: a 20-30% decrease in protein consumption was observed, although total food consumption was unaffected (Ashley et al., 1979). Though it was not recognized at the time, the data suggested that a depletion in brain serotonin may act as a stimulus for the rat to exhibit a preference for carbohydrate over protein, due to a relative deficit in brain serotonin. This interpretation has recently become more clear as a result of studies using drugs which are serotonin antagonists. Intraperitoneal injections of buspirone and 8-OH-DPAT, serotonin antagonists of various specificity, caused enhanced food intake at least in the short-term. More specifically, increased food intake was selectively obtained from the carbohydrate rich diet in a two diet choice feeding paradigm (Luo et al., 1990).

Tryptophan as a serotonin precursor and agonist has a very weak effect on food selection when given in pharmacologic doses intraperitoneally or intragastrically, even though a 30-40% reduction in food intake is brought about in the subsequent one or two hours of feeding (Morris et al., 1986). However, the serotonin agonists dexfenfluramine and fluoxetine have clear food selective actions in rats given a choice of high protein and high carbohydrate diets, but these effects are strongly dose dependent (Luo and Li, 1990). At low doses of the drugs, the decrease in food intake is clearly accounted for by a decrease in carbohydrate intake, with no change or even a small increase in protein intake. At high doses however, food intake is strongly suppressed from both diet choices.

The arguments for a physiologic role of brain serotonin in food intake regulation is based on changes in brain serotonin concentration resulting from changes in diet composition. For the past twenty years, brain serotonin synthesis has been understood to be influenced by the availability of its precursor, tryptophan. At normal brain concentrations of tryptophan, the rate limiting enzyme in serotonin synthesis (tryptophan hydroxylase) is not fully saturated. Arguments for this being of physiologic significance were provided by Fernstrom and Wurtman (1971a). They injected tryptophan intraperitoneally into fasted rats and found that brain serotonin increased linearly with tryptophan doses from 5 to 25 mg/kg and plateaued thereafter. Subsequently they allowed rats fasted overnight to eat for three hours from either a high carbohydrate or a standard laboratory chow diet, and observed an increase in brain tryptophan and serotonin only after ingestion of the high carbohydrate diet (Fernstrom and Wurtman, 1971b).

The mechanism for this increase in brain tryptophan and serotonin concentrations is straightforward. Carbohydrate increased brain serotonin by increasing tryptophan concentration in the brain. The tryptophan increase occurs as a result of the insulin release after carbohydrate consumption, which in turn increases, with the exception of tryptophan, amino acid uptake by tissues. The relative retention of tryptophan in plasma is due to its unique binding and transport by albumin. Since it shares binding sites on albumin with free fatty acids, when insulin is released and fatty acid mobilization from adipose tissue is blocked, more binding sites on albumin are available for tryptophan. Tryptophan bound to albumin is readily taken up by the brain capillary transport system for the large neutral amino acids (NAA) which also include tyrosine, phenylalanine, valine, isoleucine and leucine. Because the transport system is fully saturated at normal blood amino acid concentrations, these amino acids compete for transport sites across the blood brain barrier. Alterations in the relative concentrations of these amino acids would necessarily alter the competition for transport sites. Insulin release after carbohydrate consumption gives tryptophan an increased competitive advantage by increasing the ratio of tryptophan to NAA in plasma.

Despite the extensive evidence for the involvement of diet induced changes in brain tryptophan and serotonin in the control of feeding and diet selection, the results must be interpreted with caution. The original studies may have led to an overestimation of the physiologic significance of the relationship between food consumption and brain serotonin. In those studies, rats were allowed to eat ad libitum for three hours after a 15 hour overnight fast (Fernstrom and Wurtman, 1971b). Under such circumstances rats will eat large amounts (5 - 7 g) of food even in the first hour and the resulting modest response in brain serotonin concentration (an increase of 20-30%) may not be reflective of what would be expected to occur after normal meal eating bouts which average two grams of food (Johnson et al., 1979).

Recently, it has been shown that gavage with two grams of glucose increases, whereas gavage with one gram of casein decreases, brain tryptophan and serotonin in rats compared with saline treated controls (Teff and Young, 1988). The decrease in brain tryptophan and serotonin after gavage with protein is explained by the high content of branched chain amino acids and tyrosine and phenylalanine compared with tryptophan in the gavage solution, which results in a reduction of the plasma TRP/NAA ratio. Thus, there is little evidence that the amount of food eaten under normal feeding conditions in animals given dietary choices leads to fluctuations in serotonin synthesis or that these fluctuations account for food choices in diet selection paradigms.

The abnormal brain amino acid patterns and hyperphagia in diabetic animals could be used to argue that tryptophan and tyrosine availability to the brain is physiologically relevant to food intake regulatory mechanisms. Both tryptophan and tyrosine concentrations are reduced in brains of hyperphagic diabetic rats since in the absence of insulin, the other NAA's are not preferentially taken up into tissue and so compete effectively for transport across the blood brain barrier (Glanville and Anderson, 1985, Masiello et al., 1987). This in turn could lead to reduced levels of serotonin in the brian. In addition, at least tryptophan is more rapidly cleared from the blood stream in diabetic animals, which would also contribute to less tryptophan being available for brain uptake and subsequent serotonin synthesis (Masiello et al., 1987).

If precursor availability for synthesis of the neuro-transmitters serotonin and the catecholamines was a factor in the hyperphagia, then adding tryptophan or tyrosine to diets should reduce hyperphagia and possibly enhance the efficiency of utilization of dietary energy. Dietary tryptophan additions, however, do not reduce hyperphagia even though brain serotonin concentrations can be restored to control ranges (Woodger et al., 1979). In contrast, when tyrosine is added to the diet of diabetic rats allowed to select between high protein and high carbohydrate diets, the rats show a 40% increase in weight gain and improved feed efficiency. Of course it is also possible that the growth promoting effect of tyrosine is related to tyrosine stimulation of central and peripheral catecholamines (Glanville and Anderson, 1984b).

Thus, experimental animals studies show that rats given choices of high protein and high carbohydrate diets regulate their food selection and intake of these macronutrients. Furthermore, using the example of the neurotransmitter serotonin, it can be concluded that some central nervous system control mechanisms are involved in both the regulation of energy balance as well as food selection. This conclusion, however, is based primarily on pharmacologic and destructive manipulations of serotonergic neurons. What physiologic signal, if any, triggers the involvement of these neurons during normal feeding is unknown. While precursor availability as the physiologic signal offers an attractive hypothesis, proof for this hypothesis has been elusive.

CONCLUSION

Regulation of food intake and selection is achieved through a large array of inputs to the nervous system. A signal reflecting the link between the metabolic energy produced by individual macronutrients and food intake has not been found. Similarly, there is no evidence of a primary metabolic signal arising from one of the macronutrients which would account for the food selection/feeding behaviour of an animal. However, it is clear that the ingestion, digestion and metabolism of carbohydrate, fat and proteins influences not only the quantity of food consumed but also food selection. Identification of the mechanisms by which the brain regulates the intake of total energy and of macronutrients will no doubt be assisted by rapid developments in the neurosciences and by techniques which allow measurement of metabolic events in the brain.

REFERENCES

Anderson, G. H., 1979, Control of protein and energy intake: Role of plasma amino acids and brain neurotransmitters. Can. J. Physiol. Pharmacol., 57:1043.

Anderson, G. H., 1988, Metabolic regulation of food intake, in: "Modern Nutrition in Health and Disease, 7th ed.," M. E. Shils and V. R. Young, eds., Lea and Febiger, Philadelphia, Penn.

Anderson, G. H., and Johnston, J. L., 1983, Nutrient control of brain neurotransmitter synthesis and function. Can. J. Physiol. Pharmacol., 61:271.

Anderson, G. H., Leprohon, C., Chambers, J. H., and Coscina, D. V., 1979, Intact regulation of protein intake during the development of hypothalamic or genetic obesity in rats. Physiol. Behav., 22:777.

Anderson, G. H., Li, E. T. S., and Glanville, N.T., 1984, Brain mechanisms and the quantitative and qualitative aspects of food intake. Brain Res. Bull., 12:167.

Anil, M. H., and Forbes, J. M., 1987, Neural control and neurosensory functions of the liver. Proc. Nutr. Soc., 46:125.

Ashley, D. V. M., Coscina, D. V., and Anderson, G. H., 1979, Selective decrease in protein intake following drug induced brain serotonin depletion. <u>Life Sci.</u>, 24:973.

Collier, G., Leshner, A. I., and Squibb, R. L., 1969, Dietary self-selection in active and non-active rats. <u>Physiol. Behav.</u>, 4:79.

Campfield, L. A., Brandon, P., and Smith, F. J., 1985, On-line continuous measurement of blood glucose and meal pattern in free feeding rats: the role of glucose in meal initiation. <u>Brain Res. Bull.</u>, 14:605.

Carlisle, H. J., and Stellar, E., 1969, Caloric regulation and food preference in normal, hyperphagic and aphagic rats. <u>J. Comp. Physiol. Psych.</u>, 69:107.

Dyer, J. R., Greenwood, C. E., and McBurney, M. I., 1988, The effects of diet and duration of diabetes on hypermethionemia in streptozotocin-diabetic rats. <u>Can. J. Physiol. Pharmacol.</u>, 66:95.

Durin, J. G. V. A., 1961, "Appetite" and the relationship between expenditure and intake of calories in man. <u>J. Physiol.</u>, 156:294.

Fernstrom, J. D., 1987, Food-induced changes in brain serotonin synthesis: is there a relationship to appetite for specific macronutrients? <u>Appetite</u>, 8:163.

Fernstrom, J. D., and Wurtman, R. J., 1971a, Brain serotonin content: physiological dependence on plasma tryptophan levels. <u>Science</u>, 173:183.

Fernstrom, J. D., and Wurtman, R. J., 1971b, Brain serotonin content: increase following the ingestion of carbohydrate diet. <u>Science</u>, 174:1023.

Friedman, M. I., Tordoff, M. G., and Ramirez, I., 1986, Integrated metabolic control of food intake. <u>Brain Res. Bull.</u>, 17:855.

Glanville, N. T., and Anderson, G. H., 1984a, Altered methionine metabolism in streptozotocin diabetic rats. <u>Diabetologia</u>, 27:468.

Glanville, N. T., and Anderson, G. H., 1984b, Dietary tyrosine supplementation enhances weight gain in streptozotocin-diabetic rats. <u>Can. J. Physiol. Pharmacol.</u>, 62:781.

Glanville, N. T., and Anderson, G. H., 1985, The effect of insulin deficiency, dietary protein intake and plasma amino acid concentrations on brain amino acid levels in rats. <u>Can. J. Physiol. Pharmacol.</u>, 63:487.

Glick, Z., and Mayer, J., 1968, Hyperphagia caused by cerebral ventricular infusion of phlorizin. <u>Nature London</u>, 219:1374.

Hrboticky, N., Leiter, L. A., and Anderson, G. H., 1985, Effects of L-tryptophan on short term food intake in lean men. <u>Nutr. Res.</u>, 5:595.

Jeanningros, R., 1982, Vagal unitary responses to intestinal amino acid infusions in the anesthetized cat: a putative signal for protein induced satiety. <u>Physiol. Behav.</u>, 28:9.

Johnson, D. S., Li, E. T. S., Coscina, D. V., and Anderson, G. H., 1979, Different diurnal rhythms of protein and non-protein energy intake by rats. <u>Physiol. Behav.</u>, 21:777.

Lat, J., 1967, Self-selection of dietary components, <u>in</u>: "Handbook of Physiology, Vol. 1, Section 6", American Physiological Society, Washington, D.C.

Le Magnen, J., 1983, Body energy balance and food intake: a neuroendocrine regulatory mechanism. <u>Physiol. Rev.</u>, 63:314.

Le Magnen, J., and Devos, M., 1970, Metabolic correlates of the meal onset in the free food intake of rats. <u>Physiol. Behav.</u>, 5:805.

Leibowitz, S. F., and Shor-Posner, G., 1986, Brain serotonin and eating behavior. <u>Appetite</u>, 7 (Suppl):1.

Li, E. T. S., and Anderson, G. H., 1982, Meal composition influences subsequent food selection in the rat. <u>Physiol. Behav.</u>, 29:779.

Louis-Sylvestre, J., and Le Magnen, J., 1980, A fall in blood glucose level precedes meal onset in free feeding rats. <u>Neurosci. Biobehav. Rev.</u>, 4 (Suppl. 1):13.

Luo, S., and Li, E. T. S., 1990, Food intake and selection pattern of rats treated with dexfenfluramine, fluoxetine and RU 24969. <u>Brain Res. Bull.</u>, 24:729.

Luo, S., Ransom, T., and Li, E. T. S., 1990, Selective increase in carbohydrate intake by rats treated with B-hydroxy-2-(Di-N-propylamino)-tetraline or buspirone. <u>Life Sci.</u>, 46:1643.

Mayer, J., 1953, Glucostatic mechanism in regulation of food intake. <u>N. Engl. J. Med.</u>, 249:13.

Mayer, J., 1980, Physiology of hunger and satiety: regulation of food intake, <u>in</u>: "Modern Nutrition in Health and Disease" R. S. Godhard and M. E. Shills. eds., Lea and Febiger, Philadelphia.

Masiello, P., Balestreri, E., Bacciola, D., and Bergamini, E., 1987, Influence of experimental diabetes on brain levels of monoamine neurotransmitters and their precursor amino acids during tryptophan loading. <u>Acta. Diabetol. lat.</u>, 24:43.

Mellinkoff, S. M., Franklin, M., Boyle, D., and Geipell, M., 1956, Relationship between serum amino acid concentration and fluctuation in appetite. <u>J. Appl. Physiol.</u>, 8:535.

Morley, J. E., and Levine, A. S., 1983, The central control of appetite. <u>Lancet</u>, 1:398.

Morris, P., Li, E. T. S., MacMillan, M., and Anderson, G. H., 1986, Effect of tryptophan on food intake and selection in rats. <u>Physiol. Behav.</u>

Muller, E. E., Paneri, A., Cocchi, D., Frohman, L. A., and Mantegazza, P., 1973. Central glucoprivation: some physiological effects induced by the intraventricular administration of 2-deoxy-D-glucose. <u>Experientia</u>, 29:874.

Musten, B., Peace, D., and Anderson, G. H., 1974, Food intake regulation in the weanling rat: self-selection of protein and energy. <u>J. Nutr.</u>, 104:563.

Oomura, Y., 1976, Significance of glucose, insulin and free fatty acids in the hypothalamic feeding and satiety neurons, <u>in</u>: "Hunger: Basic Mechanism and Clinical Implications", D. Novin, W. Wyrwicka, G. Bray, eds., Raven Press, New York.

Osborne, T. B., and Mendel, L. B., 1918, The choice between adequate and inadequate diets, as made by rats, <u>J. Biol. Chem.</u>, 35:19.

Pardridge, W. M., 1984, Transport of nutrients and hormones through the blood brain barrier, <u>Federation Proc.</u>, 43:201.

Pardridge, W. M., and Oldendorf, W. H., 1975, Kenetic analysis of blood-brain barrier transport of amino acids, <u>Biochim. Biophys. Acta.</u>, 401:128.

Pardridge, W. M., and Oldendorf, W. H., 1977, Transport of metabolic substrates through the blood-brain barrier, <u>J. Neurochem.</u>, 28:5.

Parkinson, W., and Weingarten, H. P., 1989, A dissociative analysis of the ventromedial hypothalamic obesity syndrome. <u>Am. J. Physiol.</u>, in press.

Reuterving, C.-O., and Hagg, E., 1987, Circadian eating and drinking habits in alloxan diabetic rats, <u>Diabetes and Metabolisme (Paris)</u>, 13:99.

Rezek, M., Havlicek, V, and Novin, D., 1975, Satiety and hunger induced by small and large duodenal loads of isotonic glucose, <u>Am. J. Physiol.</u>, 299:545.

Richter, C. P., Holt, L. E., and Barelare, B., 1938, Nutritional requirements for normal growth and reproduction in rats studied by the self-selection method, <u>Am. J. Physiol.</u>, 122:734.

Rogers, Q. R., and Leung, P. M. B., 1973, The influence of amino acids on the neuroregulation of food intake, <u>Fed. Proc.</u>, 32:1709.

Schmitt, M., 1973, Circadian rhythmicity in responses of cells in the lateral hypothalamus, <u>Am. J. Physiol.</u>, 225:1096.

Sellers, E. A., You, R. W., and Moffat, N. W., 1954, Regulation of food consumption by calorie value of the ration in rats exposed to cold. <u>Am. J. Physiol.</u>, 177:367.

Shimazu, T., 1981, Central nervous system regulation of liver and adipose tissue metabolism. <u>Diabetologia</u>, 20(Suppl):343.

Shimomura, Y., Takahashi, M., Shimizu, H., Sato, N., Uehara, Y., Negishi, M., Inukai, T., Kobayashi, I., and Kobayashi, S., 1990, Abnormal feeding behavior and insulin replacement in STZ-induced diabetic rats, <u>Physiol. Behav.</u>, 47:731.

Smith, G. P., and Gibbs, J., 1981, Brain-gut peptides and the control of food intake. <u>in</u>: "Neurosecretion and Brain Peptides" J. B. Martin, S. Reichlin and K. L. Bick, eds., Raven Press, New York.

Sokoloff, L., FitzGerald, G. G., and Kaufman, E. E., 1977, Cerebral nutrition and energy metabolism, <u>in</u>: "Nutrition and the Brain, Vol. 1", R. W. Wurtman, J. J. Wurtman, eds., Raven Press, New York.

Stricker, E. M., Rowland, N., and Saller, C. F., 1977, Homeostasis during hypoglycemia: central control of adrenal secretion and peripheral control of feeding. <u>Science</u>, 196:79.

Teff, K., and Young, S. M., 1988, Effects of carbohydrate and protein administration on rat tryptophan and 5-hydroxytryptamine: differential effects on the brain, intestine, pineal, and pancreas. <u>Can. J. Physiol. Pharmacol.</u>, 66:683.

Wayner, M. J., Ono, T., DeYoung, A., and Barone, F. C., 1975, Effects of essential amino acids on central neurons. <u>Pharmac. Biochem. Behav.</u>, 3(Suppl.1):85.

Weingarten, H. P., Chang, P. K., and McDonald, T. J., 1985, Comparison of the metabolic and behavioral disturbances following paraventricular- and ventromedial-hypothalamic lesions. <u>Br. Res. Bull.</u>, 14:551.

Woodger, T. L., Sirek, A., and Anderson, G. H., 1979, Diabetes, dietary tryptophan, and protein intake regulation in weanling rats. <u>Am.J. Physiol.</u>, 236:R307.

Wurtman, J. J., Moses, P. L., and Wurtman, R. J., 1983, Prior carbohydrate consumption affects the amount of carbohydrate that rats choose to eat. <u>J. Nutr.</u>, 113:70.

NOVEL PEPTIDES AND ISLET FUNCTION

Kazuhiko Tatemoto

Peptide Research Laboratory, Department of Psychiatry
and Behavioral Sciences, Stanford University
School of Medicine, Stanford, California

INTRODUCTION

Hormones and neurotransmitters are known to be involved in the regulation of islet function, and in recent years, it has become clear that neuropeptides released from pancreatic neurons may also be involved in the islet hormone release (1). Several hormones and neurotransmitters are known to inhibits insulin secretion. For example, noradrenaline (2) and somatostatin (3) inhibit insulin secretion from the pancreas. Between 1980 and 1986, we isolated a number of novel peptides and found that some strongly inhibit insulin secretion. This paper briefly summarizes studies on these inhibitory peptides, and describes their relations to islet function, and the potential use of their antagonists as drugs for diabetes.

NOVEL PEPTIDES WITHIN THE NEUROENDOCRINE SYSTEMS

With the accumulation of knowledge of the processing of many peptide hormones and neuropeptides, it is now evident that these peptides are produced from their precursor proteins by post-translational processing (4). The processing of the peptides frequently includes unique modifications of the peptide molecules such as phosphorylation, sulfation, acetylation, pyroglutamation, and C-terminal amidation. The C-terminal amidation, in particular, occurs only in neuropeptides and peptide hormones in mammalian tissues (5). We therefore proposed a chemical assay in which the amide structure was used as a chemical marker to detect neuropeptides and peptide hormones (6). Such a chemical approach is radically different from the traditional approach which uses specific biological activity as a marker. When tissue extracts were subjected to this assay, it was found that they contained several previously unknown peptides with the C-terminal amide structure (6). This finding led to the isolation of a series of novel peptides including PHI-27 (7, 8), peptide YY (7, 9), neuropeptide Y (10), galanin (11,12), and pancreastatin (13). Three of these peptides, neuropeptide Y, galanin and pancreastatin, were subsequently found to inhibit insulin secretion.

PANCREASTATIN: STRUCTURE AND FUNCTION

Pancreastatin was discovered in porcine pancreatic extracts by the presence of its unique C-terminal structure, a glycine amide (13). This peptide consists of 49 amino acid residues (Fig.1) and was named pancreastatin because of its inhibitory effects on pancreatic secretion. Pancreastatin was also isolated from bovine (14) and human (15,16) tissues. The primary

Fuel Homeostasis and the Nervous System, Edited by M. Vranic *et al.*
Plenum Press, New York, 1991

```
1              5                  10                 15
Gly-Trp-Pro-Gln-Ala-Pro-Ala-Met-Asp-Gly-Ala-Gly-Lys-Thr-Gly-Ala-Glu-


              20                 25                 30
Glu-Ala-Gln-Pro-Pro-Glu-Gly-Lys-Gly-Ala-Arg-Glu-His-Ser-Arg-Gln-Glu-


35                 40                 45
Glu-Glu-Glu-Glu-Thr-Ala-Gly-Ala-Pro-Gln-Gly-Leu-Phe-Arg-Gly-NH2
```

Fig. 1. Primary structure of pancreastatin (porcine).

structures of these peptides were found to be identical to the regions of the corresponding chromogranin A molecules (17, 18). Chromogranin A is an acidic protein known to be present in a variety of endocrine cells and neurons, but its biological function is unknown (19). The discovery of pancreastatin provides evidence that chromogranin A serves as the precursor of pancreastatin.

High concentrations of pancreastatin were found in pancreas, anterior pituitary, gastrointestinal tract, adrenal gland (20, 21), and brain (22). Immunohistochemical studies revealed that pancreastatin cells are located in the islets of Langerhans, particularly in the secretory granules of insulin and somatostatin-containing cells (23, 24). Pancreastatin was also found in gut endocrine cells, adrenal chromaffin cells, and anterior pituitary cells (24, 25). Many endocrine tumors such as carcinomas and glucagonomas contain high concentrations of pancreastatin immunoreactivities (26).

Pancreastatin regulates pancreatic functions. The peptide inhibits both insulin secretion (13, 27-32) and pancreatic exocrine secretion (33, 34). It also affects extrapancreatic functions. For example, pancreastatin inhibits secretion from parietal cells (35) and parathyroid cells (36). Central administration of pancreastatin enhances memory retention (37) and elevates blood glucose, free fatty acid, and corticosterone (38).

Pancreastatin and Islet Function

Pancreastatin inhibits glucose-induced insulin secretion (13, 27) and augments arginine-induced glucagon secretion (27) from isolated perfused pancreas. Infusion of 1-10 nM pancreastatin significantly inhibits the insulin release stimulated by 16.7 mM glucose from the isolated pancreas (27). Pancreastatin inhibits glucose-induced insulin secretion from isolated rat islets (27-29) and in conscious rats or mice (30, 31). Pancreastatin also inhibits carbachol (31)- or terbutaline (32)-induced insulin secretion. This peptide lowers basal plasma insulin concentration, increases basal plasma glucagon concentration, and induces a transient hyperglycemia in mice (31). Although the mechanism is not clear, it is possible that pancreastatin acts directly on pancreatic *beta*-cells through its receptors. Pretreatment of a pancreatic *beta*-cell tumor with pertussis toxin abolishes the inhibitory effect of pancreastatin on carbachol-induced insulin release, suggesting that pancreastatin may exert its action directly on beta-cells through a pertussis toxin-sensitive G-protein pathway (39). Since pancreastatin is present in high concentrations in secretory granules of insulin cells, and co-released with insulin (40), pancreastatin may regulate insulin secretion in an autocrine fashion.

GALANIN: STRUCTURE AND FUNCTION

Galanin was originally isolated from porcine intestine (11,12). This 29 amino acid-peptide was named after its N-terminal glycine and C-terminal alanine amide structure (Fig.2). The precursor structures of galanin were deduced from the cDNA sequences of pig (41), rat (42, 43) and cow (44).

```
  1           5              10                15
Gly-Trp-Thr-Leu-Asn-Ser-Ala-Gly-Tyr-Leu-Leu-Gly-Pro-His-Ala-


       20            25
Ile-Asp-Asn-His-Arg-Ser-Phe-His-Asp-Lys-Tyr-Gly-Leu-Ala-NH2
```

Fig. 2. Primary structure of galanin (porcine).

Galanin is widely located in neurons of the central nervous system (45-47), gastro-intestinal tract (45, 48), respiratory tract (49), genitourinary tract (50), adrenal gland (51), skin (52), and thyroid gland (53). Numerous galanin-containing nerve fibers innervate the Islets of Langerhans (54). Galanin coexists with acetylcholine (55, 56), catecholamines, GABA, or 5-hyroxytryptamine (57) in certain neurones.

Galanin has a broad spectrum of biological activities. It exhibits direct contractile effects on smooth muscle preparations (12, 58) and potentiates noradrenaline-induced contraction (59). It inhibits dopamine (60), acetylcholine (61), pancreatic polypeptide (62), and gastrin release (63, 64), whereas it stimulates the release of 5-hyroxytryptamine (65), prolactin (66), vasoactive intestinal peptide (67), and growth hormone (68, 69). Central admini-stration of galanin stimulates food intake (70), and peripheral administration inhibits intestinal motility (71, 72). Galanin may also play important roles in neuropeptidergic regulation of islet hormone secretion. Studies of galanin actions on pancreatic endocrine function are summarized in the following section.

Galanin and Islet Function

Galanin inhibits basal insulin (54, 73-76) and somatostatin (54, 76) secretion, stimulates basal glucagon secretion (54, 75), and induces hyperglycemia (73-76). It impairs the insulin response to glucose (74, 75, 77), arginine (77), tolbutamide (76), and carbachol (75), but, it does not alter the glucagon responses (75-77). Galanin also inhibits insulin release induced by peptide hormones such as cholecystokinin, vasoactive intestinal peptide, glucagon, and gastric inhibitory peptide (78, 79).

Recent studies suggest that the inhibitory effect of galanin on insulin secretion is a physiological event mediated by direct actions of galanin on pancreatic *beta*-cells, which block a central step of the insulin secretory mechanism. The evidence for this is as follows: a dense network of galanin-containing nerve fibers is closely associated with pancreatic islets (54), and the amount of galanin released from the nerves is sufficient to influence islet function (80). Pancreatic *beta*-cells have enough galanin binding sites to influence insulin secretion (81, 82). Consequently, galanin directly inhibits insulin secretion from isolated pancreatic *beta*-cells (83, 84). Galanin hyperpolarizes the cell membrane and reduces cytoplasmic free calcium concentration (83-85). The galanin inhibition of glucose-induced insulin secretion from the *beta*-cells is abolished by treatment with pertussis toxin, suggesting the involvement of a pertussis toxin-sensitive G protein (84-86). The peptide activates ATP-dependent potassium channels (87-90), which is reversed by a sulfonylurea, glibenclamide (87).

In a possible mechanism of the galanin action, galanin is first released from the nerve terminals surrounding the *beta*-cells, then binds to galanin-specific receptors on the cells. This activates a pertussis toxin-sensitive G protein that, in turn, activates an ATP-dependent potassium channel either directly or via activation of a protein kinase. The activation of the ATP-dependent potassium channel hyperpolarizes the membrane and reduces intracellular calcium concentration. As a result, the exocytosis of insulin is inhibited. Glucose or a sulfonylurea, on the other hand, decreases the activity of this channel with consequent depolarization and increase in intracellular calcium concentration, and stimulates insulin secretion.

NEUROPEPTIDE Y: STRUCTURE AND FUNCTION

Neuropeptide Y (NPY) was discovered by the presence of its C-terminal tyrosine amide structure in porcine brain extracts (10). This peptide consists of 36 amino acids (Fig.3) and has structural similarities to peptide YY and pancreatic polypeptide (91). NPY was isolated from various species including pig (10), human (92), rat (93), guinea-pig, rabbit (94), cow (95), and sheep (96). The precursor structures of human and rat NPY were deduced from the cDNA structures of NPY mRNAs (97, 98) and NPY genes (99, 100).

NPY, the most abundant peptide in the central nervous system (101), is involved in a broad spectrum of brain functions including food intake, blood pressure regulation, hormone secretion, sexual behavior, and circadian rhythmicity (95,102-104). It is known to be the most potent stimulant of feeding (105-107). It also stimulates drinking (106,107). Central NPY injection changes the plasma concentrations of many pituitary hormones including leuteinizing hormone (108, 109), growth hormone, prolactin, thyrotropin (110), gonadotropin-releasing hormone (111), ACTH (112) and vasopressin (112, 113). Central administration of NPY induces hypotension and changes in heart rate (110). NPY injection into the suprachiasmatic region of the hypothalamus induces a shift in the circadian rhythm (114), while intraventricular injection of NPY suppresses copulatory behaviors (115).

NPY is also abundantly distributed throughout the peripheral nervous system including neurons of blood vessels, respiratory tract, digestive tract, genitourinary tract, spleen, adrenal gland (116, 117), thyroid gland (118), skin (119), eye (120), and pancreas (116, 121-123). Peripheral administration of NPY induces strong vasoconstriction (124). NPY may thus play important roles in sympathetic vascular controls. This peptide inhibits colonic motility (125) and also modulates epithelial ion transport (126). NPY coexists with catecholamines in some neurons (127,128), and inhibits the release of noradrenaline at the presynaptic level (129), and potentiates noradrenaline-evoked vasoconstriction at the post-synaptic level (130).

NPY and Islet Function

NPY inhibits glucose-stimulated insulin secretion in pancreatic isolated islets (131). In perfused pancreas, NPY markedly inhibits both basal and glucose-stimulated insulin secretion (131). NPY also inhibits the adrenoceptor agonist, terbutaline-induced insulin and glucagon secretion (132). In contrast, central NPY administration produces significant elevation of circulating insulin (131, 133). It is therefore suggested that NPY acts directly on the islets to inhibit insulin secretion while the central actions of NPY indirectly result in an increase in plasma insulin (131). Although it is not clear whether peripheral NPY plays significant roles for the islet hormone release under physiological conditions (134, 135), the release of hypothalamic NPY may play important roles in obesity and diabetes. It has been reported that NPY (136-138) and its mRNA (139) concentrations in the hypothalamus are considerably increased in diabetic rats. These results indicate that the NPY release may contribute to certain neuroendocrine disturbances in insulin-deficient diabetes. Insulin therapy prevents the NPY increase in the hypothalamus (140). In contrast, insulin-induced hypoglycemia increases the levels of NPY and its mRNA in chromaffin granules (141) and elevates plasma NPY concentrations (142). These results suggest that fuel homeostasis may be involved in the release of NPY that regulates food intake.

```
  1              5                  10                 15
Tyr-Pro-Ser-Lys-Pro-Asp-Asn-Pro-Gly-Glu-Asp-Ala-Pro-Ala-Glu-Asp-Leu-Ala-


  20             25                 30                 35
Arg-Tyr-Tyr-Ser-Ala-Leu-Arg-His-Tyr-Ile-Asn-Leu-Ile-Thr-Arg-Gln-Arg-Tyr-NH2
```

Fig. 3. Primary structure of neuropeptide Y (porcine).

PEPTIDE ANTAGONISTS AND AGONISTS: DRUGS FOR DIABETES ?

The receptor antagonists of peptide hormones and neuropeptides are useful for understanding the mechanisms of peptide actions, as well as elucidating structure-function relationships and introducing new pharmacological drugs. Since the peptides within the neuroendocrine systems, such as galanin, pancreastatin, and NPY, may be involved in the regulation of islet hormone secretion, receptor agonists or antagonists of these peptides are potentially useful for the development of therapeutic drugs for diabetes. However, such attempts have previously been hampered by the fact that synthesis of receptor agonists and antagonists for peptides is a difficult, time-consuming and often elusive task. Attempts at applying structural analysis and computer imaging technology have not yet produced any efficient general method for antagonist design, and this process still relies on hit or miss synthesis and testing. Therefore, we recently developed a new strategy, called "analog mixture-screening," for the synthesis of receptor antagonists and agonists (143). This new strategy relies on the synthesis and screening of peptide mixtures containing hundreds of agonist or antagonist candidates, as opposed to previous strategies that relied on the time-consuming synthesis of individual peptides. There are many ways of generating such peptide analog mixtures. For example, an analog mixture with various amino acid substitutions can be synthesized by dividing the solid-phase resin into several portions at desired positions during the synthesis, coupling a different amino acid to each portion, and then re-combining the portions for further coupling steps. An analog mixture with various chain lengths can similarly be synthesized by withdrawing portions of the resin at desired positions, and re-combining the portions for cleavage from the resin. After a series of peptide analog mixtures are designed and synthesized, a specific mixture with agonist or antagonist activity is screened using a bioassay or competitive receptor assay. The peptide analog responsible for the activity is then isolated from the selected mixture using HPLC and other separation techniques, in a manner similar to isolation of natural peptides from tissue extracts. Subsequently, the primary structure of the isolated peptide is determined and a larger amount of the peptide analog is re-synthesized for further studies. In this way, numerous analogs can be synthesized simultaneously and tested in a simple and efficient manner.

Synthesis of NPY Receptor Antagonists Using an Analog Mixture-Screening Strategy

Using this approach, two NPY analogs with NPY receptor antagonist activity, designated PYX-1 and PYX-2, were recently synthesized (143). PYX-1 is a decapeptide amide with a modified amino acid at the N-terminus and PYX-2 is a decapeptide amide with modified amino acids at both N- and C-termini. PYX-1 corresponded to Ac-[3-Cl$_2$Bzl-Tyr^{27}, D-Thr^{32}] NPY27-36, and PYX-2 to Ac-[3-Cl$_2$Bzl-Tyr^{27,36}, D-Thr^{32}] NPY 27-36 (Fig 4). These analogs strongly inhibited the NPY actions of releasing intracellular calcium in human erythroleukemia cells and inhibited the specific binding of NPY to its receptors (143).

Further studies on the use of such NPY receptor antagonists in the treatment of diabetes might introduce a generation of new drugs. The development of receptor antagonists of other peptides might also yield useful drugs. For example, receptor antagonists of galanin might be important in the treatment of diabetes, since galanin may be involved in the impairment of insulin secretion known to occur in non-insulin-dependent diabetes mellitus.

| 1 | 5 | 10 |

3-Cl$_2$Bzl

PYX-1:   Ac-Tyr-Ile-Asn-Leu-Ile-D-Thr-Arg-Gln-Arg-Tyr-NH$_2$

3-Cl$_2$Bzl   3-Cl$_2$Bzl

PYX-2:   Ac-Tyr-Ile-Asn-Leu-Ile-D-Thr-Arg-Gln-Arg-Tyr-NH$_2$

Fig. 4. Primary structures of NPY receptor antagonists.

REFERENCES

1. Miller, R. E. (1981) *Endocr. Rev.* 2: 471-494.
2. Sorenson, R. L, Elde, R., and Seybold, V. (1979) *Diabetes* 28: 899-904.
3. Efendic, S., Hökfelt, T., and Luft, R. (1978) *In Advances in Metabolic Disorders* 9; 367-424, ed. R. Luft and R. Levine, Academic Press, Yew York.
4. Andrews, P. C., Brayton, K., and Dixon, J. E. (1987) *Experientia* 43: 784-790.
5. Tatemoto, K. (1982) *Symposia Medica Hoechst* 18, pp. 507-535.
6. Tatemoto, K., and Mutt, V. (1978) *Proc. Natl. Acad. Sci. U.S.A.* 75: 4115-4119.
7. Tatemoto, K. and Mutt, V. (1980) *Nature* 285: 417-418.
8. Tatemoto, K. and Mutt, V. (1981) *Proc. Natl. Acad. Sci. USA* 78: 6603-6607.
9. Tatemoto, K. (1982) *Proc. Natl. Acad. Sci. USA* 79: 2514-2518.
10. Tatemoto, K., Carlquist, M., and Mutt, V. (1982) *Nature* 296: 659-660.
11. Tatemoto, K. (1984) *in Front. Horm. Res.* 12: 27-30, eds., M. Ratzenhofer, H. Höfler, and G. F. Walter, Karger, Basel.
12. Tatemoto, K., Rökaeus, Å., Jörnvall, H., McDonald, T. J., and Mutt, V. (1983) *FEBS Lett.* 164: 124-128.
13. Tatemoto, K., Efendic, S., Mutt, V., Makk, G., Feistner, G. J., Barchas, J. D. (1986) *Nature* 324: 476-478.
14. Nakano, I., Funakoshi, A., Miyasaka, K., Ishida, K., Makk, G., Angwin, P., Chang, D. and Tatemoto, K. (1989) *Regul. Pept.* 25: 207-213.
15. Schmidt, W. E., Siegel, E. G., Kratzin, H, Creutzfeldt, W. (1988) *Proc. Natl. Acad. Sci. USA* 85: 8231-8235.
16. Funakoshi, S., Tamamura, H., Ohta, M., Yoshizawa, K., Funakoshi, A., Miyasaka, K., Tateishi, K., Tatemoto, K., Nakano, I., Yajima, H., and Fujii, N. (1989) *Biochem. Biophys. Res. Commun.* 164: 141-148.
17. Tatemoto, K. (1989) *in Diabetes 1988*, p185-188, eds., R. Larkins, P. Zimmer, and D. Chrisholm, Elsevier Science.
18. Tatemoto, K. (1990) *in Molecular Biology of Islets of Langerhans*, pp145-152, ed., H. Okamoto, Cambridge University Press.
19. Winkler, H., Apps, D. K., and Fischer-Colbrie, R. (1986). *Neuroscience* 18: 261-290.
20. Bretherton-Watt, D., Ghatei, M. A., Bishop, A. E., Facer, P., Fahey, M., Hedges, M., Williams, G., Valentino, K.L., Tatemoto, K., Roth, K., Polak, J. M., and Bloom, S. R. (1988) *Peptides* 9: 1005-1014.
21. Schmidt, W. E., Siegel, E. G., Lamberts, R., Gallwitz, B., and Creutzfelt, W. (1988) *Endocrinology* 123: 1395-1404.
22. Kar, S., Bretherton-Watt, D., Gibson, S. J., Steel, J. H., Gentleman, S. M., Roberts, G. W., Vallentino, K., Tatemoto, K., Ghatei, M. A., Bloom, S. R., and Polak, J. M. (1989) *J. Comp. Neurol.* 288: 627-639.
23. Ravazzola, M., Efendic, S., Ostenson, C. G., Tatemoto, K., Hutton, J. C., and Orci, L. (1988) *Endocrinology* 123: 227-229.
24. Lamberts, R., Schmidt, W. E., and Creutzfelt, W. (1990) *Histochemistry* 93: 369-380.
25. Shimizu, F, Ikei, N, Iwanaga, T, and Fujita, T. (1987) *Biomed. Res.* 8: 457-462.
26. Bishop, A. E., Bretherton-Watt, D., Hamid, Q. A., Fahey, M., Shepherd, N., Valentino, K., Tatemoto, K., Ghatei, M. A., and Bloom, S. R., and Polak, J. M. (1988) *Molecular and Cellular Probes* 2: 225-235.
27. Efendic, S., Tatemoto, K., Mutt, V., Quan, C., Chang, D., Östenson, C-G. (1987) *Proc. Natl. Acad. Sci. USA* 84: 7257-7260.
28. Östenson, C.G., Sundler, S., and Efendic, S. (1989) *Pancreas* 4: 441-446.
29. Ishizuka, J., Singh, P., Greeley, G. H., Townsend, C. M., Cooper, C. W, Tatemoto, K., and Thompson, J. C. (1988) *Pancreas* 3: 77-82.
30. Funakoshi, A., Miyasaka, K., Kitani, K., and Tatemoto, K. (1989) *Regul. Pept.* 24: 225-231.
31. Ahren, B., Lindskog, S., Tatemoto, K., and Efendic, S. (1988) *Diabetes* 37: 281-285.
32. Lindskog, S., and Ahren, B. (1988) *Hormone Res.* 29: 237-240.
33. Funakoshi, A., Miyasaka, K., Nakamura, R., Kitani, K., and Tatemoto, K. (1989) *Regul. Pept.* 25: 157-166.

34. Ishizuka, J., Asada, I., Poston, G. J., Lluis, F., Tatemoto, K., Greeley, G. H., and Thompson, J. C. (1989) *Pancreas* 4: 277-281.
35. Lewis, J. J., Zdon, M. J., Adrian, T. E., and Modlin, I. M. (1988) *Surgery* 104: 1031-1036.
36. Fasciotto, B. H., Gorr, S. U., DeFranco, D. J., Levine, M. A., and Cohn, D. V. (1989) *Endocrinology* 125: 1617-1622.
37. Flood, J. F., Morley, J. F., and Tatemoto, K. (1988) *Peptides* 9: 1077-1080.
38. Gunion, M. W., Rosenthal, M. J., Tatemoto, K., and Morley, J. E. (1989) *Brain Res.* 485:251-257.
39. Lorinet, A-M., Tatemoto, K., Laburthe, M., Amiranoff, B. (1989) *Eur. J. Pharmacol.* 160:405-407.
40. Östenson, C. G., Efendic, S., and Holst, J. J. (1989) *Endocrinology* 124: 2986-90.
41. Rökaeus, Å., Brownstein, M. J. (1986) *Proc. Natl. Acad. Sci. USA* 83: 6287-6291.
42. Vrontakis, M. E., Peden, L. M., Duckworth M. L., Friesen, H. G. (1987) *J. Biol. Chem.* 262: 16755-8.
43. Kaplan, L. M., Spindel, E. R., Isselbacher, K. J., and Chin, W. W. (1988) *Proc. Natl. Acad. Sci. U. S. A.* 85: 1065-1069.
44. Rökaeus Å. and Carlquist, M. (1988) *FEBS Lett.* 234: 400-406.
45. Rökaeus, Å., Melander, T., Hökfelt, T., Lundberg, J. M., Tatemoto, K., Carlquist, M. and Mutt, V. (1984) *Neurosci. Lett.* 47: 161-166.
46. Ch'ng, J. L. C., Christofides, N.D., Anand, P., Gibson, S. J., Allen, Y. S., Su, H. C., Tatemoto, K., Morrison, J. F .B., Polak, J. M. and Bloom, S. R. (1985) *Neuroscience* 16:343-354.
47. Skofitsch, G. and Jacobowitz D. M. (1985) *Peptides* 6: 509-546.
48. Melander, T., Hökfelt, T., Rökaeus, Å., Fahrenkrug, J., Tatemoto, K. and Mutt, V. (1985) *Cell Tissue Res.* 239: 253-270.
49. Cheung A., Polak, J. M., Bauer, F. E., Cadieux, A., Christofides, N. D., Springall, D. R., Bloom, S. R. (1985) *Torax* 40: 889-896.
50. Bauer, F. E., Christofides, N. D., Hacker, G. W., Blank, M. A., Polak, J. M., Bloom, S. R. (1986) *Peptides* 7: 5-10.
51. Bauer, F. E., Hacker, G. W., Terenghi, G., Adrian, T. E., Polak, J. M., Bloom, S. R. (1986) *J. Clin. Endocr. Metab.* 63: 1372-1378.
52. Johansson, O, Vaalasti, A., Tainio, H., and Ljungberg, A. (1988) *Acta Physiol. Scand.* 132: 261-263.
53. Grundiz, T., Håkanson, R., Sundler, F., Uddman, R. (1987) *Endocrinology* 121: 575-585.
54. Dunning, B. E., Ahren, B., Veith, R. C., Böttcher, G., Sundler, F., and Taborsky, G. J. Jr. (1986) *Am. J. Physiol.* 251: E127-133.
55. Melander, T., Staines, W. A., Hökfelt, T., Rökaeus, Å., Eckenstein, F., Salvaterra, P. M., and Wainer, B. H. (1985) *Brain Res.* 360: 130-138.
56. Melander, T., and Staines, W. A. (1986) *Neurosci. Lett.* 68: 17-22.
57. Melander, T., Hökfelt. T., Rökaeus, Å., Cuello, A. C., Oertel, W. H., Verhofstad, A., and Goldstein, M. (1986) *J. Neurosci.* 6: 3640-3654.
58. Ekblad, E., Håkanson, R., Sundler, F., Wahlestedt, C. (1985) *Br. J. Pharmacol.* 86: 241-246.
59. Ohhashi, T., and Jacobowitz, D. M. (1985) *Regul. Pept.* 12: 163-171.
60. Nordström, Ö., Melander, T., Hökfelt, T., Bartfai, T., and Goldstein, M. (1987) *Neurosci. Lett.* 73: 21-26.
61. Yau, W. M., Dorsett, J. A., and Youther, M. L. (1986) *Neurosci. Lett.* 72: 305-308.
62. Silvestre, R. A., Miralles, P., Monge, L., Villanueva, M. L., and Marco, J. (1987) *Life Sci.* 40: 1829-1833.
63. Schepp, W., Prinz, C., Tatge, C., Håkanson, R., Schusdriarra, V., and Classen, M. (1990) *Am. J. Physiol.* 258: G596-602.
64. Madaus, S., Schusdziarra, V., Seufferlein, T., and Classen, M. (1988) *Life Sci.* 42: 2381-2387.
65. Sundström, E., Melander, T. (1988) *Eur. J. Pharmacol.* 146: 327-329.
66. Koshiyama, H., Kato, Y., Inoue, T., Murakami, Y., Ishikawa, Y., Yanaihara, N., and Imura, H. (1987) *Neurosci. Lett.* 75: 49-54.

67. Inoue, T., Kato, Y., Koshiyama, H., Yanaihara N., and Imura, H. (1988) *Neurosci. Lett.* 85: 95-100.
68. Ottlecz, A., Samson, W. K., McCann, S. M. (1986) *Peptides* 7: 51-53.
69. Bauer, F. E., Ginsberg, L., Venetikou, M., Mackay, D. J., Burrin, J. M., and Bloom, S. R. (1986) *Lancet* 2: 192-195.
70. Kyrkouli, S. E., Stanley, B. G., and Leibowitz, S. F. (1986) *Eur. J. Pharmacol.* 122: 159-160.
71. Fox, J.E.T., McDonald, T.J., Kostolanska, F. and Tatemoto, K. (1986) *Life Sci.* 39:103-110.
72. Bauer, F. e., Zintel, A., Kenny, M. J., Calder, D., Ghatei, M. A., and Bloom, S. R. (1989) *Gastroenterology* 97: 260-4.
73. McDonald, T. J., Dupre, J., Tatemoto, K., Greenberg, G. R., Radzuk, J., and Mutt, V. (1985) *Diabetes* 34: 192-196.
74. Manabe, T., Yoshimura, T., Kii, E., Tanabe, Y., Ohshio, G., Tobe, T., Akaji, K., and Yajima, H. (1986) *Endocr. Res.* 12: 93-98.
75. Lindskog, S., and Ahren, B. (1987) *Acta Physiol. Scand.* 129: 305-309.
76. Silvestre, R. A., Miralles, P., Monge, L., Moreno, P., Villanueva, M. L., and Marco, J. (1987) *Endocrinology* 121: 378-383.
77. McDonald, T. J., Dupre, J., Greenberg, G. R., Tepperman, F., Brooks, B., Tatemoto, K. and Mutt, V. (1986) *Endocrinology* 119: 2340-2345.
78. Hramiak, I. M., Dupre, J., and McDonald, T. J. (1988) *Endocrinology* 122: 2486-2491.
79. Miralles, P., Peiro, E., Silvestre, R. A., Villanueva, M. L., and Marco, J. (1988) *Metabolism* 37: 766-770.
80. Dunning, B. E. and Taborsky, G. J. Jr. (1989) *Am. J. Physiol.* 256: E191-198.
81. Amiranoff, B., Servin, A., Rouyer-Fessard, C., Couvineau, A., Tatemoto, K., and Laburthe, M. (1987) *Endocrinology* 121: 284-289.
82. Sharp, G. W. G., Le Marchand-Brustel, Y., Yada, T, Russo, L. L., Bliss, C. R., Cormont., M., Monge, L., and Van Obberghen, E. (1989) *J. Biol. Chem.* 264: 7302-7309.
83. Ahren, B., Arkhammar, P., Berggren, P. O., and Nilsson, T. (1986) *Biochem. Biophys. Res. Commun.* 140: 1059-1063.
84. Amiranoff, B., Lorinet, A. M., Lagny-Pourmir, I., and Laburthe, M. (1988) *Eur. J. Biochem.* 177: 147-152.
85. Nilsson, T., Arkhammar, P., Rorsman, P., and Berggren, P. O. (1989) *J. Biol. Chem.* 264: 973-980.
86. Ullrich, S., and Wollheim, C. B. (1989) *FEBS Lett.* 247: 401-404.
87. De Weille, J., Schmid-Antomarchi, H., Fosset, M., and Lazdunski, M. (1988) *Proc. Natl. Acad. Sci. USA* 85: 1312-1316.
88. Dunne, M. J., Bullett, M. J., Li, G., Wollheim, C. B., and Pettersen, O. H. (1989) *EMBO J.* 8: 413-420.
89. Ahren, B., Berggren, P-O., Bokvist, K., and Rorsman, P. (1989) *Peptides* 10: 453-547.
90. Drews, G., Debuyser, A., Nenquin, M., and Henquin , J. C. (1990) *Endocrinology* 126: 1646-1653.
91. Tatemoto, K. (1982) *Proc. Natl. Acad. Sci. U.S.A.* 79: 5485-5489.
92. Corder, R., Emson, P. C., and Lowry, P. J. (1984) *Biochem. J.* 219: 699-706.
93. Corder, R., Gaillard, R. C., and Böhlen, P. (1988) *Regul. Pept.* 21: 253-261.
94. O'Hare, M. M.T., Tenmoku, S., Aakerlund, L., Hilsted, L., Johnsen, A. and Schwartz, T. W. (1988) *Regul. Pept.* 20: 293-304.
95. Tatemoto, K. (1989) *in Neuropeptide Y*, pp13-21, eds., V. Mutt, K., Fuxe, T. Hökfelt & J. Lundberg, Raven Press.
96. Sillard, R, Agerberth, B., Mutt, V., and Jörnvall, H. (1989) *FEBS Lett.* 258: 263-265.
97. Allen, J. M., Novotny, L., Martin, J., and Heinrich, G. (1987) *Proc. Natl. Acad. Sci. U.S.A.* 84: 2532-2536.
98. Minth, C. D., Bloom, S. R., Polak, J. M., and Dixon, J. E. (1984) *Proc. Natl. Acad. Sci. U.S.A.* 81: 4577-4581.

99. Minth, C. D., Andrews, P. C., and Dixon, J. E. (1986) *J. Biol. Chem.* 261:11974-11979.
100. Larhammer, D., Ericsson, A., and Persson, H. (1987) *Proc. Natl. Acad. Sci. U.S.A.* 84: 2068-2072.
101. Allen, Y. S., Adrian, T. E., Allen, J. M., Tatemoto, K., Crow, T. J., Bloom, S. R., and Polak, J. M. (1983) *Science* 221: 877-879.
102. Allen, J. M., Bloom, S. R., Docray, G. J., Maccarrone, C., and Jarrott, B. (1986) *Neurochem. Int..* 8: 1-22.
103. Potter, E. K. (1988) *Pharmac.Ther.* 37: 251-273.
104. Wahlestedt, C. (1987) *in Neuropeptide Y: Actions and Interactions in Neurotransmission (thesis)*, Department of Pharmacology, University of Lund, Lund, Sweden.
105. Clark, J. T., Karla, P. S., Crowley, W. R., and Karla, S. P. (1984) *Endocrinology* 115: 427-429.
106. Levine, A. S., and Morley, J. E. (1984) *Peptides* 5: 1025-1029.
107. Stanley, B. G., and Leibowitz, S. F. (1984) *Life Sci.* 35: 2635-2642.
108. Karlra, S. P., and Crowley, W. R. (1984) *Life Sci.* 35: 1173-1176.
109. McDonald, J. K., Lumpkin, M. D., Samson, W. K., and McCann, S. M. (1985) *Proc. Natl. Acad. Sci. U.S.A.* 82: 561-564.
110. Fuxe, K., Agnati, L. F., Härfstrand, A., Zini, I., Tatemoto, K., Pich, E. M., Hökfelt, T., Mutt, V., and Terenius, L. (1983) *Acta Physiol. Scand.* 118: 189-192.
111. Khorram, O., Pau, K-Y-F.and Spies, H. G. (1987) *Neuroendocrinology* 45: 290-297.
112. Härfstrand, A., Fuxe, K., Agnati, L. F., Eneroth, P., Zini, I., Mutt, V., and Goldstein, M. (1986) *Neurochem. Int.* 8: 355-376.
113. Willoughby, J. O., and Blessing, W. W. (1987) *Neurosci. Lett.* 75: 17-22.
114. Albers, H. E., and Ferris, C. F. (1984) *Neurosci. Lett.* 50: 163-168.
115. Karla, S. P., Allen, L. G., Clark, J. T., Crowley, W. R., and Karla, P. S. (1986) *In Neural and Endocrine Peptides and Receptors,* pp 353-366, ed., T. W. Moody, Plenum, New York.
116. Lundberg, J. M., Terenius, L., Hökfelt., T., Martling, C. R., Tatemoto, K., Mutt, V., Polak, J., Bloom, S., and Goldstein, M. (1982) *Acta Physiol. Scand.* 116: 477-480.
117. Lundberg, J. M., Terenius, L., Hökfelt, T., and Goldstein, M. (1983) *Neurosci. Lett.* 42: 167-172.
118. Grunditz, T., Håkanson, R., Rerup, C., Sundler, F., and Uddman, R. (1984) *Endocrinology* 115: 1537-1542.
119. Johanson, O. (1986) *Acta Physiol. Scand.* 128: 147-153.
120. Terenghi, G., Polak, J. M., Allen, J. M., Zhang, S. Q., Unger, W. G., Bloom, S. R. (1983) *Neurosci. Lett.* 42: 33-38.
121. Sundler, F., Moghimzadeh, E., Håkanson, R., Ekelund, M., and Emson, P. (1983) *Cell Tissue Res.* 230: 487-493.
122. Carlei, F., Allen, J. M., Bishop, A. E., Bloom, S. R., Polak, J. M. (1984) *Experientia* 41: 1554-1557.
123. Sundler, F., Håkanson, R., Ekblad, E., Uddman, R., and Wahlestedt, C. (1986) Ann. Rev. Cytol. 102: 234-269
124. Lundberg, J. M., and Tatemoto, K. (1982) *Acta Physiol. Scand.* 116: 393-402.
125. Hellström, P. M., Olerup, O., and Tatemoto, K. (1984) *in Gastrointestinal Motility,* pp 433-440., ed., C. Roman, MTP Press, Lancaster.
126. Fried, D. D., Miller, R. J., and Walker, M. W. (1986) *Br. J. Pharmac.* 88:425-128.
127. Hökfelt, T., Lundberg, J. M., Tatemoto, K., Mutt, V., Terenius, L., Polak, J., Bloom, S. R., Sasek, C., Elde, R., and Goldstein, M. (1983) *Acta Physiol. Scand.* 117: 315-318.
128. Hökfelt, T., Lundberg, J. M., Lagercrantz, H., Tatemoto, K., Mutt, V., Lindberg, J., Terenius, L., Everitt, B. J., Fuxe, K., Anganti, L., and Goldstein, M. (1983) *Neurosci. Lett.* 36; 217-222.
129. Lundberg, J. M., and Stärne, L. (1984) *Acta Physiol. Scand.* 120: 477-479.
130. Wahlestedt, C., Edvinsson, L., Ekblad, E., and Håkanson, R (1985) *J. Phamacol. Exp. Therap.* 234: 735-741.

131. Moltz, J. H., and McDonald, J. K. (1985) *Peptides* 6: 1155-1159.
132. Petersson, M., Lundquist, I., Ahren, B. (1987) *Endor. Res.* 13: 407-417.
133. Kuenzel, W. J., and McMurtry, J. (1988) *Physiol. Behav.* 44: 669-678.
134. Dunning, B. E., Ahren, B., Böttcher, G., Sundler, F., and Taborsky, G. J. Jr. (1987) *Regul. Pept.* 18: 253-265.
135. Holst, J. J., Orskov, C., Knuhtsen, S., Sheikh, S., Nielsen, O. V. (1989) *Acta Physiol. Scand.* 136; 519-526.
136. Williams, G., Steel, J. H., Cardoso, H., Ghatei, M. A., Lee, Y. C., Gill. J. S., Burrin, J. M., Polak, J. M., and Bloom, S. R. (1988) *Diabetes*, 37: 763-772.
137. Williams, G., Gill. J. S., Lee. Y. C., J. H., Cardoso, H., Okpere, B. E., and Bloom, S. R. (1989) *Diabetes*, 38: 321-327.
138. Williams, G., Lee, Y. C., Ghatei, M. A., Cardoso, H. M., Ball, A. J., Bone, A. J., Baird, J. D., and Bloom, S. R. (1989) *Diabet. Med.* 6: 601-607.
139. White, J. D., Olchovsky, D., Kershaw, M., and Berelowitz, M. (1990) *Endocrinology* 126: 765-772.
140. Sahu, A., Sninsky, C. A., Kalra, P. S., and Kalra, S. P. (1990) *Endocrinology* 126; 192-198.
141. Laslop, A., Wohlfarter, T., Fischer-Colbrie, R., Steiner, H. J., Humpel, Ch., Saria, A., Schmid, K. W., Sperk, G., and Winkler, H. (1989) *Regul. Pept.* 26: 191-202.
142. Takahashi, K., Mouri, T., Murakami, O., Itoi, K., Sone, M., Ohneda, M., Nozuki, M., Yoshinaga, K. (1988) *Peptides* 9: 433-435.
143. Tatemoto, K. (1990) *in Ann. New York Acad. Sci.* in press.

RELATIONSHIPS BETWEEN THE HYPOTHALAMUS AND ADIPOSE TISSUE MASS

Daniel A. K. Roncari

Institute of Medical Science and
Departments of Medicine, University of Toronto and
Sunnybrook Health Science Centre
2075 Bayview Avenue, Toronto, ON, Canada, M4N 3M5

INTRODUCTION

A system influencing the availability of energy to the entire organism, expectedly, depends on a widespread network directed by the central nervous system. Indeed, there are complex interactions between "controllers" in the brain and peripheral mechanisms for energy consumption and release, as well as the complex machinery mediating chemical energy storage (predominantly as triglyceride) and mobilization. Critical afferent and efferent systems connect the hypothalamic and other brain centres which control food intake and energy disposal, with the relevant target sites. This brief review will summarize four mechanisms whereby the brain, particularly the hypothalamus along with its interconnecting neural structures, can influence body fat content.

1. Hypothalamic-Autonomic Nervous System Influences on Food Intake and Energy Expenditure

Recent comprehensive reviews detail the anatomy and function of hypothalamic nuclei, their neural afferent, inter-connecting, and efferent pathways, as well as the possible role of these networks in obesity.[1,2]

Initially perceived as centres controlling only food intake, the classic, interdependent ventromedial nucleus and lateral hypothalamus also play a major role in energy disposal.[1,2] In fact, the ventromedial nucleus, through its (beta)-adrenergic efferents, stimulates adaptive thermogenesis, decreases insulin levels (recent evidence indicates that the sympathetic neuro-transmitter, galanin, has a critical role in suppressing insulin secretion[3]), and leads to depletion of chemical energy stores, i.e., triglyceride and glycogen, through augmented lipolysis and glycogenolysis, respectively.

The lateral and dorsomedial hypothalamic centres are connected to distal sites through vagal efferents. Important influences include stimulation of insulin secretion and, consequently, promotion of triglyceride and glycogen stores.[1,2] Augmented lateral and dorsomedial hypothalamus-vagal activity results in stimulation of both food intake and insulin secretion (primary and elicited by nutrients). In turn, insulin is the predominant hormone promoting storage of chemical energy.[4] Indeed, insulin plays an overriding role in triglyceride accretion in adipocytes through multiple effects.[4] In addition, by inhibiting the ventromedial nucleus, and through the parasympathetic efferents, the lateral hypothalamus dampens adaptive thermogenesis.

Experimental and clinical lesions of the paraventricular nucleus and/or its connecting ventral noradrenergic bundle, result in hyperphagia, while more extensive damage including the ventromedial hypothalamic nucleus cause, in addition, dampened adaptive thermogenesis and promotion of triglyceride storage.[1,2] In these states, the lateral hypothalamus and dorsomedial nucleus are unrestricted, and dominantly trigger food intake, and through unrestrained vagal overactivity (in the face of decreased galanin levels), lead to marked hyper-insulinemia.[1,2] These abnormalities result in classic hypothalamic corpulence. It has been proposed that certain genetic types of rodent obesity, as well as human obesity, might be due to dysfunctions analogous to those produced by structural lesions.[1,2]

2. Neuropeptides and Gut Peptides

Several neuropeptides appear to influence food intake by interacting with hypothalamic nuclei, some increasing ingestion (notably through the paraventricular nucleus), while several by suppressing it.[1,2,5] These peptides, which frequently reside in the same neurones with neurotransmitters (e.g., serotonin, dopamine, γ-aminobutyric acid), apparently adjust the size of meals. While the physiological role of many of these peptides remains to be established, corticotropin-releasing factor (CRF) may turn out to have particular significance. Produced by the paraventricular nucleus, CRF both decreases food intake and augments energy expenditure.[1,2,5,6] Indeed, its levels are elevated in states associated with weight loss, e.g., anorexia nervosa and primary adrenocortical insufficiency. Conversely, glucocorticoid hormones are essential, but not sufficient for the development of obesity.[1,2] Cortisol may act by two different mechanisms.[1,2] First, it regulates CRF levels. Secondly, it is apparently permissive for the α_2-adrenergic action on the paraventricular nucleus resulting in increased food intake. Glucocorticoids are also important peripherally because they may be critical inducers of adipose differentiation, as will be described in section 3.

As yet another example of coordinated, concerted influences of brain and adipose tissue, some peptides both depress food intake centrally and suppress triglyceride accretion in adipocytes. For example, acidic fibroblast growth factor, cachectin, and interleukin-Iß attenuate food intake.[7,8] Basic fibroblast growth factor (and probably other members of the heparin-binding growth factor family), cachectin, and interleukin-Iß inhibit adipose differentiation, and have other effects which oppose triglyceride accretion.[9] While these peptides might reach adipocytes through the endocrine pathway, e.g., a heparin-binding growth factor from the adenohypophysis, paracrine/autocrine mechanisms may well be important.[10-12] A fundamentally important question asks how the central and peripheral mechanisms are coordinated. The answer should lie in the poorly understood interrelationships between neuroendocrine, endocrine, and paracrine/autocrine mechanisms. The peptides inhibiting food intake and triglyceride accretion probably act in concert with the other coordinated system facilitating energy depletion, i.e. activation of the ventromedial nucleus.

A number of gastrointestinal-pancreatic peptides suppress food intake.[13] The small intestinal hormone, cholecystokinin, and its biologically active form, the sulfated octapeptide, have been studied most extensively, and may be illustrative of other peptides.[13] Cholecystokinin acts through vagal afferents projecting onto the nucleus tractus solitarius, in turn possibly relaying messages to the paraventricular nucleus. Moreover, as exemplified by cholecystokinin, the neuropeptide acting centrally is structurally the same as the gut peptide, which influences the same centres via vagal afferents, a complex interactive loop controlling food intake.[5,13] The gut peptides also appear to augment adaptive thermogenesis.

3. Hypothalamic-Pituitary-Adipose Cell Axis

The possible existence of a hypothalamic-pituitary-adipose axis was originally proposed[10,11,14,15] on the basis of two findings: a) Early studies indicated that hypophysectomy in rats virtually abolished the *in vivo* replication of adipose tissue stromal cells (which turned out to be at least partly preadipocytes), as determined by labelling the DNA of cells with injections of radioactive thymidine[16], and b) Adenohypophysial basic fibroblast growth factor, which is produced by folliculo-stellate cells, as well as related higher molecular mass pituitary heparin-binding growth factors, were mitogenic on cultured human and rat preadipocytes[10,11,15]. The involvement of the hypothalamus is postulated because of its known interactions with all anterior pituitary endocrine axes.

We have also found that basic fibroblast growth factor is a potent inhibitor of adipose differentiation.[9] It is probable that the larger, related mitogenic pituitary proteins are also inhibitory. We have thus proposed that the adenohypophysis modulates the size of preadipocyte (and other mesenchymal cell) pools throughout the body through the action of heparin-binding growth factors. By being both mitogenic and inhibitory of differentiation, an appropriate complement of potentially replicative preadipocytes would be maintained.[9] Then, specific nutritional and hormonal (e.g., glucocorticoids, insulin) factors would trigger the process of differentiation. The resulting mature fat cells would then expand adipose depots. A possible clinical reflection of the loss of pituitary heparin-binding polypeptide growth factors is the mild adiposity of panhypopituitary subjects even after target gland hormone replacement.

We have recently discovered that cultured human preadipocytes themselves express genes encoding proteins belonging to the heparin-binding (fibroblast) growth factor family, and that these mitogenic proteins are released by preadipocytes into the medium. Preadipocytes from massively obese persons produce significantly greater quantities than cells with lesser degrees of adiposity. The excessive production of these mitogenic proteins acting through paracrine/autocrine mechanisms, may account, at least partly, for the adipocyte hyperplasia characteristic of massive obesity. Thus, the heparin-binding growth factors may act on fat tissue through both endocrine and paracrine/autocrine (local) pathways. While the endocrine mechanisms would involve the adenohypophysis, and thus presumably the hypothalamus, the possible relationship between the pituitary heparin-binding hormones and the cognate local principles remains to be elucidated.

Glucocorticoid hormones turn out to be critical for the induction of differentiation in preadipocytes.[9] The effects of physiologic levels should be distinguished from the dramatic differential effects of cortisol hypersecretion on different adipose depots in Cushing's syndrome. Thus, under physiologic conditions, glucocorticoids, which are the effectors of the hypothalamic-pituitary-adrenocortical axis, have important roles in the regulation of food intake, energy expenditure, and through stimulation of adipose differentiation, triglyceride storage.

Somatotropin (growth hormone) induces adipose differentiation in murine fetal preadipocyte lines.[17] After induction of early steps of differentiation by growth hormone, the fetal cells become particularly susceptible to the mitogenic influence of insulin-like growth factor I, resulting in clonal expansion of committed, developing fat cells.[18] We, and others, have found that growth hormone does not induce the differentiation of preadipocytes from post-natal subjects.[9] Thus, somatotropin, whose production and secretion is controlled by hypothalamic growth hormone-releasing hormone and somatostatin, might only contribute to preadipocyte development and growth during prenatal life.

4. Putative Hypothalamic-Neural Regulation of Cytoskeletal Activity

In mediating the diverse biomechanical functions of cells, the cytoskeleton (microtubules, microfilaments, intermediate filaments and related structures) consumes prodigious quantities of energy, whose precise magnitude remains to be defined. A few years ago, the author postulated a novel concept to explain the genesis of obesity.[4,19] The hypothesis invokes interindividual genetic differences in cytoskeletal activity, and hence, energy consumption for biochemical processes. A major portion of the unused energy would eventually be stored as chemical energy, predominantly triglycerides in adipocytes.[4,19] There would be an inverse relationship between energy utilized for cytoskeletal function, e.g., various motile processes, and triglyceride depositon. Variations in degree of cytoskeletal activity in the general population would be due to genetic polymorphism. At the extreme of massive obesity, a discrete mutation in a gene encoding a structural or regulatory cytoskeletal protein would result in appreciable dampening of biomechanical processes; consequently, rather large quantities of energy would remain available for triglyceride storage.[4,19] The early evidence obtained in our laboratory is entirely supportive of this hypothesis.

The endocrine and nervous systems mediate communication between diverse cells widespread through the body. I propose that a unique, until now unrecognized hypothalamic-efferent neural pathway modulates the activity of the cytoskeleton in various cell types including preadipocytes and other motile cells. I also propose the term "N" (capital form of the Greek letter ν - "nu") for this new pathway. It may well be related to the hypothalamic-efferent adrenergic pathway, but with distinct receptor and intracellular machinery influencing selectively cytoskeletal activity. Thus, structural, functional, or pharmacologic (e.g., phenothiazines, butyrophenones, lithium, cyproheptadine, valproic acid) abnormalities of this pathway would decrease the quantity of energy utilized for cytoskeletal functions, channeling more energy to chemical storage, and thus predisposing to obesity.[4,19] One would also speculate that there is integration between effects on the hypothalamus and on the cytoskeleton, mediated by neural and neuroendocrine mechanisms. For example, when the organism requires more energy, the stimuli acting normally on the hypothalamus to increase food intake would act in concert with those decreasing cytoskeletal activity, thus promoting energy conservation. In subjects predisposed to obesity, a putative pathogenetic mechanism would involve decreased hypothalamic responsiveness to afferent nutritional stimuli. Another possibility would invoke constitutive dampening of target cell cytoskeletal activity, with partial unresponsiveness to stimulation by the "N" pathway.

SUMMARY

The brain, particularly certain nuclei of the hypothalamus and their neural connections, have a major influence on energy balance, through effects on both food intake and energy expenditure. As summarized in Table 1, there are indeed extensive interactions between the hypothalamus and adipose tissue, the predominate site of storage of chemical energy. Structural, and possibly functional, abnormalities of the neural structures facilitate the development of obesity. This review has described four components of the interactive system. Two of these components are still partly conjectural; while we have increasing experimental support, the hypothalamic-pituitary-adipose axis and the hypothalamic-efferent neural-cytoskeletal pathway are the subject of continuing intense investigation. More complete knowledge of the pathophysiology of obesity will, in turn, facilitate prevention and treatment of corpulence, as well as such frequent associations as non-insulin dependent diabetes mellitus.

ACKNOWLEDGEMENTS

We thank Mr. Bradford Hamilton and Dr. Krystyna Teichert-Kuliszewska in our laboratory, as well as Dr. Mervyn Deitel, St. Joseph's Health Centre - University of Toronto, for their invaluable assistance. For the research conducted by the author, the support of the Medical Research Council of Canada (Grant MT-8460), the Canadian-Ontario Heart and Stroke Foundation, as well as the Sunnybrook Health Science Centre Foundation are acknowledged gratefully.

Table 1. Central-Peripheral Connections

Pathway	Functions	
Hypothalamus-Adrenergic Efferents	Increased Adaptive Thermogenesis Decreased Food Intake and Triglyceride Storage	
Hypothalamic-Vagal Efferents	Increased Food Intake and Triglyceride Storage	
Hypothalamic-Neuro-peptides-Gut Peptides	Increased or Decreased Food Intake	
Hypothalamic-Pituitary-Adrenocortical Axis	CRF:	Decreased Food Intake Increased Adaptive Thermogenesis
	Cortisol:	Increased Food Intake Increased Adipose Differentiation
Hypothalamic-Pituitary-Adipose Axis	Development and Maintenance of Immature Preadipocytes	
Putative Hypothalamic-Efferent Neural Cytoskeletal ("N") Pathway	Modulation of Cellular Biomechanical Activity and Associated Energy Utilization (with Reciprocal Relationship to Triglyceride Storage)	

References

1. G. A. Bray, D. A. York, and J. S. Fisler, Experimental obesity: A homeostatic failure due to defective nutrient stimulation of the sympathetic nervous system, Vitamins and Hormones 45:1 (1989).
2. G. A. Bray, J. Fisler, and D. A. York, Neuroendocrine control of the development of obesity: Understanding gained from studies of experimental animal models, Frontiers in Neuroendocrinology 11:0 (1990).
3. G. J. Taborsky, Jr. This issue.
4. D. A. K. Roncari, Obesity and lipid metabolism, in "Clinical Medicine", J. A. Spittell, Jr., ed., Harper & Rowe, Philadelphia (1986).
5. J. E. Morley, Neuropeptide regulation of appetite and weight, Endocr. Rev. 8:256 (1987).
6. A. S. Levine, B. Rogers, J. Kneip, M. Grace, and J. E. Morley, Effect of centrally administered corticotropin-releasing factor (CRF) on multiple feeding paradigms, Neuropharmacol. 22:337 (1983).
7. K. Hanai, Y. Oomura, Y. Kai, K. Nishikawa, N. Shimizu, H. Morita, and C. R. Plata-Salaman, Central action of acidic fibroblast growth factor in feeding regulation, Am. J. Physiol. 256:R217 (1989).
8. C. R. Plata-Salaman, Y. Oomura, and Y. Kai, Tumor necrosis factor and interleukin-Iß: suppression of food intake by direct action in the central nervous system, Brain Res. 448:106 (1988).

9. D. A. K. Roncari and P. E. Le Blanc, Inhibition of rat perirenal preadipocyte differentiation, Biochem. Cell Biol. 68:238 (1990).

10. D. C. W. Lau, D. A. K. Roncari, D. K. Yip, S. Kindler, and S. G. E. Nilsen, Purification of a pituitary polypeptide that stimulates the replication of adipocyte precursors in culture. FEBS Lett. 153:395 (1983).

11. D. A. K. Roncari, Pre-adipose cell replication and differentiation, Trends Biochem. Sci. 9:486 (1984).

12. D. C. W. Lau, D. A. K. Roncari, and C. H. Hollenberg, Release of mitogenic factors by cultured preadipocytes from massively obese human subjects, J. Clin. Invest. 79:632 (1987).

13. J. E. Morley, Appetite regulation by gut peptides, Annu. Rev. Nutr. 10:383 (1990).

14. D. A. K. Roncari, Obesity and lipid metabolism, in "Systematic Endocrinology", C. Ezrin, J. O. Godden, and R. Volpe, eds., Harper and Rowe, Hagerstown (1979).

15. D. A. K. Roncari, Hormonal influences on the replication and maturation of adipocyte precursors, Int. J. Obes. 5:547 (1981).

16. C. H. Hollenberg and A. Vost, Regulation of DNA synthesis in fat cells and stromal elements from rat adipose tissue, J. Clin. Invest. 47:2485 (1968).

17. M. Morikawa, T. Nixon, and H. Green, Growth hormone and the adipose conversion of 3T3 cells, Cell 29:783 (1982).

18. K. M. Zezulak and H. Green, The generation of insulin-growth factor-I sensitive cells by growth hormone action, Science 233:551 (1986).

19. D. A. K. Roncari, Individual variations in energy utilized for biomechanical processes and molecular mobility account for diverse susceptibility to obesity, Medical Hypotheses 23:11 (1987).

NEURAL CONTROL OF ISLET FUNCTION BY NOREPINEPHRINE AND SYMPATHETIC NEUROPEPTIDES

Beth E. Dunning and Gerald J. Taborsky, Jr.

Diabetes Depart., Sandoz Research Institute, East Hanover, NJ 07936 and Division of Metabolism, Endocrinology & Nutrition, Depart. of Medicine, Seattle VA Medical Center and University of Washington Seattle, WA 98108.

INTRODUCTION

Stress activates the sympathoadrenal system increasing the circulating levels of epinephrine (EPI) and norepinephrine (NE). Both the magnitude of the plasma catecholamine response and the resultant hyperglycemia are dependent on the severity and type of stress. Severe stress also alters the secretion of the metabolically-active hormones, insulin (impaired relative to the hyperglycemia) and glucagon (enhanced despite the hyperglycemia). Most evidence links the activation of the sympathoadrenal system to these changes of pancreatic hormone secretion, suggesting that either the increase of adrenomedullary EPI or increased activity of the sympathetic nerves of the pancreas mediates these changes of insulin and glucagon secretion.

Previous work has focused on EPI as the mediator. The role of the pancreatic sympathetic nerves has received less attention probably because of the technical difficulties involved in such studies. Thus, although it has long been known that direct electrical stimulation of these nerves reproduces the inhibition of insulin and stimulation of glucagon secretion characteristic of stress, it has only recently been demonstrated that these nerves are, in fact, activated during certain stresses (45). Since NE is the classical sympathetic neurotransmitter, it has generally been assumed that NE mediates the effects of sympathetic nerve activation on islet function. However, experimental tests of that assumption, coupled with the discovery of peptidergic neurotransmitters, have led to the proposal that neuropeptide co-transmitters may mediate a significant portion of the sympathetic neural effects on islet function. Thus, the first goal of this article is to summarize the evidence in support of the "neuropeptide hypothesis." The second goal is to address the broader question of the physiologic role of the sympathetic nerves of the pancreas, regardless of the particular neurotransmitter.

To accomplish these goals, the chapter will be divided into three parts. The first section reviews the effects of electrical stimulation of sympathetic nerves on insulin (IRI), glucagon (IRG) and somatostatin (SLI) secretion, as well as on pancreatic blood flow. The second section reviews evidence that three different neurotransmitters may combine to mediate these sympathetic neural effects on the islet: the classical sympathetic neurotransmitter, NE, and the putative peptide co-transmitters, galanin and Neuropeptide Y (NPY). This section will include evidence of their localization in and actions on the endocrine pancreas, the presence of their receptors on islet cells and their release during pancreatic sympathetic nerve stimulation, and finally the influence of blocking their actions on the hormonal response to such neuronal activation. The third section will summarize evidence for a physiologic role of the pancreatic sympathetic nerves in regulating pancreatic hormone secretion during stress, pointing out both the problems and promise in that area of investigation.

Fuel Homeostasis and the Nervous System, Edited by M. Vranic *et al.*
Plenum Press, New York, 1991

The Effect of Sympathetic Neural Activation on the Endocrine Pancreas

Methods: There are two methods for directly electrically stimulating the sympathetic nerves of the pancreas: mixed pancreatic nerve stimulation (MPNS) and splanchnic nerve stimulation (SpNS). These exploit the differences between the neuroanatomy of the sympathetic and parasympathetic branches of the autonomic nervous system illustrated in Figure 1. The SpNS technique relies on the fact that splanchnic nerves contain predominantly sympathetic fibers. However, they innervate not only the pancreas but also the liver and adrenal, as well as the rest of the splanchnic bed. The resulting hepatic and adrenal activation elicits hyperglycemia which offsets the neural impairment of **basal** IRI release. More importantly, the high levels of circulating EPI do not allow one to determine the effect of the pancreatic sympathetic nerves *per se*. This drawback can be avoided by stimulating the splanchnic nerves still attached to an isolated buffer perfused pancreas or by performing SpNS in adrenalectomized animals. However, the former approach may bias the results, since NE produces primarily a-adrenergic effects on IRI release *in vitro* (52) as opposed to the mixed α-and ß-adrenergic effects seen *in vivo* (85). The latter approach may also mislead since adrenalectomy can impair the functioning the sympathetic nerves which can be dependent on the $ß_2$-adrenergic stimulation provided by circulating EPI (90).

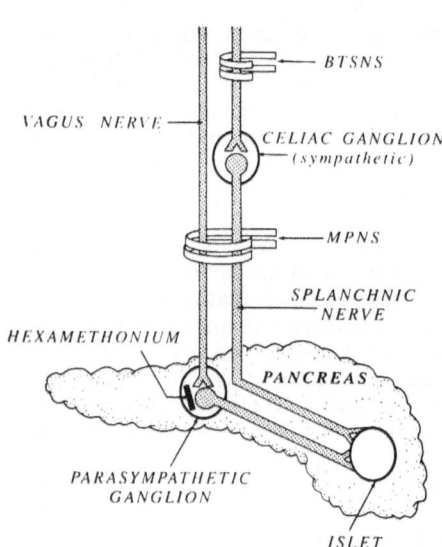

Figure 1. Two techniques for electrically stimulating the sympathetic nerves of the pancreas. Bilateral Thoracic Splanchnic Nerve Stimulation (BTSNS; also SpNS in text) activates the sympathetic nerves to the pancreas, but also those to the liver and adrenal medulla (see text). Mixed Pancreatic Nerve Stimulation (MPNS) activates the sympathetic nerves of the pancreas but also produces an activation of parasympathetic nerves which can be blocked by Hexamethonium (see text).

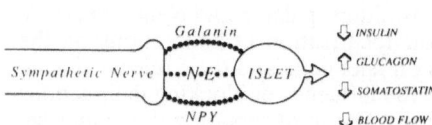

Figure 2. The sympathetic nerves of the canine pancreas release the neuropeptides galanin and NPY, in addition to the classical sympathetic neurotransmitter norepinephrine (NE). Together these transmitters produce a decrease of pancreatic blood flow, an inhibition of insulin and pancreatic somatostatin secretion and a stimulation of glucagon secretion.

MPNS activates the fibers coursing with the major arterial supply to the pancreas. This mixed autonomic nerve contains pre-ganglionic parasympathetic fibers as well as post-ganglionic sympathetic fibers (see Figure 1). The parasympathetic influence can be eliminated by using hexamethonium to block the ganglionic action of acetylcholine leaving a local, sympathetic neural activation.

Effects on Pancreatic Hormones and Blood Flow: Direct neural effects on pancreatic hormone secretion were first demonstrated by Porte, *et al.* who reported that MPNS following atropine treatment in dogs, markedly inhibited the IRI response to glucose (84, Figure 2). The effect of sympathetic neural activation has since been confirmed in many species and experimental systems including SpNS *in vivo* in cats (13), pigs (48) , dogs (54), calves and sheep (17). Additionally, an inhibition of **basal** IRI secretion is elicited by MPNS following atropine and hexamethonium in dogs (6), during SpNS *in vitro* in puppies (88), and a transient inhibition of basal IRI output is seen during SpNS *in vivo* in dogs, prior to the onset of hyperglycemia (35).

Sympathetic neural activation also markedly stimulates IRG secretion (Figure 2). This occurs during MPNS following cholinergic blockade (6), during SpNS *in vitro* (50) and even during SpNS *in vivo*(54), despite marked hyperglycemia.

Sympathetic neural effects on pancreatic SLI secretion have been studied less extensively. Since pancreatic somatostatin makes little contribution to systemic SLI levels (98), such studies require sampling of the pancreatic venous effluent. SpNS in puppy and pig pancreas preparations inhibits pancreatic SLI output (87,50, Figure 2), as does MPNS during cholinergic blockade *in vivo* (7). SpNS does not inhibit pancreatic SLI output *in vivo* (unpublished observation) perhaps because of the offsetting stimulus of hyperglycemia. Since SpNS still stimulates IRG secretion *in vivo*, inhibition of pancreatic SLI release, a potential paracrine inhibitor of IRG secretion (97), apparently does not mediate the IRG response seen during this type of pancreatic nerve stimulation.

MPNS during cholinergic blockade transiently reduces pancreatic blood flow (Figure 2). In constant flow perfusion systems, SpNS increases perfusion pressure (50) reflecting pancreatic vasoconstriction. SpNS *in vivo* increases mean arterial blood pressure which overcomes the local vasoconstriction, resulting in increased pancreatic blood flow (35). Thus, activation of pancreatic sympathetic nerves consistently produces vasoconstriction despite variable effects on pancreatic blood flow.

In summary, electrical stimulation of the sympathetic nerves of the pancreas produces a local vasoconstriction, inhibits IRI and SLI secretion and markedly stimulates IRG secretion (Figure 2).

Mediators of Sympathetic Neural Effects on the Endocrine Pancreas

Traditionally, NE was thought to mediate all effects of sympathetic nerves. Recent evidence of the localization, release and action of peptides in peripheral sympathetic nerves suggest that NE may not function alone. The following section reviews the evidence that two sympathetic neuropeptides, galanin and NPY, may, with NE, mediate the effects of pancreatic sympathetic nerves on islet hormone secretion (Figure 2). We review here their 1) presence in pancreatic nerves, 2) effects on islet hormone secretion, 3) receptors on islet cells, 4) release during sympathetic neural activation as well as 5) the effect of their selective antagonism on the islet response to sympathetic neural activation.

Sympathetic Innervation of The Pancreas: Most of the sympathetic input to the pancreas leaves the thoracic spinal cord, courses in the splanchnic nerves and synapses in the celiac ganglia. The post-ganglionic fibers then follow the arterial blood supply into the pancreas, as illustrated in figure 1. There are, however, descriptions of sympathetic fibers arising from the cervical and lumbar spinal cord that directly innervate the pancreas (70) as well as sympathetic fibers in the vagus nerve (103). The sympathetic fibers innervate not only pancreatic arterioles but also the islets themselves. Islet nerves do not form classical synapses but rather, terminate blindly, releasing neurotransmitter which presumably

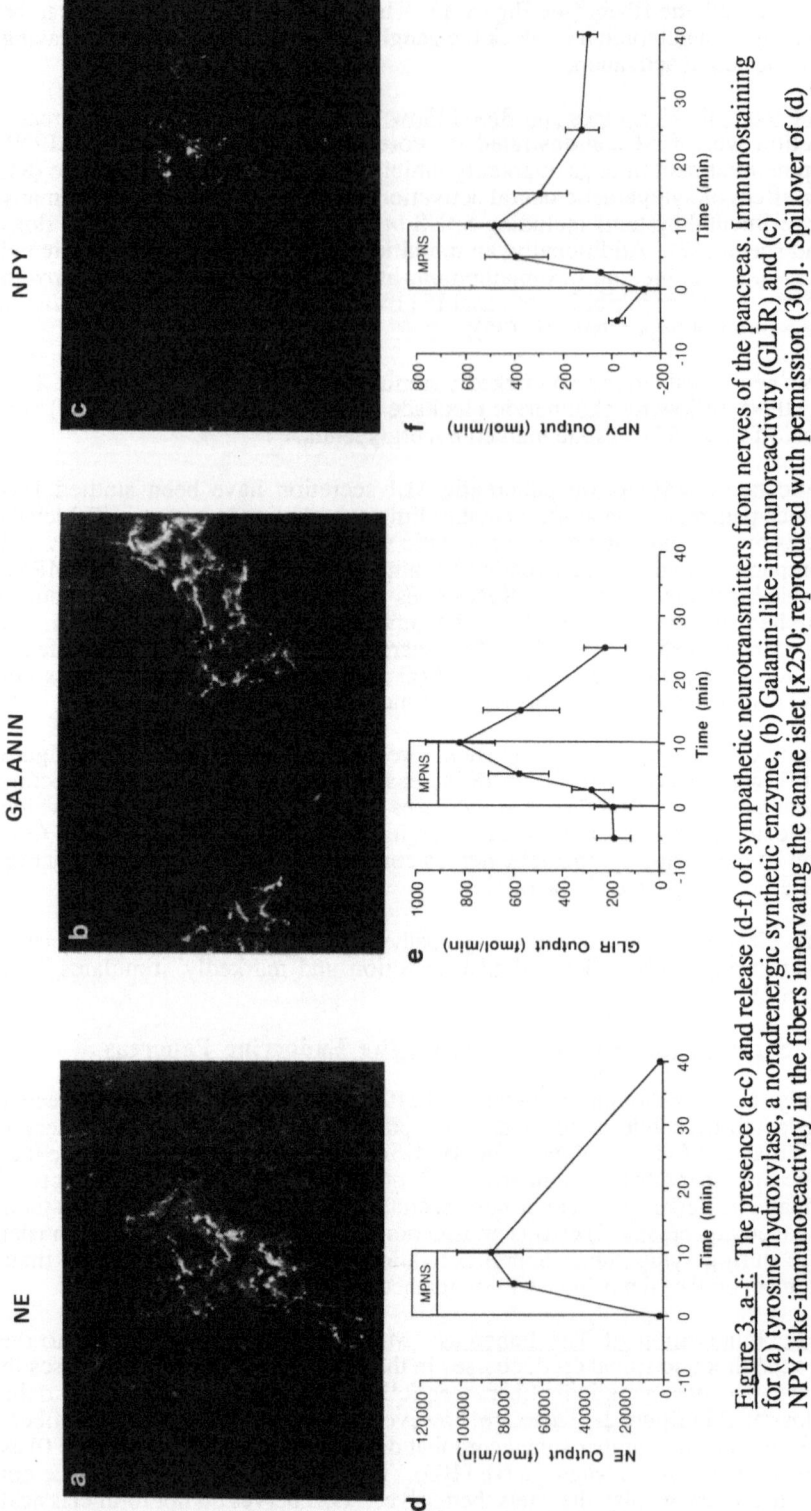

Figure 3, a-f: The presence (a-c) and release (d-f) of sympathetic neurotransmitters from nerves of the pancreas. Immunostaining for (a) tyrosine hydroxylase, a noradrenergic synthetic enzyme, (b) Galanin-like-immunoreactivity (GLIR) and (c) NPY-like-immunoreactivity in the fibers innervating the canine islet [x250; reproduced with permission (30)]. Spillover of (d) norepinephrine, (e) galanin-like-immunoreactivity (GLIR), or (f) NPY-like-immunoreactivity into pancreatic venous plasma of halothane-anesthetized dogs during mixed pancreatic nerve stimulation [MPNS; 8 Hz, 10 mA, 1msec, 10 min.; reproduced with permission (34) or unpublished observation].

diffuses to many adjacent target cells (95). Although there is some species variation in the adrenergic innervation of the islet, the species variation in the sympathetic peptidergic innervation of the islet is considerable (see below), making generalizations inadvisable.

NE: The presence of NE-containing islet nerves has been demonstrated in many species using formaldehyde-induced autofluorescence or electron-microscopy (103) as well as the difficult technique of immunostaining for the catecholamine synthetic or degradative enzymes (Figure 3a) or even for NE itself (69). Noradrenergic nerves preferentially innervate islets (vs exocrine tissue) and there is also abundant noradrenergic innervation of blood vessels in both the islets and exocrine parenchyma (69). Additionally, noradrenergic nerves innervate intrapancreatic ganglia (61), providing an anatomical basis for sympathetic modulation of parasympathetic neural activity.

Galanin: The presence of galanin-containing nerves in pancreatic tissue has been demonstrated by immunofluorescent staining of pancreatic tissue of the dog (31, Figure 3b), rat (96), mouse (67), pig (73), and recently, human (28). In dogs, fibers containing galanin-like immunoreactivity (GLIR) have been found to densely and preferentially innervate islets, with few fibers in the intrapancreatic ganglia, blood vessels and acinar tissue, in contrast to other species. In dog islets, nearly all GLIR positive fibers also stain for tyrosine hydroxylase, a noradrenergic marker (Figure 3a), suggesting colocalization with NE. Galanin staining in mouse pancreas appears to be less dense than that in the dog but is more dense than that in rats, consistent with measurements of pancreatic GLIR content (unpublished observation). However, both techniques depend on antibody recognition of the galanin molecule, the primary structure of which is known to vary across species. Neuronal cell bodies containing GLIR are not found in pancreatic tissue but rather in sympathetic ganglia (3), again colocalized with tyrosine hydroxylase, suggesting that galanin is a post-ganglionic sympathetic neurotransmitter.

NPY: The presence of the putative sympathetic neurotransmitter, NPY, has been demonstrated in pancreatic nerves of several species by immunofluorescent staining. In the dog, NPY fibers are found in islets (Figure 3c) as well as in the exocrine parenchyma, with preferential innervation of the blood vessels supplying these structures. In the dog, NPY fibers also innervate intrapancreatic ganglia: occasionally NPY-positive cell bodies are also present there (30). More often, NPY-positive cell bodies are found in the celiac ganglia, usually co-localized with galanin and NE, as is the NPY in the fibers supplying canine islets (3). Thus, in the dog, most pancreatic NPY fibers are extrinsic and therefore probably sympathetic. In the rat (21) and mouse (83) pancreas, NPY fibers only occasionally innervate the islets; the majority innervate the vasculature. In the pig, NPY fibers innervate pancreatic blood vessels, but there is also dense NPY innervation of intrapancreatic ganglia as well as many NPY-positive cell bodies, suggesting that, in the pig, NPY may serve both a sympathetic and a parasympathetic function (49).

Effects of Exogenous NE, Galanin and NPY on Pancreatic Hormone Secretion and Blood Flow:

NE: The classical sympathetic neurotransmitter, NE, is a mixed adrenergic agonist capable of activating of α-and ß-adrenergic receptors. Whereas ß-adrenergic activation stimulates the release of insulin IRI, IRG and SLI, α-adrenergic activation inhibits IRI and SLI release in several species in experimental systems, both in vivo and in vitro (51,89,57,32). The effects of α-adrenergic activation on IRG release are more controversial in that stimulatory (57), inhibitory (41), and more frequently, weak (89) or no effects (32) have been reported. In general, available evidence suggests that α_2 receptor mediation of the inhibitory effects and β_2 receptor mediation of the stimulatory effects on islet secretion. Thus, a mixed agonist like NE can produce two opposite and offsetting adrenergic effects on IRI, SLI and even IRG secretion.

Since the ratio of α to ß-adrenergic receptors and their effectiveness can vary between species and experimental systems, the net effect of NE on IRI, SLI and IRG secretion is difficult to predict. Dogs tend to be ß-adrenergically sensitive thus, infusions of NE tend to stimulate basal IRI secretion (6). Rats tend to be more sensitive to α-adrenergic effects and NE infusions either inhibit or have no effect on basal IRI release in this species (82). Further, *in vitro* systems seem to be predominantly a-adrenergic, thus even in a buffer perfused dog pancreas NE inhibits IRI release (52). The decreased effectiveness of the ß-adrenergic receptor *in vitro* may be due to the low oxygen carrying capacity of the buffer perfusing the pancreas since hypoxia is known to impair the function of this receptor (15). Finally, the effect of combined α- and ß-adrenergic activation on the pancreatic B-cell differs depending on the type of IRI secretion examined. For example, intrapancreatic infusion of NE in the dog elicits a mild stimulation of **basal** IRI secretion, but markedly impairs arginine-**stimulated** IRI release (5). Thus, the ability of the classical sympathetic neurotransmitter, NE, to account for all of the effects of sympathetic nerve activation may depend on the species and experimental circumstances in which this question is examined.

In dogs, attempts to answer this question have led to the conclusion that NE cannot reproduce all of the effects of local sympathetic nerve stimulation on islet function. For example, pancreatic arterial infusion of NE at a wide range of doses, including that which reproduces the decrease of blood flow elicited by MPNS (Figure 4a) did not inhibit basal IRI secretion and only partially reproduced the inhibition of SLI and stimulation of IRG that are induced by MPNS (6). A second, independent approach led to the same conclusion: complete and combined α-and ß-adrenergic blockade demonstrated at the level of the islet only modestly reduced the degree of inhibition of IRI and SLI and did not affect the stimulation of IRG induced by MPNS (32). Studies of this type, together with the presence of peptides in islet nerves and their colocalization with NE, suggest a role for neuropeptides in this sympathetic control of pancreatic hormone secretion.

NE has well known and readily demonstrable vasoconstrictor actions: pancreatic administration of NE reduces blood flow *in vivo* (Figure 4d), and increases perfusion pressure in constant flow systems *in vitro* (50). Yet, α_1 adrenergic antagonism, sufficient to block pancreatic vasoconstriction during local NE infusion, reduced, but did not eliminate the decrease of pancreatic venous blood flow during MPNS (unpublished observations). Therefore although NE mediates part of the vasoconstriction, some other sympathetic neurotransmitter, perhaps NPY (see below), may contribute.

Galanin: The first clue that the 29 amino acid peptide, galanin, impaired IRI release was a report of hyperglycemia during galanin infusion in conscious dogs (102). Thereafter, synthetic porcine galanin was shown to markedly inhibit basal (Figure 4b) and stimulated IRI secretion in several species including dogs (31,72,46), rats (66,94), and mice (65). The data in pigs and man are mixed, with some finding no inhibition (73,42), while others report impairment of IRI release (68,28,62). Since galanin is without effect on pancreatic blood flow (Figure 4e) and since its inhibition of IRI release persists during a-adrenergic blockade (46), it is unlikely that galanin's inhibitory actions on IRI release are mediated either via an interaction with an α_1 adrenergic receptor on the pancreatic vasculature or via activation on the inhibitory α_2 adrenergic receptor on the B-cell itself. Since galanin inhibits IRI release from isolated islets (1) and monolayer cultures of neonatal pancreatic cells (33) and cultured B-cell lines (10), the inhibitory effect of galanin is almost certainly direct.

The intracellular mechanism by which galanin inhibits IRI release is controversial. There exists electrophysiologic and/or pharmacologic evidence to suggest that galanin acts via at least 3 different mechanisms, each possibly involving G-proteins: 1) opening ATP-dependent K+ channels (26,27,2) 2) inhibiting adenylate cyclase and reducing cAMP levels (9) and 3) blocking a step distal to both depolarization and adenylate cyclase, perhaps the process of exocytosis itself (78,91,104).

Although the bioactive site of the galanin molecule has not yet been precisely defined, structure-activity studies using galanin fragments suggest that the N-terminus of the molecule is critical while the C-terminus may not be essential. For example, elimination of

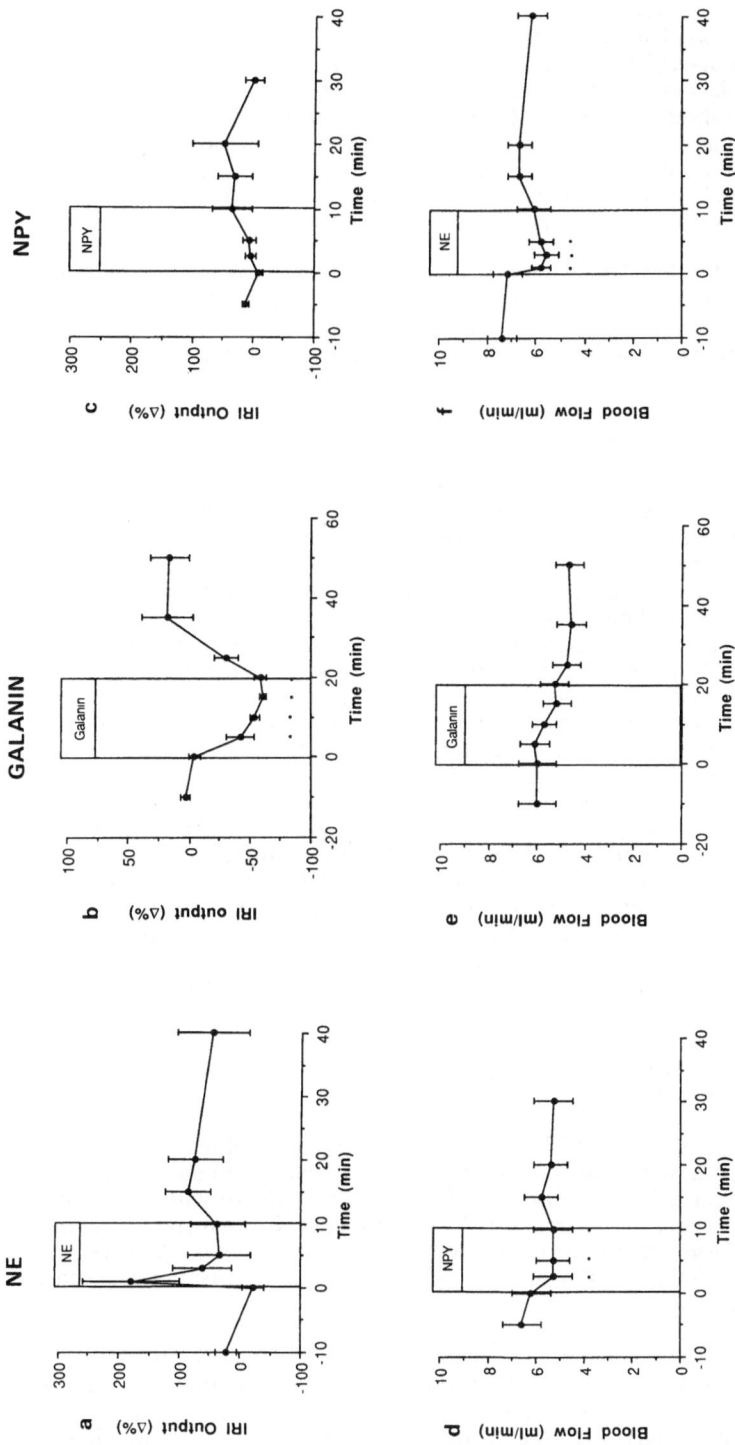

<u>Figure 4 a-f:</u> The actions of locally infused synthetic neurotransmitters on insulin (IRI) output (a-c) or blood flow (d-f) from the right lobe of the canine pancreas. (a) Norepinephrine (NE) was infused at 700 pmol/min for 10 min., (b) porcine galanin was infused at 3 pmol/min for 20 min. and (c) human NPY was infused at 50 pmol/min for 10 min. into the superior pancreatico-duodenal artery; insulin output was measured in the superior pancreatico-duodenal vein of anesthetized dogs [reproduced with permission (6,31)]. (d) Norepinephrine (NE), (e) porcine galanin and (f) human NPY were infused as above. Blood flow was measured using an electromagnetic flow probe in an extracorporeal shunt between the superior pancreatico-duodenal vein and the portal vein of anesthetized dogs [reproduced with permission (6,30,31)].

the two N-terminal amino acids eliminates IRI-inhibiting activity, whereas deletion of 14 C-terminal amino acids only modestly reduces the potency of galanin (47,10). These bioactivity studies are in complete agreement with the binding of galanin fragments and analogues to ß-cell membranes (60), suggesting that all galanin-related peptides examined thus far act as weak agonists rather than antagonists. Although the primary structure of galanin is known for only three species (pig (102), rat (55), and cow (86)), the structural differences occur within the C-terminal 7 amino acids. Thus, these known differences would be expected to have minor effects on the potency of galanin between species. Nonetheless, testing of the homologous (native) form of galanin is needed to definitively rule out this explanation for certain negative findings (73,42).

Galanin markedly inhibits basal pancreatic SLI secretion in dogs (46,31) and modestly inhibits SLI release from the perfused rat pancreas in some reports (81), but not others (59). Galanin has also been found to inhibit somatostatin release from gastrointestinal D-cells (59). This is consistent with reports that iv galanin infusion decreases circulating SLI (31), which is thought to be of gastrointestinal origin (98,37). Galanin may even impair somatostatin release from hypothalamic neurons, a mechanism proposed to explain the stimulatory effect of galanin on growth hormone (25).

The effects of galanin on IRG secretion are less pronounced and more varied than those on IRI or SLI release. In dogs, galanin modestly stimulates pancreatic IRG output to a degree that has little effect on peripheral IRG levels (31). In perfused rat pancreas, the rat, but not porcine, form of galanin also stimulates IRG output (81,94). These stimulatory effects on IRG secretion may be mediated through inhibition of pancreatic somatostatin secretion since pancreatic somatostatin tonically restrains pancreatic glucagon secretion (97) and since IRG release seems to increase only when galanin is effective in suppressing pancreatic SLI release (31,81). When somatostatin tone is low, galanin administration seems to inhibit IRG secretion as illustrated by the effect of galanin on basal IRG levels in the rat (66) and high doses of galanin in the buffer-perfused dog pancreas (46). This weak inhibitory effect of galanin may reflect a direct action on the A-cell, since inhibition is also seen in monolayer cultures of neonatal rat pancreas (33).

NPY: The 36 amino acid peptide, NPY, was originally isolated from extracts of porcine brain (100) using a method for detection of the C-terminal amide structure characteristic of many biological active peptides. NPY modestly inhibits glucose-stimulated IRI in vivo in mice (83) and rats (82) and in vitro from isolated rat islets and perfused rat pancreas (75), but has little effect on basal IRI secretion in these systems. In dogs, high doses are required to produce minor islet effects: NPY modestly inhibits pancreatic SLI output, with no effect on basal (Figure 4c) or glucose-stimulated IRI release (30). At very high doses, NPY modestly stimulates IRI and IRG release and inhibits SLI output (30). In the buffer-perfused pig pancreas, NPY modestly inhibits IRG release with no effect on SLI or IRI secretion (49). In general, NPY does not have a major or consistent effect on islet function.

Perhaps a more important role for NPY is found in the regulation of pancreatic blood flow, analogous to its hypothesized role in other organs. For example, in the isolated pig pancreas, NPY increases perfusion pressure, an effect that was additive with that of NE (49). In dogs, pancreatic arterial infusion of NPY modestly decreases pancreatic venous blood flow (Figure 4f). These findings are consistent with the innervation of blood vessels by NPY fibers and the interaction of NPY and NE to increase vascular tone in many other sympathetic-innervated organs.

In summary, we suggest that while **NE** is capable of influencing pancreatic hormone secretion via both α-adrenergic inhibitory and ß-adrenergic stimulatory actions, effects of electrical activation of sympathetic nerves on pancreatic hormone secretion and blood flow **cannot** be explained solely by NE. **Galanin** exerts a potent and direct sympathomimetic effect on IRI and SLI release, as well as a modest, probably indirect, sympathomimetic effect on IRG secretion. Although **NPY's** effect on pancreatic hormone secretion is modest, it may augment NE's effect to increase pancreatic vascular resistance.

Receptors for NE, Galanin and NPY on Endocrine Pancreatic Tissue

Studies of tissue "receptors" usually involve either isolated membranes or tissue autoradiography. Both techniques use the binding and displacement of radiolabelled ligands to estimate the number and affinity as well as specificity of the receptor. The difficulty in obtaining large quantities of pure, isolated islets to prepare membranes, and the low density of islets *vs* acinar tissue for autoradiography, has limited the receptor work done on islet cells.

Recently a few studies have demonstrated directly the binding and competitive displacement of radiolabelled α- and ß-adrenergic ligands to isolated and semi-purified endocrine cells. While there is general agreement on the affinity of these receptors (Kd ~ 1nM), there are apparently conflicting reports about the relative abundance of α- and ß-adrenergic receptors on islet cells (22,39,40).

For galanin, membranes for binding studies have been prepared from a hamster B-cell tumor line. This cell line contains a single population of specific, high affinity (Kd = 1.5 nM) and low capacity galanin binding sites with an apparent molecular weight for the "receptor complex" of 54 KD (11). Receptors for galanin having an IC50 of 0.3 nM have also been demonstrated on the rat B-cell line, Rin m 5F (60). The specificity of this galanin binding has been also demonstrated utilizing fragments and analogues of galanin. These studies indicate 1) the critical importance of the N-terminal amino acid for receptor binding, 2) a requirement for an aromatic amino acid in position 2 and 3) a modest influence of the C-terminal half of the galanin molecule (60).

Although there are NPY binding studies utilizing brain tissue, we know of no analogous studies on the membranes of pancreatic islet cells. In other tissues, NPY receptors have been classified as Y_1 (92), exemplified by the post-synpatic receptor mediating vasoconstriction (105), and Y_2 (92), exemplified by the presynpatic receptor restraining neuronal NE release (105). Activation of the Y_1 receptor also markedly potentiates the post-synaptic actions of NE (105). Thus it is likely, but unproven, that NPY's effect on pancreatic vasculature is mediated by a Y_1 receptor. The report that the α-adrenergic antagonist, phentolamine, competitively blocks NPY's inhibition of glucose-stimulated IRI release in isolated islets (8), suggests that NPY's effect on IRI release may also be via potentiation of NE's inhibitory effect on the B-cell. However, the inhibitory action of NE, itself, on IRI release is mediated by α_2 adrenergic receptors, originally thought to be solely presynaptic in location. By analogy, it is premature to rule out mediation via post-synaptic Y_2 receptors. The development of specific antagonists needed to resolve such questions is just starting (101,74,38).

Release of NE, Galanin and NPY during Sympathetic Nerve Stimulation. Peripheral plasma concentrations of NE have been used as an index of whole-body sympathetic neural activity. Such measurements do not, however, reflect the activity of pancreatic sympathetic nerves, since the pancreas makes a negligible contribution to systemic NE levels (45). Therefore, estimates of pancreatic NE release require measurements of pancreatic venous NE concentration and blood flow. If the neural activation is local, e.g. MPNS, those measurements are sufficient to calculate pancreatic NE spillover. During more generalized neural activation, such as SpNS or stress, measurements of arterial NE levels and estimates of pancreatic NE extraction are also needed to determine the contribution of the circulating neurotransmitter to that measured in the pancreatic vein and thus to correctly calculate pancreatic NE spillover (Figure 5). Similar considerations apply for the calculation of pancreatic galanin (35) and NPY (unpublished observations) release.

Using these calculations, one finds little basal spillover of NE from the pancreas. During local stimulation of sympathetic nerves (MPNS) pancreatic venous NE concentrations increase by 25nM and pancreatic spillover increases in excess of 80 pmol/min (34, Figure 3d). During SpNS, pancreatic venous NE concentrations increase by 4nM and pancreatic spillover increases by 20 pmol/min (35). The smaller amount of pancreatic NE released during SpNS is probably due to stimulation of only part of the sympathetic input to the pancreas; i.e. some fibers innervating the pancreas may enter the sympathetic trunk below the SpNS electrodes (Figure 1). Thus, as expected, electrical stimulation of the sympathetic input to the pancreas produces a large spillover of NE into the pancreatic venous effluent. However, as described above, direct infusion of NE into the pancreatic artery, at a rate that reproduces these pancreatic venous NE concentrations, does not reproduce the inhibition of IRI secretion that occurs during the nerve stimulation (6).

As with NE, there is little galanin spillover in the basal state. However, during MPNS, pancreatic venous concentrations of GLIR increase by ~0.2nM and pancreatic GLIR spillover increases by 0.8 pmol/min (34, Figure 3e). During SpNS, pancreatic venous GLIR concentrations increase by ~0.1nM. However, SpNS also increases arterial levels of GLIR, and only part of that increase contributes to the pancreatic venous GLIR levels (35). Using a formula analogous to that developed to calculate pancreatic NE spillover (see above, Figure 5), pancreatic galanin spillover during SpNS was calculated as 0.5 pmol/min. It should be noted that these calculations assume that the increase of both arterial and pancreatic venous GLIR are due to real galanin. Support for that assumption comes from gel filtration studies showing that the increases of GLIR are due to galanin-sized molecules (34,35).

On a molar basis, the amount of galanin released during sympathetic nerve stimulation is only 1-2% of the NE released, however, galanin is a very potent inhibitor of IRI secretion, at least in the dog. Thus, direct infusion of galanin at a rate that reproduces the nerve-stimulated GLIR output, elicits a marked inhibition of IRI secretion (Figure 4b) equal to that produced by nerve stimulation.

MPNS in dogs also increased pancreatic NPY spillover (Figure 3f, unpublished observation, in collaboration with M.R. Brown). The molar amount of NPY released is only marginally less than galanin, and approximately 1% of the NE released. In contrast to galanin, pancreatic infusion of NPY had no significant effect on IRI secretion (Figure 4c), although it did reduce pancreatic venous blood flow (Figure 4f). Since α_1 adrenergic blockade does not eliminate the decrease of pancreatic blood flow elicited by sympathetic nerve stimulation (unpublished), NPY may contribute to the regulation of pancreatic blood flow.

In pigs, MPNS also releases pancreatic NPY (4) and SpNS releases NPY from the perfused pig pancreas (49). However, vagal nerve stimulation induces a 10-fold greater NPY release than SpNS (93), calling into question the exclusive sympathetic role of NPY in this species.

The Effect of Antagonizing the Action of Released Neurotransmitters on Pancreatic Hormone Secretion. Several studies describe the effect of α- or ß-adrenergic antagonists on the pancreatic response to sympathetic nerve stimulation. However, since NE is a dual agonist and since the effect of α receptor activation opposes the effect of ß receptor activation on IRI, SLI and perhaps even IRG secretion (see above), such "single" adrenergic blockade will simply reveal the opposing effect. For example, ß-adrenergic blockade during NE infusion reveals an α adrenergic inhibition of IRI release, even though NE itself has little net effect on basal IRI release (32). Thus, the net contribution of endogenous NE can only be discerned during complete and combined adrenergic blockade, and the few studies in which this was done fail to produce a consistent picture, perhaps because of differences between species or *in vivo vs in vitro* conditions.

For example, combined adrenergic blockade *in vitro* in pigs (50) and rats (58) nearly eliminated the IRG response to SpNS, whereas *in vivo* in dogs (32), cats (13) and calves (18) IRG response was virtually unaffected by combined blockade. Combined adrenergic blockade modestly reduced (but did not block) the neural inhibition of IRI and SLI in some studies (32,58) but virtually abolished these responses in others (50,13). Further complicating the issue, in one study, ß-adrenergic blockade alone (which would be expected to **augment** neural inhibition of IRI) markedly **reduced** the IRI inhibition (13).

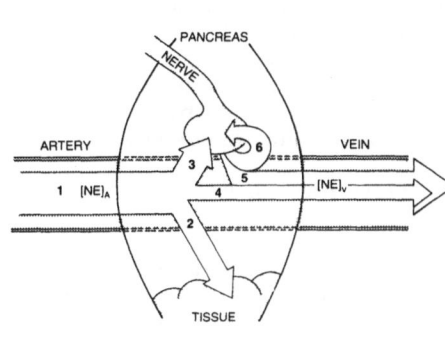

Figure 5. Calculation of pancreatic norepinephrine spillover. Circulating norepinephrine (NE) (1) enters the pancreas by the pancreatic artery. Approximately 75% of this incoming NE is extracted during transit through the pancreas, either by uptake into tissues (2), or nerves (3). Only 25% of the arterial NE escapes pancreatic extraction (4), and contributes to the pancreatic venous NE level (4+5). The nerves within the pancreas release NE (5 + 6) but the vast majority is taken back up by the nerves (6), leaving the minority (5) to spillover into pancreatic venous plasma. This spillover is therefore an attenuated index in neuronal NE release. Pancreatic venous NE spillover (5) is calculated as the venous NE level (4 + 5) minus the arterial contribution (4) multiplied by the pancreatic venous plasma flow [reproduced with permission (Havel, P.J., G.J. Taborsky, Jr. *Endo. Rev.* 10:332-350, 1989)].

In another, the α-blocker that was effective to block neural inhibition of IRI is considered selective for α_1 receptors (50), and since adrenergic inhibition of IRI is thought to be mediated by α_2 receptors, this antagonist would not be expected to be effective. In some studies, the decrease of pancreatic blood flow or increase of perfusion pressure was reduced in magnitude (32,58), and in another (50), was eliminated by combined adrenergic blockade. In summary, endogenous NE appears to mediate at least part of the pancreatic response to sympathetic nerve activation. On the other hand, combined adrenergic blockade does not abolish the effect of sympathetic nerve stimulation, suggesting that non-adrenergic neurotransmitters, such as the neuropeptides galanin and NPY, may contribute to the pancreatic response.

Since there are currently no known antagonists for galanin, there are no studies for galanin analogous to those cited above for NE. An alternative approach is to use high titer antibodies to neutralize neurally-released galanin before it activates its receptor on islet B-cells. Such experiments require antisera capable of very rapid and high affinity binding *in vivo* as well as effective penetration of the large molecular weight antibodies through the pancreatic vasculature to the target cells. Our preliminary findings with this approach provide suggestive, but not conclusive, evidence for galanin's contribution to neural inhibition of IRI secretion. During the intrapancreatic infusion of normal rat serum, SpNS significantly inhibited basal IRI output, whereas the later increase of IRI output due to developing hyperglycemia, was not statistically significant. In contrast, during the intrapancreatic infusion of galanin antiserum, the early inhibition of basal IRI output during SpNS was not significant, whereas the later stimulation was (unpublished observation). These data suggest, but by no means prove, that endogenous galanin contributes both to the early absolute inhibition and the later retraint of the IRI response to hyperglycemia.

We know of no studies attempting to immunoneutralize the local action of NPY on the pancreatic hormonal or vascular responses to sympathetic nerve stimulation. NPY antagonists are not yet commercially available, although there are preliminary reports of two different compounds with some NPY antagonism (101,74).

In summary, experiments using combined adrenergic antagonists demonstrate a role for NE as well as suggest a role for other neurotransmitters in mediating the pancreatic response to sympathetic nerve stimulation. Definitive proof that galanin and NPY play such an auxillary role awaits the development of specific, high affinity neuropeptide antagonists. Development of peptide antagonists may be accelerated by using information derived from the structure-activity and structure-binding studies mentioned above. Alternatively, several naturally-occurring non-peptide antagonists, e.g. to the neuropeptide CCK, have been discovered by screening fermentation broths. However, development of antagonists solves only half the problem because definitively demonstrating the role of sympathetic neuropeptides requires careful experimental design and several types of control experiments. For example, demonstrating that an antagonist does not eliminate a response requires proof that an effective blockade was achieved, and showing that an antagonist alters or eliminates a response requires proof that it did not do so by enhancing an offsetting response or altering the amount of neurotransmitter/cotransmitter released.

Evaluation of Experimental Approaches to Determine the Role of Pancreatic Sympathetic Nerves During Stress

Here we discuss three approaches to determining the importance of the sympathetic innervation of the pancreas in the regulation of islet function. They are: 1) blocking the actions of sympathetic neurotransmitters; 2) denervating the pancreas; and 3) measuring pancreatic neural activity. We will discuss the pros and cons of each, suggesting potential solutions to current problems.

The Influence of Sympathetic Blockers on Islet Function During Stress Surprisingly, studies of the effect of combined adrenergic blockade on the pancreatic responses during systemic stress are relatively scarce. One recent study reported that combined adrenergic blockade eliminated stress-induced inhibition of glucose-stimulated insulin IRI levels in swimming mice but did not block the stress-induced stimulation of IRG levels (56). Conversely, combined adrenergic blockade in baboons blocked neuroglucopenia-induced stimulation of IRG, but did not affect the concurrent inhibition of IRI (14). Combined adrenergic blockade has also been reported to have no influence on the pancreatic response to hypoglycemic stress (80). These varied findings may reflect differences in the role of the sympathetic nerves during different types and grades of stress, or in different species.

The major problem with this approach is that systemic adrenergic blockade will also block the actions of circulating EPI on the islet. Thus, unless the effect of combined blockade was also studied adrenalectomized animals, it would be difficult to ascribe the effects to blockade of the activity of pancreatic noradrenergic nerves. However, adrenalectomy itself may cause problems, since EPI apparently facilitates neuronal NE release via tonic stimulation of pre-synaptic ß-adrenergic receptors (90). Another problem with systemic adrenergic blockade is its potential to increase both neuronal and adrenal medullary repsonses by 1) centrally increasing sympathetic outflow (43); 2) decreasing pre-synaptic inhibition of NE release (106); and 3) decreasing clearance of circulating EPI (23). Finally, adrenergic blockade may not block the effect of sympathetic co-transmitters. Therefore, a failure of adrenergic antagonists to block the pancreatic response to a stress may not rule out an important role for the sympathetic nerves, but rather indicate the importance of non-adrenergic neurotransmitters.

We know of only one preliminary study which blocked neuropeptide action on the pancreas during stress: immunoneutralization of galanin blocked the inhibition of glucose-stimulated IRI levels normally induced by swimming stress in mice (36). Although the study suggested a role for galanin in mediating this stress-induced impairment of IRI secretion, again this methodology cannot distinguish between the effects of local, neurally-released galanin and those of circulating galanin which was also increased in response to this stress (36).

Pancreatic Denervation A more direct approach to assessing the importance pancreatic sympathetic nerves is to either surgically interrupt the pancreatic innervation, or to chemically ablate the sympathetic nerves. Two elegant studies compared the pancreatic response to either simulated hypotension (carotid sinus denervation) (53) or hemorrhagic hypotension (12) in sympathetically-intact cats with those in cats that underwent selective pancreatic sympathetic denervation. In the control animals this stress markedly increased IRG levels and decreased IRI levels in the portal vein despite a rapid doubling of plasma glucose. Pancreatic sympathectomy reduced the IRG response by half and converted the inhibition of IRI to a modest stimulation, despite a marked reduction in the hyperglycemic response. These findings indicate that pancreatic sympathetic nerves can mediate the hormonal and metabolic response to this stress. Moreover, since adrenalectomy had a similar effect, it appears that the adrenals and pancreatic sympathetic nerves can act redundantly. In a separate study, the IRG response to moderate hypoxia in conscious calves was blocked by splanchnic nerve section. Since adrenal catecholamine output was not stimulated by this degree of hypoxia in calves, the IRG response can be attributed to activation of the pancreatic sympathetic nerves (19).

Another interesting study compared the effects of adrenal demedullation and immunosympathectomy produced by administration of anti-nerve growth factor serum to neonatal rats, on the pancreatic response to 60 minutes of forced swimming. The exercise-induced increase of IRG levels was present in adrenal demedullated rats but was absent in chemically sympathectomized rats (71), suggesting a role for the sympathetic nerves of the pancreas.

Pancreatic sympathetic nerves may also have a tonic inhibitory influence on IRI secretion since CNS ablation increases IRI output from the neurally-intact but vascularly isolated perfused rat pancreas (24) and since selective pancreatic denervation augments the IRI response to glucose in dogs (16).

Although we have emphasized the simple interpretation of these technically-difficult and relatively scarce studies, these approaches also have drawbacks. First, unless the sympathetic denervation of the pancreas is selective, effects ascribed to these nerves may be mediated by other, unmeasured factors. Second, denervation or destruction of pancreatic sympathetic nerves may up-regulate the adrenergic receptors on the islets and allow increased effectiveness of circulating EPI to substitute for the original contribution of locally released NE.

In summary, certain denervation studies suggest that the pancreatic sympathetic nerves regulate islet function during stress and in other states. However, the relative scarcity of these studies and the conflicting evidence that for example, basal and glucose-stimulated IRI levels (20) and the IRG response to hypoglycemia (79) are normal in quadriplegic (sympathectomized) man, mandates further investigation in this area.

Measurement of Pancreatic Sympathetic Neural Activity The hypothesis of a physiologic role for the pancreatic sympathetic nerves requires evidence that their activity increases during stress. Such evidence comes from studies using 3 different techniques to estimate such activity: 1) measurement of pancreatic NE turnover, 2) measurement of neuronal firing rate and 3) measurement of pancreatic neurotransmitter release. Thus, using the first technique it has been demonstrated that the stress of cold exposure increased pancreatic NE turnover in rats (108). Even non-stress states may change pancreatic sympathetic tone for diet-induced (63) or genetically obese (64) rats have reduced pancreatic NE turnover relative to appropriate lean controls. Preliminary data (36) suggest that swimming stress in mice increases the activity of pancreatic galaninergic nerves since pancreatic GLIR content is reduced. Likewise, preliminary data show that genetically obese (ob/ob) mice have lower pancreatic GLIR content consistent with the reduction of chronic galaninergic tone (29). These data support a physiologic role for the pancreatic sympathetic nerves to regulate IRI secretion. However, findings that a 2 day fast in rats reduced, and sucrose overfeeding increased pancreatic NE turnover (108) suggest that the role of pancreatic sympathetic nerves may not be limited to impairing the IRI response during stress. However, NE turnover is an indirect measurement of pancreatic neural activity which should be confirmed by more direct approaches.

The electrophysiological recording of nerve activity, done primarily by Niijima et al., is such a direct approach. However, most of their studies measure electrical activity of either the pancreatic branch of the vagus or the adrenal nerve in rabbits. These authors have clearly demonstrated that hyperglycemia increases pancreatic vagal activity and decreases adrenal nerve activity, whereas hypoglycemia or neuroglucopenia decreases pancreatic vagal and increases adrenal nerve activity (76,77). Their studies have generally assumed that activity of the pancreatic sympathetic nerves would parallel that of the adrenal nerves. This assumption has not been directly tested but different studies (see below) suggest otherwise. This group has also recorded from the pancreatic branch of the splanchnic nerve in rats, and report that lesions of the VMH, DMH and PVN reduce pancreatic sympathetic nerve activity whereas LHA lesion often increased this activity (107). Unfortunately, the recording of pancreatic sympathetic activity during stress apparently has yet to be done.

The third approach is to directly measure pancreatic venous and arterial neurotransmitter concentrations and pancreatic neurotransmitter extraction using the formula and concepts discussed above and illustrated in Figure 5 to calculate pancreatic neurotransmitter spillover during stress. Using such an approach, in halothane-anesthetized, laparotomized dogs, pancreatic NE spillover increased markedly during neuroglucopenic stress (45, Figure 6). In contrast, pancreatic NE spillover did not increase during the stress of hemorrhagic hypotension or hypoxia (Figure 6), despite increases of arterial EPI equivalent to those produced by neuroglucopenic stress. Thus, the activation of the sympathetic nerves of the pancreas appear to be stress specific. Further, the increases of arterial EPI suggest that the adrenal nerve was activated by all three stresses, only one of which apparently activates the pancreatic sympathetic nerves, suggesting that adrenal nerve activity is not likely to be an index of pancreatic sympathetic nerve activity. Finally, since arterial NE increased in all three states, yet pancreatic NE spillover increased only during neuroglucopenia, stresses can apparently produce a regionally-selective activation of sympathetic nerves rather than generalized activation, the older, more traditional view.

Although measurement of neurotransmitter spillover is an attractive and direct way to assess pancreatic sympathetic activity, at present the technique has two major drawbacks. First, the levels of peptidergic neurotransitters, usually 1-2 % of NE levels, are frequently too low to reliably determine if pancreatic galaninergic nerves are activated during stress. Second, the need to sample pancreatic venous blood and measure pancreatic blood flow currently require acute major surgery with the attendent anesthesia. Certain anesthetics, e.g. pentobarbitol, markedly suppress the activity of the sympathetic nervous system normally produced by stress (99). Low dose halothane is used, on the other hand, because it allows a near normal catecholamine response to neuroglucopenia suggesting a less suppressive effect (44). However, its effects on other types and grades of stress are unknown. Finally, surgical stress itself or the attendent nociceptive activation may significantly modify the pancreatic sympathetic responses. Development of a chronically-instrumented model to allow similar direct measurement of pancreatic neurotransmitter release in conscious animals would circumvent these potential drawbacks and would allow not only measurement of pancreatic sympathetic responses to stress uncontaminated by surgical stress or anesthesia but also investigations of the effect of changes in basal sympathetic tone upon islet function in a variety of non-stress states.

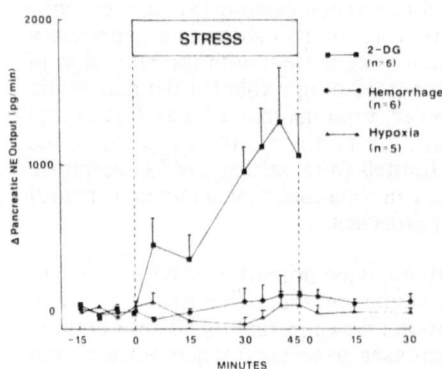

Figure 6. The change of pancreatic NE output in response to 2-DG- induced neuroglucopenia, hemorrhagic hypotension, or hypoxia in halothane-anesthetized, laporatomized dogs [reproduced with permission (45)].

SUMMARY

It is clear that the sympathoadrenal system has a role in the regulation of endocrine pancreatic function and that the sympathetic nerves of the pancreas can change pancreatic hormone secretion to increase the availability of metabolic fuels. It seems likely that the classical sympathetic neurotransmitter, NE, acts in concert with peptide co-transmitters, such as galanin and NPY. Each is released during the stimulation of pancreatic sympathetic nerves and each is capable of influencing either islet function or pancreatic blood flow. There is considerable indirect evidence that the sympathetic innervation of the pancreas is activated during acute stress and influences the endocrine pancreas. However, proving such a physiologic role is difficult because of redundant mechanisms that influence the secretion of the metabolically-crucial hormones, insulin and glucagon. Such definitive proof therefore awaits the development of new techniques to dissect and dissociate these mechanisms.

ACKNOWLEDGEMENTS

The authors wish to acknowledge our collaborators Dr. Bo Ahren, Dr. Richard C. Veith, Peter Havel, Tom Mundinger, Dr. Gerhard Bottcher, Dr. Frank Sundler, Dr. Marvin Brown, Dr. Stephan Kowalyk and Dr. Daniel Porte, Jr. for their contribution to the concepts presented in this review. This work was supported by the Department of Veterans Affairs, National Institutes of Health grants DK-07526, DK-12829 and DK-17047, and the American Diabetes Association.

REFERENCES

1. Ahren, B., P. Arkhammar, P.-O. Berggren and T. Nilsson. Galanin inhibits glucose-stimulated insulin release by a mechanism involving hyperpolarization and lowering of cytoplasmic free Ca^{2+} concentration. *Biochem. Biophys. Res. Comm.* 140(3): 1059-1063, 1986.

2. Ahren, B., P-O. Berggren, K. Bokvist and P. Rorsman. Does galanin inhibit insulin secretion by opening of the ATP-regulated K+ channel in the ß-cell? *Peptides* 10: 453-457, 1989.

3. Ahren, B., G. Bottcher, S. Kowalyk, B. E. Dunning, F. Sundler and G. J. Taborsky, Jr. Galanin is co-localized with noradrenaline and neuropeptide Y (NPY) in dog pancreas and celiac ganglion. *Cell Tissue Res.* 261:49-58, 1990.

4. Ahren, B., H. Martensson and R. Ekman. Pancreatic nerve stimulation releases neuropeptide Y- but not galanin- or calcitonin gene-related peptide-like immunoreactivity from the pig pancreas. *J. Auton. Nerv. Syst.* 27: 11-16, 1989.

5. Ahren, B. and G. J. Taborsky, Jr. Effects of pancreatic noradrenaline infusion on basal and stimulated islet hormone secretion in the dog. *Acta Physiol. Scand.* 132: 143-150, 1988.

6. Ahren, B., R. C. Veith and G.J. Taborsky, Jr. Sympathetic nerve stimulation versus pancreatic norepinephrine infusion in the dog. 1. Effects on basal release of insulin and glucagon. *Endocrinology* 121: 323-331, 1987.

7. Ahren, B., R. C. Veith, T. L. Paquette and G.J. Taborsky, Jr. Sympathetic nerve stimulation versus pancreatic norepinephrine infusion in the dog: 2) Effects on basal release of somatostatin and pancreatic polypeptide. *Endocrinology* 121: 332-339, 1987.

8. Alwmark, A. and B. Ahren. Phentolamine reverses NPY-induced inhibition of insulin secretion in isolated rat islets. *Eur. J. Pharm.* 135: 307-311, 1987.

9. Amiranoff, B., A.-M. Lorinet, I. Lagny-Pourmir and M. Laburthe. Mechanism of galanin-inhibited insulin release. *Eur. J. Biochem.* 177: 147-152, 1988.

10. Amiranoff, B., A.-M. Lorinet, N. Yanaihara and M. Laburthe. Structural requirement for galanin action in the pancreatic ß cell line Rin m 5F. *Eur. J. Pharm.* 163: 205-207, 1989.

11. Amiranoff, B., A. L. Servin, C. Rouyer-Fessard, A. Couvineau, K. Tatemoto and M. Laburthe. Galanin receptors in a hamster pancreatic ß-cell tumor: Identification and molecular characterization. *Endocrinology* 121(1): 284-289, 1987.

12. Andersson, P.-O., L.-O. Farnebo, B. B. Fredholm, B. Hamberger, J. Holst and J. Jarhult. Metabolic and hormonal adjustments during hemorrhage in cats after interference with the sympatho-adrenal system. *Acta Physiol. Scand.* 114: 111-119, 1982.

13. Andersson, P.-O., J. Holst and J. Jarhult. Effects of adrenergic blockade on the release of insulin, glucagon and somatostatin from the pancreas in response to splanchnic nerve stimulation in cats. *Acta Physiol. Scand.* 116: 403-409, 1982.

14. Asplin, C. M., P. L. Werner, J. B. Halter, P. M. Hollander and J. P. Palmer. Autonomic nervous system control of glucagon secretion during neuroglucopenia. *Endocrinology* 112(5): 1585-1589, 1983.

15. Baum, D., D. Porte, Jr. and J. Ensinck. Hyperglucagonemia and α-adrenergic receptor in acute hypoxia. *Am. J. Physiol.* 237(5): E404-E408, 1979.

16. Bewick, M., A. R. Mundy, B. Eaton and F. Watson. Endocrine function of the heterotopic pancreatic allotransplant in dogs III. The cause of hyperinsulinemia. *Transplantation* 31(1): 23-25, 1981.

17. Bloom, S. R. and A. V. Edwards. The release of pancreatic glucagon and inhibition of insulin in response to stimulation of the sympathetic innervation. *J. Physiol.* 253: 157-173, 1975.

18. Bloom, S. R. and A. V. Edwards. Characteristics of the neuroendocrine responses to stimulation of the splanchnic nerves in bursts in the conscious calf. *J. Physiol.* 346: 533-545, 1984.

19. Bloom, S. R., A. V. Edwards and R. N. Hardy. The role of the autonomic nervous system in the control of glucagon, insulin and pancreatic polypeptide release from the pancreas. *J. Physiol.* 280: 9-23, 1978.

20. Brodows, R. G., F. X. Pi-Sunyer and R. G. Campbell. Insulin secretion in adrenergic insufficiency in man. *J. Clin. Endocrinol. Metab.* 38:1103-1108, 1974.

21. Carlei, F., J. M. Allen, A. E. Bishop, S. R. Bloom and J. M. Polak. Occurrence, distribution and nature of neuropeptide Y in the rat pancreas. *Experientia* 41: 1554-1557, 1985.

22. Cherksey, B., N. Altszuler and J. Zadunaisky. Preponderance of ß-adrenergic binding sites in pancreatic islet cells of the rat. *Diabetes* 30: 172-174, 1981.

23. Cryer, P. E., R. A. Rizza, M. W. Haymond and J. E. Gerich. Epinephrine and norepinephrine are cleared through beta-adrenergic, but not alpha-adrenergic mechanisms in man. *Metabolism* 29(Suppl. 1): 1114-1118, 1980.

24. Curry, D. L. Direct tonic inhibition of insulin secretion by central nervous system. *Am. J. Physiol.* 244(Endocrinol. Metab. 7): E425-E429, 1983.

25. Davis, T. M. E., J. M. Burrin and S. R. Bloom. Growth hormone (GH) release in response to GH-releasing hormone in man is 3-fold enhanced by galanin. *J. Clin. Endocrinol. Metab.* 65: 1248-1252, 1987.

26. DeWeille, J., H. Schmid-Antomarchi, M. Fosset and M. Lazdunski. ATP-sensitive K+ channels that are blocked by hypoglycemia-inducing sulfonylureas in insulin-secreting cells are activated by galanin, a hyperglycemia-inducing hormone. *Proc. Natl. Acad. Sci.* 85: 1312-1316, 1988.

27. Dunne, M. J., M. J. Bullett, G. Li, C. B. Wollheim and O. H. Petersen. Galanin activates nucleotide-dependent K+ channels in insulin-secreting cells via a pertussis toxin-sensitive G-protein. *EMBO J.* 8(2): 413-420, 1989.

28. Dunning, B., A. Ar'Rajab, G. Bottcher, F. Sundler and B. Ahren. Galanin (GAL) is present in human pancreatic nerves and inhibits insulin (IRI) release from isolated human islets. *Diabetes* 39(Suppl 1): 136A, 1990.

29. Dunning, B. E. and B. Ahren.   Reduced pancreatic content of the inhibitory neuropeptide, galanin, in genetically obese, hyperinsulinemic mice. *Diabetologia* 33(Suppl):A27, 1990.

30. Dunning, B. E., B. Ahren, G. Bottcher, F. Sundler and G.J. Taborsky, Jr.   The presence and actions of NPY in the canine endocrine pancreas. *Reg. Peptides* 18: 253-265, 1987.

31. Dunning, B. E., B. Ahren, R. C. Veith, G. Bottcher, F. Sundler and G.J. Taborsky, Jr. Galanin: a novel pancreatic neuropeptide. *Am. J. Physiol.* 251(Endocrinol. Metab. 18): E127-E133, 1986.

32. Dunning, B. E., B. Ahren, R. C. Veith and G. J. Taborsky, Jr.   Nonadrenergic sympathetic neural influences on basal pancreatic hormone secretion. *Am. J. Physiol.* 255(Endocrinol. Metab. 18): E785-E792, 1988.

33. Dunning, B. E., W. Y. Fujimoto, K. Tatemoto and G. J. Taborsky, Jr.   Comparison of the effects of galanin and pancreastatin on basal insulin release. *69th Ann. Mtg. Endocrine Soc.* Abstract # 292: 94, 1987.

34. Dunning, B. E. and G.J. Taborsky, Jr.   Galanin release during pancreatic nerve stimulation is sufficient to influence islet function. *Am. J. Physiol.* 256(Endocrinol. Metab. 19): E191-E198, 1989.

35. Dunning, B. E., P. J. Havel, R. C. Veith and G.J. Taborsky, Jr.   Pancreatic and extrapancreatic galanin release during sympathetic neural activation. *Am. J. Physiol.* 258 (Endocrinol. Metab. 21): E436-E444, 1990.

36. Dunning, B. E., S. Karlsson and B. Ahren.   Galanin contributes to stress-induced inhibition of insulin secretion in swimming mice. *Diabetes* 38(Suppl. 2): 48A, 1989.

37. Dunning, B. E. and G. J. Taborsky, Jr.   Calcitonin gene-related peptide: a potent and selective stimulator of gastrointestinal somatostatin secretion. *Endocrinology* 120(5): 1774-1781, 1987.

38. Fuhlendorff, J., U. Gether, L. Aakerlund, N. Langeland-Johansen, H. Thogersen, S. G. Melberg, U. B. Olsen, O. Thastrup and T. W. Schwartz.   [Leu31, Pro34]neuropeptide Y: a specific Y1 receptor agonist. *Proc. Natl. Acad. Sci. USA* 87(1): 182-186, 1990.

39. Fyles, J. M., M. A. Cawthorne and S. L. Howell.   The characteristics of beta-adrenergic binding site on pancreatic islets of Langerhans. *J. Endocrinol.* 111(2): 263-70, 1986.

40. Fyles, J. M., M. A. Cawthorne and S. L. Howell.   The determination of alpha-adrenergic receptor concentration on rat pancreatic islet cells. *Biosci. Rep.* 7(1): 17-22, 1987.

41. Gerich, J. E., M. Langlois, C. Noacco, V. Schneider and P. H. Forsham.   Adrenergic modulation of pancreatic glucagon secretion in man. *J. Clin. Invest.* 53: 1441-1446, 1974.

42. Gilbey, S. G., J. Stephenson, D. J. O'Halloran, J. M. Burrin and S. R. Bloom. High-dose porcine galanin infusion and effect on intravenous glucose tolerance in humans. *Diabetes* 38(9): 1114-1116, 1989.

43. Goldberg, M. R. and D. Robertson.   Yohimbine: A pharmacologic probe for study of the alpha2-adrenoceptor. *Pharm. Rev.* 35(2): 143-180, 1983.

44. Havel, P. J., D. E. Flatness, J. B. Halter, J. D. Best, R. C. Veith and G.J. Taborsky, Jr.   Halothane anesthesia does not suppress sympathetic activation produced by neuroglucopenia. *Am. J. Physiol.* 252 (Endocrinol. Metab. 15): E667-E672, 1987.

45. Havel, P. J., R. C. Veith, B. E. Dunning and G.J. Taborsky, Jr.   Pancreatic noradrenergic nerves are activated by neuroglucopenia but not by hypotension or hypoxia in the dog. *J. Clin. Invest.* 82: 1538-1545, 1988.

46. Hermansen, K.   Effects of galanin on the release of insulin, glucagon and somatostatin from the isolated, perfused dog pancreas. *Acta Endo.* 119: 91-98, 1988.

47. Hermansen, K., N. Yanaihara and B. Ahren. On the nature of the galanin action on the endocrine pancreas: studies with six galanin fragments in the perfused dog pancreas. *Acta Endo.* 121: 545-550, 1989.

48. Holst, J. J., R. Gronholt, O. B. Schaffalitzky de Muckadell and J. Fahrenkrug. Nervous control of pancreatic endocrine secretion in pigs. *Acta Physiol. Scand.* 113: 279-283, 1981.

49. Holst, J. J., C. Orskov, S. Knuhtsen, S. Sheikh and O. V. Nielsen. On the regulatory functions of neuropeptide Y (NPY) with respect to vascular resistance and exocrine and endocrine secretion in the pig pancreas. *Acta Physiol Scand* 136: 519-526, 1989.

50. Holst, J. J., T. W. Schwartz, S. Knuhtsen, S. L. Jensen and O. V. Nielsen. Autonomic nervous control of the endocrine secretion from the isolated, perfused pig pancreas. *J. Auton. Nerv. Sys.* 17: 71-84, 1986.

51. Itoh, M. and J. E. Gerich. Adrenergic modulation of pancreatic somatostatin, insulin, and glucagon secretion: evidence for differential sensitivity of islet A, B, and D cells. *Metabolism* 31(7): 715-720, 1982.

52. Iverson, J. Adrenergic receptors and the secretion of glucagon and insulin from the isolated, perfused canine pancreas. *J. Clin. Invest.* 52: 2102-2116, 1973.

53. Jarhult, J. and J. J. Holst. Reflex adrenergic control of endocrine pancreas evoked by unloading of carotid baroreceptors in cats. *Acta Physiol. Scand.* 104: 188-202, 1978.

54. Kaneto, A., H. Kajinuma and K. Kosaka. Effect of splanchnic nerve stimulation on glucagon and insulin output in the dog. *Endocrinology* 96: 143-150, 1975.

55. Kaplan, L. M., E. R. Spindel, K. J. Isselbacher and W. W. Chin. Tissue-specific expression of the rat galanin gene. *Proc. Natl. Acad. Sci. USA* 85: 1065-1069, 1988.

56. Karlsson, S. and B. Ahren. Insulin and glucagon secretion in swimming mice. Effect of autonomic receptor antagonism. *Metabolism* : 1990, in press.

57. Knudtzon, J. Adrenergic effects on plasma levels of glucagon, insulin, glucose and free fatty acids in rabbits - influences of selective blocking drugs. *Acta. Physiol. Scand.* 120: 353-361, 1984.

58. Kurose, T., Y. Seino, S. Nishi, K. Tsuji, T. Taminato, K. Tsuda and H. Imura. Mechanism of sympathetic neural regulation of insulin, somatostatin and glucagon secretion. *Am. J. Physiol.* 258(Endocrinol. Metab. 21): E220-E227, 1990.

59. Kwok, Y. N., C. B. Verchere, C. H. S. McIntosh and J. C. Brown. Effect of galanin on endocrine secretions from the isolated perfused rat stomach and pancreas. *Eur. J. Pharm.* 145: 49-54, 1988.

60. Lagny-Pourmir, I., A. M. Lorinet, N. Yanaihara and M. Laburthe. Structural requirements for galanin interaction with receptors from pancreatic beta cells and from brain tissue of the rat. *Peptides* 10: 757-761, 1989.

61. Larsson, L.-I. and J. F. Rehfeld. Peptidergic and adrenergic innervation of pancreatic ganglia. *Scand. J. Gastroent.* 14: 433-437, 1979.

62. Leonhardt, U., E. G. Siegel, H. Kohler, M. Barthel, A. Tytko, K. Nebendahl and W. Creutzfeldt. Galanin inhibits glucose-induced insulin release *in vitro*. *Horm. Metab. Res.* 21: 100-101, 1989.

63. Levin, B. E., J. Triscari, S. Hogan and A. C. Sullivan. Resistance to diet-induced obesity: food intake, pancreatic sympathetic tone, and insulin. *Am. J. Physiol.* 252(Regulatory Integrative Comp. Physiol. 21): R471-R478, 1987.

64. Levin, B. E., J. Triscari and A. C. Sullivan. Studies of origins of abnormal sympathetic function in obese Zucker rats. *Am. J. Physiol.* 245(Regulatory Integrative Comp. Physiol.): R364-R371, 1983.

65. Lindskog, S. and B. Ahren. Galanin: effects on basal and stimulated insulin and glucagon secretion in the mouse. *Acta Physiol. Scand.* 129: 305-309, 1987.

66. Lindskog, S. and B. Ahren. Effects of galanin on insulin and glucagon secretion in the rat. *Int. J. Pancreatology* 4: 335-344, 1989.

67. Lindskog, S., B. Ahren, B. E. Dunning and F. Sundler. Distribution of galanin in the mouse and rat pancreas. *Diabetologia* 33(Suppl):A110, 1990.

68. Lindskog, S., B. E. Dunning, H. Martensson, A. Ar'Rajab, G. J. Taborsky, Jr. and B. Ahren. Galanin of the homologous species inhibits insulin secretion in the rat and in the pig. *Acta Physiol. Scand.* 139:591-596, 1990.

69. Luiten, P. G. M., G. J. terHorst, R. M. Buijs and A. B. Steffens. Autonomic innervation of the pancreas in diabetic and non-diabetic rats. A new view on intramural sympathetic structural organization. *J. Auton. Nerv. Syst.* 15: 33-44, 1986.

70. Luiten, P. G. M., G. J. terHorst, S. J. Coopmans, M. Rietberg and A. B. Steffens. Preganglionic innervation of the pancreas islet cells in the rat. *J. Auton. Nerv. Syst.* 10: 27-42, 1984.

71. Luyckx, A. S., A. Dresse, A. Cession-Fossion and P. J. Lefebvre. Catecholamines and exercise-induced glucagon and fatty acid mobilization in the rat. *Am. J. Physiol.* 229(2): 376-383, 1975.

72. McDonald, T. J., J. Dupre, K. Tatemoto, G. R. Greenberg, J. Radziuk and V. Mutt. Galanin inhibits insulin secretion and induces hyperglycemia in dogs. *Diabetes* 34: 192-196, 1985.

73. Messel, T., H. Harling, G. Bottcher, A. H. Jonsen and J. J. Holst. Galanin in the porcine pancreas. *Regul. Pept.* 28(2): 161-176, 1990.

74. Michel, M. C. and H. J. Motulsky. He 90481: A competitive non-peptidergic antagonist at NPY receptors. *Program of Mtg: "Central and peripheral significance of neuropeptide Y and its related peptides* : PI-22, 1990.

75. Moltz, J. H. and J. K. McDonald. Neuropeptide Y: Direct and indirect action on insulin secretion in the rat. *Peptides* 6: 1155-1159, 1985.

76. Niijima, A. An electrophysiological study on the regulatory mechanism of blood sugar level in the rabbit. *Brain Res.* 87: 195-199, 1975.

77. Niijima, A. Neural mechanisms in the control of blood glucose concentration. *J. Nutr.* 119: 833-840, 1989.

78. Nilsson, T., P. Arkhammar, P. Rorsman and P.-O. Berggren. Suppression of insulin release by galanin and somatostatin is mediated by a G-protein. *J. Biol. Chem.* 264(2): 973-980, 1989.

79. Palmer, J. P., D. P. Henry, J. W. Benson, D. G. Johnson and J. W. Ensinck. Glucagon response to hypoglycemia in sympathectomized man. *J. Clin. Invest.* 57: 522-525, 1976.

80. Patel, D. G. Role of sympathetic nervous system in glucagon response to insulin hypoglycemia in normal and diabetic rats. *Diabetes* 33:1154-1159, 1984.

81. Peiro, E., R. A. Silvestre, P. Miralles, P. Degano and J. Marco. Diabetogenic effect of homologous galanin in the rat pancreas. *Diabetologia* 32(Suppl): 528A, 1989.

82. Pettersson, M. and B. Ahren. Insulin secretion in rats: effects of neuropeptide Y and noradrenaline. *Diabetes Res.* , 1990, in press.

83. Pettersson, M., B. Ahren, I. Lundquist, G. Bottcher and F. Sundler. Neuropeptide Y: Intrapancreatic neuronal localization and effects on insulin secretion in the mouse. *Cell. Tiss. Res.* 248: 43-48, 1987.

84. Porte, D., Jr., L. Girardier, J. Seydoux, Y. Kanazawa and J. Posternak. Neural regulation of insulin secretion in the dog. *J. Clin. Invest.* 52: 210-214, 1973.

85. Ribes, G., J. P. Blayac and M. M. Loubatieres-Mariani. Differences between the effects of adrenaline and noradrenaline on insulin secretion in the dog. *Diabetologia* 24: 107-112, 1983.

86. Rokaeus, A. and M. Carlquist. Nucleotide sequence analysis of cDNAs encoding a bovine galanin precursor protein in the adrenal medulla and chemical isolation of bovine gut galanin. *FEBS Lett.* 234: 400-406, 1988.

87. Roy, M. W., M. S. Jones and R. E. Miller. Pancreatic somatostatin secretion is suppressed by splanchnic nerve stimulation. *Diabetologia* 20: 102-105, 1981.

88. Roy, M. W., K. C. Lee, M. S. Jones and R. E. Miller. Neural control of pancreatic insulin and somatostatin secretion. *Endocrinology* 115(2): 770-775, 1984.

89. Samols, E. and G. C. Weir. Adrenergic modulation of pancreatic A, B, and D cells. *J. Clin. Invest.* 63: 230-238, 1979.

90. Scheurink, A. J. W., A. B. Steffens, H. Bouritius, G. H. Dreteler, R. Bruntink, R. Remie and J. Zaagsma. Adrenal and sympathetic catecholamines in exercising rats. *Am. J. Physiol.* 256: R155-R160, 1989.

91. Sharp, G. W. G., Y. L. Marchand-Brustel, T. Yada, L. L. Russo, C. R. Bliss, M. Cormont, L. Monge and E. V. Obberghen. Galanin can inhibit insulin release by a mechanism other than membrane hyperpolarization or inhibition of adenylate cyclase. *J. Biol. Chem.* 264(13): 7302-7309, 1989.

92. Sheikh, S. P., R. Hakanson and T. W. Schwartz. Y1 and Y2 receptors for neuropeptide Y. *FEBS Lett.* 245(1-2): 209-214, 1989.

93. Shiekh, S. P., J. J. Holst, T. Skak-Nielsen, U. Knigge, J. Warberg, E. Theodorsson-Norheim, T. Hokfelt, J. M. Lundberg and T. W. Schwartz. Release of NPY in pig pancreas: dual parasympathetic and sympathetic regulation. *Am. J. Physiol.* 255: G46-G54, 1988.

94. Silvestre, R. A., P. Miralles, L. Monge, P. Moreno, M. L. Villanueva and J. Marco. Effects of galanin on hormone secretion from the *in situ* perfused rat pancreas on glucose production in rat hepatocytes *in vitro. Endocrinology* 121(1): 378-383, 1987.

95. Smith, P. H. and B. J. Davis. Morphological and functional aspects of pancreatic islet innervation. *J. Auton. Nerv. Syst.* 9: 53-66, 1983.

96. Su, H. C., A. E. Bishop, R. F. Power, Y. Hamada and J. M. Polak. Dual intrinsic and extrinsic origins of CGRP- and NPY-immunoreactive nerves of rat gut and pancreas. *J. Neurosci.* 7(9): 2674-2687, 1987.

97. Taborsky, G. J., Jr. Evidence of a paracrine role for pancreatic somatostatin in vivo. *Am. J. Physiol.* 245 (Endocrinol. Metab. 8): E598-E603, 1983.

98. Taborsky, G. J., Jr. and J. W. Ensinck. Contribution of the pancreas to circulating somatostatin-like immunoreactivity in the normal dog. *J. Clin. Invest.* 73: 216-223, 1984.

99. Taborsky, G. J., Jr., J. B. Halter, D. Baum, J. D. Best and D. Porte, Jr. Pentobarbital anesthesia suppresses basal and 2-deoxy-D-glucose-stimulated plasma catecholamines. *Am. J. Physiol.* 247 (Regulatory Integrative Comp. Physiol. 16): R905-R910, 1984.

100. Tatemoto, K. Neuropeptide Y: complete amino acid sequence of the brain peptide. *Proc. Natl. Acad. Sci. USA* 79: 4585-5489, 1982.

101. Tatemoto, K. Neuropeptide Y and its related peptides. *Program of mtg: Central and peripheral significance of neuropeptide Y and its related peptides* 1: 1990.

102. Tatemoto, K., A. Rokaeus, H. Jornvall, T. J. McDonald and V. Mutt. Galanin-a novel biologically active peptide from porcine intestine. *FEBS Lett.* 164: 124-128, 1983.

103. Tiscornia, O. M. The neural control of exocrine and endocrine pancreas. *Am. J. Gastroent.* 67: 541-560, 1977.

104. Ullrich, S. and C. B. Wollheim. Galanin inhibits insulin secretion by direct interference with exocytosis. *FEBS Lett.* 247(2): 401-404, 1989.

105. Wahlestedt, C., N. Yanaihara and R. Hakanson. Evidence for different pre- and post-junctional receptors for neuropeptide Y and related peptides. *Regul. Pept.* 13(3-4): 307-318, 1986.

106. Westfall, T. C. Evidence that noradrenergic transmitter release is regulated by presynaptic receptors. *Fed. Proc.* 43: 1352-1357, 1984.

107. Yoshimatsu, H., A. Niijima, Y. Oomura, K. Yamabe and T. Katafuchi. Effects of hypothalamic lesion on pancreatic autonomic nerve activity in the rat. *Brain Res.* 303: 147-152, 1984.

108. Young, J. B. and L. Landsberg. Effect of diet and cold exposure on norepinephrine turnover in pancreas and liver. *Am. J. Physiol.* 236(5): E524-E533, 1979.

NEUROPEPTIDES IN THE REGULATION OF ISLET HORMONE SECRETION - LOCALIZATION, EFFECTS AND MODE OF ACTION

Bo Ahrén[1,2], Per-Olof Berggren[3], Patrik Rorsman[4], Claes-Göran Östenson[3], and Suad Efendic[3]

Departments of [1]Pharmacology and [2]Surgery, Lund University, Lund, [3]Endo- crinology, Karolinska Institute, Stockholm, and [4]Medical Physics, Gothenburg University, Gothenburg, Sweden

Islet Innervation

It has been known since long that the pancreatic islets are richly innervated by autonomic nerves (1-3). These nerves have been subdivided in sympathetic-adrenergic and parasympathetic-cholinergic nerves. The sympathetic nerve fibers pass in the splanchnic nerves, synapse in the celiac ganglion and enter the pancreas along the arteries as post-ganglionic nerves (3). They innervate intrapancreatic vessels and some fibers enter the islets and terminate close to the endocrine cells (3-5). Sympathetic fibers also enter the pancreas as pre-ganglionic fibers and terminate in intrapancreatic sympathetic ganglia (6). The parasympathetic nerve fibers pass in the vagus and enter the pancreas along its arteries and terminate in intrapancreatic ganglia (3,7), with the postganglionic fibers innervating the islets. During recent years, evidence has also presented that the pancreatic islets are innervated by also peptidergic nerves (8). This review will present our present knowledge on the peptidergic transmitters of these nerves, their localization, effects and mode of action, with regard to islet hormone secretion.

Nerve activation and islet hormone secretion

To study the influence of the autonomic nerves on islet hormone secretion, electrical activation has been undertaken. Sympathetic nerve stimulation is usually performed by activation of either the splanchnic nerves or of the mixed pancreatic autonomic nerves after blockade of cholinergic receptors with atropine or hexamethonium. Such activation results in inhibition of insulin and somatostatin secretion and activation of glucagon secretion (2,3,8-12). Parasympathetic activation, performed by vagal activation, results in stimulation of insulin and glucagon secretion (2,3,8,9,13-15), whereas the effect on somatostatin secretion is variable, and both stimulation (16) and inhibition (17) have been demonstrated.

It was previously assumed that the classical neurotransmitters noradrenaline and acetylcholine mediated the above mentioned effects on islet hormone secretion. However, recent studies have challenged this view. First, it has been studied whether the effect of sympathetic nerve activation on islet hormone secretion is blocked by complete α- and β-adrenoceptor blockade. In the dog, adrenergic blockade did not affect either inhibition of insulin and somatostatin secretion or stimulation of glucagon secretion evoked by sympathetic activation (18). Likewise, in the perfused rat pancreas, adrenergic blockade did not fully antagonize the inhibition of splanchnic nerve activation on insulin and somatostatin secretion (19). Furthermore, in the calf, sympathetically induced activation of glucagon secretion was not affected by complete adrenergic blockade (10). Second, noradrenaline, infused directly into the pancreatic artery in the dog could not reproduce the sympathetically induced inhibition of basal insulin and somatostatin secretion, whereas stimulation of glucagon secretion was only partially reproduced (11,12). These studies therefore suggest that non-adrenergic mechanisms participate in the mediation of the effects on islet hormone secretion evoked by sympathetic activation.

Studies on the possible mediation by acetylcholine of vagally induced changes in islet hormone secretion have used atropine to antagonize the islet action of this neurotransmitter. Thus, vagally induced stimulation of insulin secretion was largely inhibited by atropine in the dog (13) but not in the pig (15). Furthermore, atropine did not affect vagally induced glucagon secretion in the dog or pig (13,15), and somatostatin secretion in the dog was only partially antagonized by atropine (16). Thus, non-cholinergic effects exist in the vagal regulation of islet function.

Thus, non-adrenergic, non-cholinergic mechanisms exist in the neural regulation of islet function. Likely candidates for mediating these effects are the neuropeptides. Further evidence for neuropeptides as mediators of the influences are that the neuropeptides are localized to intrapancreatic nerve fibers, that they are released from the pancreas by autonomic nerve activation and that they affect islet hormone secretion (8).

Neuropeptides and islet function

Neuropeptide Y (NPY)

In several different species it has been observed that some nerve fibers within the pancreas react with antibodies directed against the 36 amino acid peptide NPY (5,20-23). These NPY fibers seem to innervate both vessels and endocrine cells. Further characterization of the NPY fibers in the dog has demonstrated that most of them react also with antibodies directed against tyrosine hydroxylase, the marker for adrenergic nerves and that the dog celiac ganglion contains numerous cell bodies positive for NPY-immunoreactivity (24). This suggests that NPY and noradrenaline are co-localized to intrapancreatic, postganglionic, sympathetic nerves. In addition, NPY is released from the pancreas upon sympathetic nerve activation (23,25) and inhibits insulin secretion in mice and in the perfused rat pancreas (5,26-28). Furthermore, in the dog, NPY stimulates glucagon secretion and inhibits insulin secretion (22). Hence, NPY is a candidate to mediate the inhibition of insulin secretion evoked by sympathetic nerve activation. However, in the mouse, most NPY fibers are non-adrenergic (5), and in the perfused pig pancreas, NPY has been shown to be without inhibitory influence on islet hormone secretion (29). In the perfused pig pancreas, the neuropeptide causes pancreatic vasoconstriction, suggesting a possible role in the regulation of pancreatic blood flow (27). Thus, the role of NPY for islet hormone secretion is not yet established.

Galanin

The 29 amino acid peptide, galanin, has been demonstrated in intra-pancreatic nerves in several species (30,31). In the dog, these fibers are associated with the islets (30). Furthermore, a recent study has shown that galanin fibers in the dog pancreas also store NPY- and tyrosine hydroxyla-se-like immunoreactivity and numerous cell bodies in the celiac ganglion contain galanin-like immunoreactivity (24). This shows that galanin, like NPY, is localized to intrapancreatic sympathetic nerve fibers, suggesting that galanin is an intrapancreatic sympathetic neurotransmitter. Galanin is released from the dog pancreas upon sympathetic nerve activation (32) and potently inhibits insulin secretion in several experimental species (33). However, no effect of galanin on insulin secretion has been observed upon its intravenous infusion in man (34,35). Furthermore, galanin inhibits somatostatin secretion, whereas its influence on glucagon is variable, since both stimulatory and inhibitory effects have been reported (33). Nevertheless, the localization to intrapancreatic nerves, the pancreatic release upon sympathetic nerve activation and the influence on islet hor-mone secretion suggest that galanin is a neuropeptide that mediates changes in islet hormone secretion induced by sympathetic activation.

Studies in a hamster pancreatic β-cell tumour and in clonal insulin producing RINm5F cells have identified the occurrence of galanin receptors (36,37). This suggests that the inhibitory action of galanin on insulin secretion is exerted through a direct action on the β-cells. Although it might be argued that the effects of galanin on the insulin secretory pro-cess are accounted for by interference with glucose metabolism, we have not observed any such effect when this parameter was evaluated from experiments of glucose oxidation (Eizirik D and Berggren PO, unpublished). In a pre-vious study, we have demonstrated that galanin induced a rapid decrease in membrane potential and the free cytoplasmic concentration of calcium, $[Ca^{2+}]_i$, in isolated β-cells stimulated with 20 mM glucose, without affec-ting membrane potential or $[Ca^{2+}]_i$ under non-stimulatory conditions (38,39). In contrast, there was no interference of the peptide with the rise in $[Ca^{2+}]_i$ induced by depolarizing the cells with high concentration of K^+ (38,39). However, also under these conditions, insulin release was suppressed, indicating a dissociation between changes in $[Ca^{2+}]_i$ and the secretory machinery. This inhibitory action of galanin on insulin secretion is partially reversed by increasing the concentration of glucose in the incubation medium, indicating that the sugar is able to promote secretion by a mechanism not related to changes in $[Ca^{2+}]_i$ (39). Moreover, studies in permeabilized cells have revealed that galanin directly inhibits exocytosis in a Ca^{2+}-independent manner (40). It is unlikely that galanin directly affects the voltage-activated Ca^{2+} channels, since the peptide was without effect on $[Ca^{2+}]_i$ subsequent to stimulation with high concentrations of K^+ (38,39).

Since changes in K^+ permeability have profound effects on β-cell mem-brane potential, the mechanism underlying galanin-induced repolarization is most likely explained in terms of an increased K^+ conductance. Accordingly, it was demonstrated that neither an increased influx of Cl^- nor a reduced inflow of Ca^{2+} was responsible for agonist-evoked repolarization (40). Furthermore, galanin has recently been shown to slightly increase the efflux of $^{86}Rb^+$ from prelabelled normal islets (41). Moreover, based on experiments in clonal insulin-producing RINm5F cells, the repolarizing action of galanin has been suggested to involve activation of the glucose- and ATP-regulated K^+-channels (40,42-45). However, these effects are not consistently obtained when studying normal pancreatic β-cells (46), a fact that may be accounted for by washing out intracellular coupling factors during conventional whole-cell recordings. By taking the advantage of the perforated-patch whole-cell configuration of the patch-clamp technique, a novel method which permits the recording of whole-cell membrane currents from metabolically intact cells (47), we have recently demonstrated that galanin instead activated a sulphonylurea-insensitive and low-conductance K^+-channel (48,49; Fig. 1).

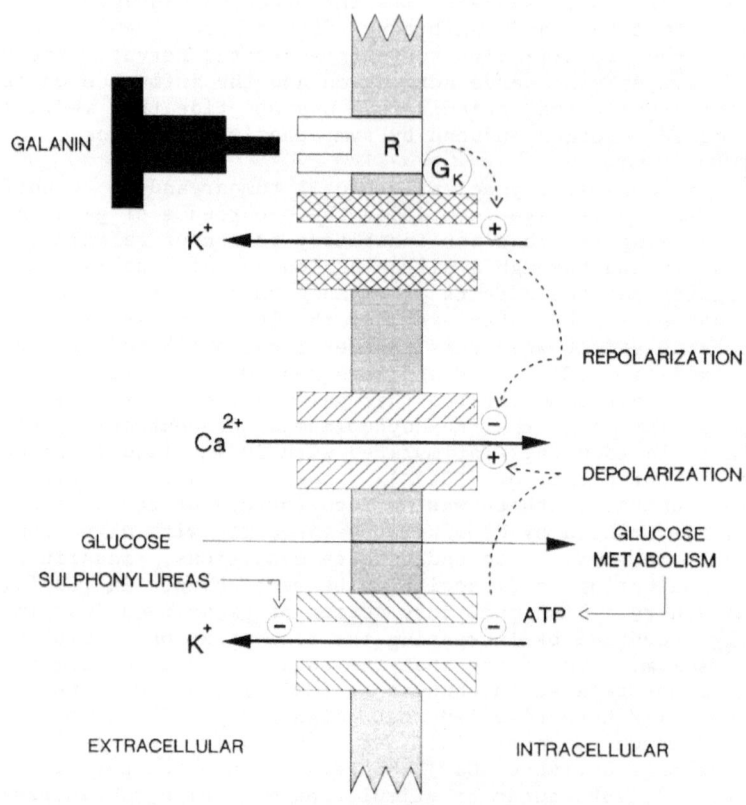

FIG. 1. Regulation of β-cell electrical activity by fuel secretagogues, sulphonylureas and galanin. Metabolism of glucose leads to the production of ATP. An increased ATP/ADP ratio induces closure of specific ATP-regulated K^+-channels in the plasma membrane. This leads to membrane depolarization with subsequent opening of voltage-dependent L-type Ca^{2+}-channels, increase in $[Ca^{2+}]_i$ and stimulation of insulin release. Also the hypoglycemic sulphonylureas stimulate insulin release by directly closing ATP-regulated K^+-channels. Galanin, however, inhibits insulin release by activating a K^+-channel distinct from the ATP-regulated K^+-channel. This K^+-channel is not sensitive to sulphonylureas. The effect of galanin on the actual K^+-channel is mediated by a G-protein (G_K).

The fact that the effects of galanin are reversed if the cells are pretreated with pertussis toxin (39,50) indicates the participation of a G-protein (G_K) in the inhibitory signal-transduction pathway (51). It is not likely that pertussis toxin exerts its effects only by increasing the cellular content of cyclic AMP, since galanin induced a lowering in $[Ca^{2+}]_i$ despite the presence of the adenylate cyclase activator forskolin (39). We have also found that the ability of galanin to activate the low-conductance and sulphonylurea-insensitive K^+-channels was abolished by pretreatment of the cells with pertussis toxin (48,49), suggesting that these K^+-channels are operated by a similar type of G-protein. Since the repolarization-induced decrease in $[Ca^{2+}]_i$ was not solely responsible for the inhibitory effects of galanin on insulin release, there may exist either several galanin-associated G-proteins with sensitivity to the toxin or the putative G-protein has more than one effector system (52). Interestingly, studies performed on both permeabilized neutrophils, RINm5F cells and dialysed mast cells have demonstrated that the non-hydrolysable GTP analogue guanosine 5'-(3-0-thio) triphosphate stimulates secretion even at vanishingly low Ca^{2+} concentrations, implying a Ca^{2+}-independent mechanism for secretion regulated by a G-protein (53). It is tempting to speculate that galanin exerts its effects also by interacting with a similar G-protein-regulated pathway for exocytosis in the normal β-cell.

Vasoactive intestinal peptide (VIP)

Nerve fibers containing VIP-immunoreactivity have been demonstrated in intrapancreatic ganglia and in islets in several species including man (54-57). The islet-associated VIP fibers form a peri-insular net work and some fibers enter the islets and terminate on islet endocrine cells (55,56). Pancreatic denervation in the dog has been shown not to alter the VIP-innervation (56). Therefore, it is likely that the intrapancreatic VIP-nerves are intrinsic, postganglionic nerves, originating from intrapancreatic ganglia. VIP has been shown to be released from the pancreas upon vagal nerve activation in the pig (58). Furthermore, VIP stimulates both insulin and glucagon secretion in man as well as in the dog, the pig, the rat, and the mouse (56,59-63). Moreover, somatostatin secretion is stimulated by VIP in the perfused dog pancreas (63). Hence, the localization of VIP to intra-pancreatic, possibly postganglionic, nerve fibers, its release from the pancreas upon vagal nerve activation, and its stimulation of islet hormone secretion make this 28 amino acid peptide a candidate for mediating non-cholinergic vagal influences on islet function.

Gastrin releasing peptide (GRP)

Nerve fibers displaying GRP-immunoreactivity were initially demon-strated in the pig pancreas (64,65). Recently, pancreatic GRP-nerves have been demonstrated also in other species (31). In the pig, vagal nerve activation stimulates pancreatic release of GRP (66), and exogenous GRP stimulates the secretion of insulin and inhibits somatostatin secretion (67). This suggests that at least in the pig, GRP might be involved in the vagal control of islet function. Also in other species, GRP stimulates secretion of insulin and glucagon (68-70). In contrast, in the perfused dog pancreas, GRP stimulates insulin secretion without affecting glucagon or somatostatin secretion (71). In the mouse, GRP-stimulated insulin secretion is inhibited by atropine (69). The intrapancreatic localization and effects on insulin and glucagon secretion make this 27 amino acid peptide a poten-tial candidate for being an intrapancreatic neuropeptide.

Calcitonin gene-related peptide (CGRP)

CGRP-immunoreactive nerve fibers have been demonstrated within the pancreas in several species (72-76). The fibers run in and around the islets and in close connection to blood vessels. Furthermore, fibers in close apposition to islet endocrine cells have been demonstrated (74,75). In the rat, studies on the nature of these fibers have revealed that capsaicin treatment reduces the CGRP innervation, whereas 6-hydroxydopamine has no effect, which suggest that the fibers are sensory (74-76). Moreover, since no CGRP nerve cell bodies have been demonstrated within the pancreas, it is likely that the CGRP nerves are extrinsic (76). Functionally, the peptide has been shown to inhibit insulin secretion in pigs, rats and mice (73,77,78), whereas CGRP in the perfused dog pancreas exerts inhibition at low doses and stimulation at high doses (79). In man, however, no effect of CGRP on insulin secretion has been observed (35). The effect of CGRP on glucagon secretion seems to be variable and dependent on the experimental system, whereas the peptide does not influence somatostatin secretion (75). Thus, CGRP seems to be localized to intrapancreatic nerves and to inhibit insulin secretion. One study on the possible release of this 37 amino acid peptide from the pancreas has failed to demonstrate any elevation of plasma CGRP levels in the pancreatic vein during electrical nerve activation in the pig (25). Hence, it is still hypothetical whether this neuropeptide participates in the neural regulation of insulin secretion.

Although CGRP binding sites have been identified in various tissues, its receptor has not been well characterized (80). However, in peripheral tissues the effects of the peptide seem to be mediated by cyclic AMP, which makes it conceivable that the CGRP receptor is a GTP-protein-coupled receptor (81). In this context, it is of interest to note that CGRP has been reported to stimulate DNA synthesis and elevate intracellular cyclic AMP levels without inducing phosphoinositide hydrolysis in umbilical vein endothelial cells (82). A similar CGRP-operated cyclic AMP-based effector system is not likely to exist in the pancreatic β-cell, in view of the inhibitory effect of the peptide on insulin release. Indeed, CGRP has been shown to inhibit glucose-induced elevation of cyclic AMP levels in normal islets (83).

Cholecystokinin (CCK)

CCK-nerves have been demonstrated within the pancreas (54,84). These nerve fibers innervate intrapancreatic ganglia and islets, and the fibers have been suggested to be extrinsic (54). Sofar, no studies on the possible release of CCK from the pancreas during nerve activation have been undertaken. However, several studies have dealt with the effects of the different CCK-forms on islet hormone secretion. Generally, as for example in studies in the perfused pig and dog pancreas and in vivo in the rat and pig (84-87), these studies have demonstrated stimulation by CCK of insulin, glucagon, and somatostatin secretion. The C-terminal tetrapeptide, CCK-4, and the C-terminal octapeptide, CCK-8, seem to be equipotent in these actions (85-87). Hence, CCK might be an intrapancreatic neuropeptide regulating islet function.

Specific CCK receptors have been demonstrated on the β-cell (88). Furthermore, it has been demonstrated that CCK-8, concomitantly with its insulin releasing action, activates phospholipase C and thereby promotes a receptor-linked polyphosphoinositide (PI) hydrolysis in normal rat islets (89, Karlsson and Ahrén unpublished). However, CCK-8 only promotes insulin release at a stimulatory concentration of glucose. This has been attributed to the fact that there is no formation of diacylglycerol and thereby no activation of the protein kinase C branch of the Ca^{2+} messenger system at basal glucose concentration (89). If protein kinase C is to be involved in the regulation of insulin release, a lack of diacylglycerol formation will of course interfere with the secretory response. However, recent studies do not favour this concept, since pretreatment of the β-cells with phorbol ester, in order to downregulate protein kinase C activity, made them refractory to the phorbol ester whereas the insulin response to glucose remained unchanged (90,91). To what extent the potentiating effect of CCK on glucose-induced insulin release can be fully explained by the fact that the peptide activates the phospholipase C system merits further investigations. It is not clear to what extent CCK also activates the adenylate cyclase system in the pancreatic β-cells. Whereas CCK can activate adenylate cyclase in broken cell membrane preparations, there was no accumulation of cyclic AMP in intact pancreatic acinar cells subsequent to stimulation by the peptide (92-95). Similarly, CCK does not affect glucose-stimulate accumulation of cyclic AMP in rat islets (96). However, this latter observation can be explained by protein kinase C-mediated desensitization of the adenylate cyclase system to CCK, as proposed for glucagon-stimulated adenylate cyclase in the hepatocyte (97).

Peptide histidine-isoleucine (PHI)

The 27 amino acid peptide PHI shows structural similarity to VIP and it has been suggested that these two peptides are processed from the same precursor molecule (57,98). PHI has, as VIP, been localized to nerve fibers close to the pancreatic islets (57) and appears to exert the same biological effects as VIP. Thus, PHI stimulates the secretion of insulin, glucagon and somatostatin in the perfused rat pancreas (99) and stimulates insulin secretion in vivo in the rat (60) and the mouse (100). In contrast, however, PHI seems not to exert any effects on islet hormone secretion in man (57). Whether PHI is released from the pancreas during nerve stimulation is not yet known. Therefore, PHI is a candidate to be a pancreatic neuropeptide of functional importance for islet hormone secretion, though this has not yet been established.

Substance P

The 11 amino acid peptide substance P has been localized to intrapancreatic nerve fibers and ganglia in the cat (54), and in the perfused dog pancreas, substance P stimulates the secretion of insulin, glucagon and somatostatin (63). In contrast, in the perfused rat pancreas and in vivo in the mouse, substance P inhibits insulin secretion (101,102). Therefore, its possible role in islet function is not clear.

Neurotensin

The 13 amino acid peptide neurotensin has been demonstrated to occur in nerves in the rat pancreas (101). Studies in mice, calves, and dogs have shown that neurotensin stimulates insulin secretion (102-105). Furthermore, in isolated rat islets, the peptide stimulates secretion of insulin, glucagon and somatostatin at low glucose levels, whereas the peptide inhibits the secretion of these hormones at high glucose concentrations (106). Therefore, more studies are needed to establish a potential role of neurotensin in the control of islet function.

Conclusion

Non-adrenergic, non-cholinergic mechanisms are involved in the neural regulation of islet hormone secretion. Most likely, these are mediated by the intrapancreatic neuropeptides. Table I summarizes our knowledge on pancreatic release and effects of intrapancreatic neuropeptides and effects of electrical nerve activation. NPY and galanin, have both been shown to be localized to intrapancreatic sympathetic nerves and to be released from the pancreas during sympathetic nerve activation. Furthermore, in most studies these peptides also mimic the effects of sympathetic nerve stimulation, viz. inhibition of insulin and somatostatin secretion and activation of glucagon secretion. This may suggest that any, or both, of these neuropeptides mediate non-adrenergic, sympathetic influences on islet hormone secretion. In that case, NPY and/or galanin are involved in mediating metabolic responses to stress, with the goal to elevate blood glucose levels for adequate glucose delivery to the central nervous system by minimizing peripheral glucose uptake and maximizing hepatic glucose release. Conversely, VIP and GRP are released from the pancreas during vagal nerve activation and stimulate secretion of insulin and glucagon, i.e., exert the same effects as vagal nerve activation. Hence, these two neuropeptides are candidates for mediating non-cholinergic, vagal actions. Also the other intrapancreatic neuropeptides are potential candidates for being of physiological importance in neural regulation of islet function. The final establishment of involvement of neuropeptides in neural control of islet hormone secretion needs studies with specific neuropeptide receptor antagonists. In addition, the functional importance of co-localization of different neurotransmitters in the same nerve terminals should be explored. The mode of action underlying effects of neuropeptides on islet hormone secretion is sofar, with the exception of galanin, largely unknown.

TABLE I

Occurrence, Pancreatic Release and Effects on Islet Hormone Secretion of Neuropeptides

Neuropeptide	Intrapancreatic Neuronal Localization	Pancreatic Release During Nerve Activation	Effect on the Secretion of		
			Insulin	Somatostatin	Glucagon
NPY	yes	yes	–	–	+
Galanin	yes	yes	–	–	+ or –
VIP	yes	yes	+	+	+
GRP	yes	yes	+	0 or –	0 or +
CGRP	yes	no	–	0	variable
CCK	yes	not studied	+	+	+
PHI	yes	not studied	+	+	+
Substance P	yes	not studied	+ or –	+	+
Neurotensin	yes	not studied	+ or –	+ or –	+ or –
Sympathetic nerve activation			–	–	+
Parasympathetic nerve activation			+	+ or –	+

+ indicates stimulation, – indicates inhibition, and 0 indicates no effect. For detailed information see text.

References

1. Langerhans P. Beiträge zur mikroskopischen Anatomie der Bauchspeichel-drÜse. Inaugural-Dissertation. Friedrich-Wilhelms-Universität, Berlin, pp. 1-32 (1869).
2. Woods SC, Porte Jr D. Neural control of the endocrine pancreas. Physiol Rev 54:596-619 (1974).
3. Miller RE. Pancreatic neuroendocrinology: Peripheral neural mechanisms in the regulation of the islets of Langerhans. Endocr Rev 4: 471-494 (1981).
4. Ahrén B, Ericson LE, Lundquist I, Lorén I, Sundler F. Adrenergic inner-vation of pancreatic islets and modulation of insulin secretion by the sympatho-adrenal system. Cell Tissue Res 216:15-30 (1981).
5. Pettersson M, Ahrén B, Lundquist I, Böttcher G, Sundler F. Neuropeptide Y: intrapancreatic neuronal localization and inhibitory effects on glucose-stimulated insulin secretion in the mouse. Cell Tissue Res 248: 43-48 (1987).
6. Luiten PGM, ter Horst GJ, Buijs RM, Steffens AB. Autonomic innervation of the pancreas in diabetic and non-diabetic rats; a new view on intra-mural sympathetic structural organization. J Auton Nerv Syst 15:33-44 (1986).
7. Coupland RE. The innervation of the pancreas of the rat, cat and rabbit as revealed by the cholinesterase technique. J Anat 92:143-149 (1958).
8. Ahrén B, Taborsky Jr GJ, Porte Jr D. Neuropeptidergic versus choli-nergic and adrenergic regulation of islet hormone secretion. Diabetolo-gia 29:827-836 (1986).
9. Holst JJ, Gronholt R, Schaffalitzky de Muckadell OB, Fahrenkrug J. Nervous control of pancreatic endocrine secretion in pigs. V. In-fluences of the sympathetic nervous system on the pancreatic secretion of insulin and glucagon, and on the insulin and glucagon response to vagal stimulation. Acta Physiol Scand 113:278-283 (1981).
10. Bloom SR, Edwards AV. Certain pharmacological characteristics of the release of pancreatic glucagon in response to stimulation of the splan-chnic nerves. J Physiol 280:25-35 (1978).
11. Ahrén B, Veith RC, Taborsky Jr GJ. Sympathetic nerve stimulation versus pancreatic norepinephrine infusion in the dog: 1)Effects on basal release of insulin and glucagon. Endocrinology 121:323-331 (1987).
12. Ahrén B, Veith RC, Paquette TL, Taborsky Jr GJ. Sympathetic nerve stimulation versus pancreatic norepinephrine infusion in the dog: 2)Ef-fects on basal release of somatostatin and pancreatic polypeptide. Endocrinology 121:323-339 (1987).
13. Ahrén B, Taborsky Jr GJ. The mechanism of vagal nerve stimulation of glucagon and insulin secretion in the dog. Endocrinology 118:1551-1557 (1986).
14. Bloom SR, Edwards AV. Pancreatic endocrine responses to stimulation of the peripheral ends of the vagus nerves in conscious calves, J Physiol 315:31-41 (1981).
15. Holst JJ, Gronholt R, Schaffalitzky de Muckadell OB, Fahrenkrug J. Nervous control of pancreatic endocrine secretion in pigs. 2. The effect of pharmacological blocking agents on the response to vagal stimulation. Acta Physiol Scand 111:9-14 (1981).
16. Ahrén B, Paquette TL, Taborsky Jr GJ. Effect and mechanism of vagal nerve stimulation on somatostatin secretion in dogs. Am J Physiol 250: E212-E217 (1986).
17. Holst JJ, Jensen SL, Knuhtsen S, Nielsen OV. Autonomic nervous control of pancreatic somatostatin secretion. Am J Physiol 245:E542-E548 (1983).
18. Dunning BE, Ahrén B, Veith RC, Taborsky Jr GJ. Nonadrenergic sympa-thetic neural influences on basal pancreatic hormone secretion. Am J Physiol 255:E785-E792 (1988).

19. Kurose T, Seino Y, Nishi S, Tsuji K, Taminato T, Tsuda K, Imura H. Mechanism of sympathetic neural regulation of insulin, somatostatin, and glucagon secretion. Am J Physiol 258:E220-E227 (1990).

20. Sundler F, Moghimzadeh E, Håkanson R, Ekelund M, Emson P. Nerve fibers in the gut and pancreas of the rat displaying neuropeptide Y-immuno-reactivity. Intrinsic and extrinsic origin. Cell Tissue Res 230:487-493 (1983).

21. Carlei F, Allen JM, Bishop AE, Bloom SR, Polak JM. Occurrence, distribution and nature of neuropeptide Y in the rat pancreas. Experientia 41:1554-1557 (1985).

22. Dunning BE, Ahrén B, Böttcher G, Sundler F, Taborsky Jr GJ. The presence and actions of NPY in the canine endocrine pancreas. Regul Pept 18:253-265 (1987).

23. Sheikh SP, Holst JJ, Skak-Nielsen T, Knigge U, Warberg J, Theodorsson-Norheim E, Hökfelt T, Lundberg JM, Schwartz TW. Release of NPY in pig pancreas: dual parasympathetic and sympathetic regulation. Am J Physiol 255:G46-G54 (1988).

24. Ahrén B, Böttcher G, Kowalyk S, Dunning BE, Sundler F, Taborsky Jr GJ. Galanin is co-localized with noradrenaline and neuropeptide Y in dog pancreas and celiac ganglion. Cell Tissue Res 261:49-58 (1990).

25. Ahrén B, Mårtensson H, Ekman R. Pancreatic nerve stimulation releases neuropeptide Y- but not galanin- or calcitonin gene-related peptide-like immunoreactivity from the pig pancreas. J Autonom Nerv Syst 27:11-16 (1989).

26. Moltz JH, McDonald JK. Neuropeptide Y: direct and indirect action on insulin secretion in the rat, Peptides 6:1155-1159 (1985).

27. Pettersson M, Ahrén B: Insulin secretion in rats: effects of neuropeptide Y and noradrenaline. Diabet Res in press (1990).

28. Skoglund G, Gross R, Ahrén B, Loubatières-Mariani MM: Synergism between neuropeptide Y and noradrenaline on insulin secretion in the isolated perfused rat pancreas. Diabetologia 32:542A (1989).

29. Holst JJ, Örskov C, Knuhtsen S, Sheikh S, Nielsen OV. On the regulatory functions of neuropeptide Y (NPY) with respect to vascular resistance and exocrine and endocrine secretion in the pig pancreas. Acta Physiol Scand 136:519-526 (1989).

30. Dunning BE, Ahrén B, Veith RC, Böttcher G, Sundler F, Taborsky Jr GJ. Galanin: a novel pancreatic neuropeptide. Am J Physiol 251:E127-133 (1986).

31. Sundler F, Böttcher G. Islet innervation, with special reference to neuropeptides. In: Samols E (ed.): The Endocrine Pancreas. Raven Press, New York in press (1991).

32. Dunning BE, Taborsky Jr GJ. Galanin release during pancreatic nerve stimulation is sufficient to influence islet function. Am J Physiol 256:E191-E198 (1989).

33. Ahrén B, Rorsman P, Berggren PO. Galanin and the endocrine pancreas. FEBS Lett 229:233-237 (1988).

34. Gilbey SG, Stephenson J, O'Halloran DJ, Burrin JM, Bloom SR. High-dose porcine galanin infusion and effect on intravenous glucose tolerance in humans. Diabetes 38:1114-1116 (1989).

35. Ahrén B: Effects of galanin and calcitonin gene-related peptide on insulin and glucagon secretion in man. Acta Endocrinol 123:in press (1990).

36. Amiranoff B, Servin AL, Rouyer-Fessard C, Couvineau A, Tatemoto K, Laburthe M: Galanin receptors in a hamster pancreatic β-cell tumor: identification and molecular characterization. Endocrinology 121: 284-289 (1987).

37. Sharp GWG, LeMerchaud-Brustel Y, Yada T, Russo LL, Bliss CR, Cormont M, Monge L, van Obberghen E: Galanin can inhibit insulin release by a mechanism other than membrane hyperpolarization or inhibition of adenylate cyclase. J Biol Chem 264:7302-7309 (1989).

38. Ahrén B, Arkhammar P, Berggren PO, Nilsson T: Galanin inhibits glucose-stimulated insulin release by a mechanism involving hyperpolarization and lowering of cytoplasmic free Ca^{2+} concentration. Biochem Biophys Res Commun 140:1059-1063 (1986).

39. Nilsson T, Arkhammar P, Rorsman P, Berggren PO: Suppression of insulin release by galanin and somatostatin is mediated by a G-protein: an effect involving repolarization and reduction in cytoplasmic free Ca^{2+} concentration. J Biol Chem 264:973-980 (1989).

40. Ullrich S, Wollheim CB: Galanin inhibits insulin secretion by direct interference with exocytosis. FEBS Lett 247:401-404 (1989).

41. Lindskog S, Ahrén B: Galanin and the regulation of insulin secretion. In: Hökfelt T (ed.): Wenner-Gren International Symposium Series, in press (1990).

42. de Weille JR, Schmid-Antomarchi H, Fosset M, Lazdunski M: ATP-sensitive K^+ channels that are blocked by hypoglycemia-inducing sulfonylureas in insulin-secreting cells are activated by galanin, a hyperglycemia-inducing hormone. Proc Natl Acad Sci USA 85:1312-1316 (1988).

43. Fosset M, Schmid-Antomarchi H, deWeille JR, Lazdunski M: Somatostatin activates glibenclamide-sensitive and ATP-regulated K^+-channels in insulinoma cells via a G-protein. FEBS Lett 242:94-96 (1988).

44. Dunne MJ, Bullet MJ, Quodong L, Wollheim CB, Petersen OH: Galanin activates nucleotide-dependent K^+-channels in insulin-secreting cells via a pertussis toxin-sensitive G-protein. EMBO J 8:413-420 (1989).

45. de Weille JR, Schmid-Antomarchi H, Fosset M, Lazdunski M: Regulation of ATP-sensitive K^+ channels in insulinoma cells: activation by somatostatin and protein kinase C and the role of cAMP. Proc Natl Acad Sci USA 86:2971-2975 (1989).

46. Ahrén B, Berggren PO, Bokvist K, Rorsman P: Does galanin inhibit insulin secretion by opening of the ATP-regulated K^+-channel in the B-cell? Peptides 10:453-457 (1989).

47. Horn R, Marty A: Muscarinic activation of ionic currents measured by a new whole-cell recording mode. J Gen Physiol 92:145-169 (1988).

48. Rorsman P, Bokvist K, Ämmälä C, Berggren PO. Larsson O, Wåhlander K: Adrenaline activates a low-conductance G-protein-regulated K^+-channel in mouse pancreatic B-cells. Nature in press (1990).

49. Wåhlander K, Ämmälä C, Berggren PO. Bokvist K, Juntti-Berggren L, Rorsman P: Galanin inhibits B-cell electrical activity by a G-protein-regulated sulphonylurea-insensitive mechanism. In: Hökfelt T (ed.): Wenner-Gren International Symposium Series, in press (1990).

50. Amiranoff B, Lorinet AM, Lagny-Pourmir I, Laburthe M: Mechanism of galanin-inhibited insulin release: occurrence of a pertussis-toxin-sensitive inhibition of adenylate cyclase. Eur J Biochem 177:147-152 (1988).

51. Ui M: Islet-activating protein, pertussis toxin: a probe for functions of the inhibitory guanine nucleotide regulatory component of adenylate cyclase. Trends Biochem Sci 7:277-279 (1984).

52. Brown AM, Birnbaumer L: Direct G protein gating of ion channels. Am J Physiol 254:H401-H410 (1988).

53. Vallar L, Biden TJ, Wollheim CB: Guanine nucleotides induce Ca^{2+}-independent insulin secretion from permeabilized RINm5F cells. J Biol Chem 262:5049-5056 (1987).

54. Larsson LI. Innervation of the pancreas by substance P, enkephalin, vasoactive intestinal polypeptide and gastrin-CCK immunoreactive nerves. J Histochem Cytochem 27:1283-1284 (1979).

55. Bishop AE, Polak JM, Green IC, Bryant MGT, Bloom SR. The location of VIP in the pancreas of man and rat. Diabetologia 18:73-78 (1980).

56. Prinz RA, El Sabbagh H, Adrian TE, Bloom SR, Gardner I, Polak JM, Inokuchi H, Bishop AE, Welbourn RB. Neural regulation of pancreatic polypeptide release. Surgery 94:1011-1018 (1983).

57. Fahrenkrug J, Holst Pedersen J, Yamashita Y, Ottesen B, Hökfelt T, Lundberg JM: Occurrence of VIP and peptide HM in human pancreas and their influence on pancreatic endocrine secretion in man. Regul Pept 18:51-61 (1987).

58. Holst JJ, Fahrenkrug J, Knuhtsen S, Jensen SL, Poulsen SS, Nielsen OV. Vasoactive intestinal polypeptide (VIP) in the pig pancreas: Role of VIPergic fibers in control of fluid and bicarbonate secretion. Regul Pept 8:245-249 (1984).

59. Jensen SL, Fahrenkrug J, Holst JJ, Nielsen OV, Schaffalitzky de Mucka-dell OB. Secretory effect of VIP on isolated perfused porcine pancreas. Am J Physiol 235:E387-E391 (1978).

60. Szecowka J, Lins PE, Tatemoto K, Efendic S. Effects of porcine intestinal heptacosapeptide and vasoactive intestinal polypeptide on insulin and glucagon secretion in rats. Endocrinology 112:1469-1473 (1983).

61. Ahrén B, Lundquist I. Effects of vasoacive intestinal polypeptide (VIP), secretin and gastrin on insulin secretion in the mouse. Diabetologia 20:1-6 (1981).

62. Ahrén B, Lundquist I. Interaction of vasoactive intestinal peptide (VIP) with cholinergic stimulation of glucagon secretion. Experientia 38:405-406 (1982).

63. Hermansen K. Effects of substance P and other peptides on the release of somatostatin, insulin and glucagon in vitro. Endocrinology 107:256-261 (1980).

64. Moghimzadeh E, Ekman R, Håkanson R, Yanaihara N, Sundler F. Neuronal gastrin-releasing peptide in the mammalian gut and pancreas. Neuroscience 10:553-563 (1983).

65. Knuhtsen S, Holst JJ, Baldissera FGA, Skak-Nielsen T, Poulsen SS, Jensen SL, Nielsen OV. Gastrin releasing peptide in the porcine pancreas. Gastroenterology 92:1153-1158 (1987).

66. Knuhtsen S, Holst JJ, Jensen SL, Nielsen OV. Gastrin releasing peptide: effect on exocrine secretion, and release from isolated perfused pig pancreas. Am J Physiol 248:G281-G287 (1985).

67. Knuhtsen S, Holst JJ, Schwartz TW, Jensen SL, Nielsen OV. The effect of gastrin-releasing peptide on the endocrine pancreas. Regul Pept 17:269-276 (1987).

68. Bloom SR, Edwards AV, Ghatei MA. Endocrine responses to exogenous bombesin and gastrin-releasing peptide in conscious calves. J Physiol 344:37-48 (1983).

69. Pettersson M, Ahrén B. Gastrin releasing peptide (GRP): Effects on basal and stimulated insulin and glucagon secretion in the mouse. Peptides 8:55-60 (1987).

70. Pettersson M, Ahrén B. Insulin and glucagon secretion in the rat: Effects of gastrin releasing peptide. Neuropeptides 12:159-163 (1988).

71. Hermansen K, Ahrén B. Gastrin releasing peptide stimulates the secretion of insulin, but not that of glucagon or somatostatin, from the isolated perfused dog pancreas. Acta Physiol Scand 138:175-179 (1990).

72. Rosenfeld MG, Mermod JJ, Amara SG, Swanson LWQ, Sawchenko PE, Rivier J, Vale WW, Evans RM. Production of a novel neuropeptide encoded by the calcitonin gene via tissue-specific RNA processing. Nature 304: 129-135 (1983).

73. Pettersson M, Ahrén B, Böttcher G, Sundler F. Calcitonin gene-related peptide: occurrence in pancreatic islets in the mouse and the rat and inhibition of insulin secretion in the mouse. Endocrinology 119:865-869 (1986).

74. Sternini C, Brecha N. Immunocytochemical identification of islet cells and nerve fibers containing calcitonin gene-related peptide-like immunoreactivity in the rat pancreas. Gastroenterology 90:1155-1163 (1986).

75. Ahrén B, Pettersson M. Calcitonin gene-related peptide (CGRP) and amylin and the endocrine pancreas. Int J Pancreatol 6:1-15 (1990).

76. Su HC, Bishop AE, Power RF, Hamada Y, Polak JM. Dual intrinsic and extrinsic origins of CGRP- and NPY-immunoreactive nerves of rat gut and pancreas. J Nerurosci 7:2674-2687 (1987).

77. Ahrén B, Mårtensson H, Nobin A. Effects of calcitonin gene-related peptide (CGRP) on islet hormone secretion in the pig. Diabetologia 30: 354-359 (1987).

78. Pettersson M, Ahrén B. Insulin and glucagon secretion in rats: Effects of calcitonin gene-related peptide. Regul Pept 23:37-50 (1988).

79. Hermansen K, Ahrén B. Dual effects of calcitonin gene-related peptide on insulin secretion in the perfused dog pancreas. Regul Pept 27:149-157 (1990).

80. Niski M, Sanke T, Nagamatsu S, Bell GI, Steiner DF: Islet amyloid polypeptide. J Biol Chem 265:4173-4176 (1990).

81. Kobilka BK, Frielle T, Collins S, Yang-Fend T, Kobilka TS, Francke U, Lefkowitz RJ, Caron MG: An intronless gene encoding a potential member of receptors coupled to guanine nucleotide regulatory proteins. Nature 329:75-79 (1987).

82. Haegerstrand A, Dalsgaard CJ, Jonzon B, Larsson O, Nilsson J: Calcitonin gene-related peptide stimulates proliferation of human endothelial cells. Proc Natl Acad Sci USA 87:3299-3303 (1990).

83. Pettersson M, Ahrén B: Calcitonin gene-related peptide inhibits insulin secretion. Studies on ion fluxes and cyclic AMP in isolated rat islets. Diab Res in press (1990).

84. Rehfeld JF, Larsson LI, Goltermann NR, Schwartz TW, Holst JJ, Jensen SL, Morley JS. Neural regulation of pancreatic hormone secretion by the C-terminal tetrapeptide of CCK. Nature 284:33-38.

85. Hermansen K. Effects of cholecystokinin (CCK)-4, nonsulfated CCK-8, and sulfated CCK-8 on pancreatic somatostatin, insulin and glucagon secretion in the dog: studies in vitro. Endocrinology 114:1770-1775 (1984).

86. Szecowka J, Lins PE, Efendic S. Effects of cholecystokinin, gastric inhibitory polypeptide, and secretin on insulin and glucagon secretion in rats. Endocrinology 110:1268-1272 (1982).

87. Ahrén B, Mårtensson H, Nobin A. Cholecystokinin (CCK)-4 and CCK-8 stimulate islet hormone secretion in vivo in the pig. Pancreas 3: 279-284 (1988).

88. Verspohl EJ, Ammon HPT, Williams JA, Goldfine ID: Evidence that cholecystokinin interacts with specific receptors and regulates insulin release in isolated rat islets of Langerhans. Diabetes 35: 38-43 (1986).

89. Zawalich WS, Diaz VA, Zawalich KC: Cholecystokinin-induced alterations in B-cell sensitivity. Diabetes 36:1420-1424 (1987).

90. Hii CST, Jones PM, Persaud SJ, Howell SL: A re-assessment of the role of protein kinase C in glucose-stimulated insulin secretion. Biochem J 246:489-493 (1987).

91. Arkhammar P, Nilsson T, Welsh M, Welsh N, Berggren PO: Effects of protein kinase C activation on the regulation of stimulus-secretion coupling in pancreatic B-cells. Biochem J 264:207-215 (1989).

92. Rutten WJ, DePont JJHHM, Bonting SL: Adenylate cyclase in the rat pancreas. Properties and stimulation by hormones. Biochim Biophys Acta 274:201-213 (1972).

93. Schulz I, Stoltze HH: The exocrine pancreas: the role of secretagogues, cyclic nucleotides and calcium in enzyme secretion. Ann Rev Physiol 42: 127-156 (1980).

94. Long BW, Gardner JD: Effects of cholecystokinin on adenylate cyclase activity in dispersed pancreatic acinar cells. Gastroenterology 73: 1008-1014 (1977).

95. Willems PHGM, Tilly RHJ, DePont JJHHM: Pertussis toxin stimulates cholecystokinin-induced cyclic AMP formation but is without effect on secretoguge-induced calcium mobilization in exocrine pancreas. Biochim Biophys Acta 928:179-185 (1987).

96. Verspohl EJ, Breuning I, Ammon HPT, Mark M: Significance of Ca^{2+}, Rb^+ fluxes, of cAMP and cGMP for the CCK_8-modulated insulin release. Regul Pept 17:229-241 (1987).

97. Wakelam MJO, Murphy GJ, Hruby VJ, Houslay MD: Activation of two signal-transduction systems in hepatocytes by glucagon. Nature 324: 68-70 (1986).

98. Itoh N, Obata K, Yanaihara N, Okamoto H. Human preprovasoactive intestinal polypeptide contains a novel PHI-27-like peptide, PHM-27. Nature 304:547-549 (1983).

99. Szecowka J, Tendler D, Efendic S. Effects of PHI on hormonal secretion from perfused rat pancreas. Am J Physiol 245:E313-E317 (1983).

100. Ahrén B, Lundquist I. Effects of peptide HI on basal and stimulated insulin and glucagon secretion in the mouse. Neuropeptides 11:159-162 (1988).

101. Efendic S, Luft R, Pernow B. Effect of substance P on arginine-induced insulin and glucagon release from the isolated perfused rat pancreas. In: v Euler US, Pernow B (eds.) Substance P. Raven Press, New York, pp 241-245 (1977).

102. Lundquist I, Sundler F, Ahrén B, Alumets J, Håkanson R. Somatostatin, pancreatic polypeptide, substance P, and neurotensin: Cellular distribution and effects on stimulated insulin secretion in the mouse. Endocrinology 104:832-838 (1979).

103. Feurle GE, Reinecke M. Neurotensin interacts with carbachol, secretin, and caerulein in the stimulation of the exocrine pancreas of the rat in vitro. Regul Pept 7:137-143 (1983).

104. Ishida T. Stimulatory effect of neurotensin on insulin and gastrin secretion in dogs. Endocrinol Japon 24:335-342 (1977).

105. Blackburn AM, Bloom SR, Edwards AV. Pancreatic endocrine responses to exogenous neurotensin in the conscious calf. J Physiol 314:11-21 (1981).

106. Iguchi A, Sakamoto N, Burleson PD. The effects of neuropeptides on glucoregulation. Adv Metab Disorders 10:421-434 (1983).

PROGLUCAGON-DERIVED PEPTIDES IN THE NEUROENDOCRINE SYSTEM

P.L.Brubaker[*], K.M.Stobie, J.N.Roberge, E.Y.T.Lui and D.J.Drucker[@]

Departments of Physiology[*] and Medicine[@]
University of Toronto
Toronto, Ontario M5S 1A8 Canada

INTRODUCTION

The proglucagon gene is expressed in a diversity of cell types, most notably the pancreatic A cell, the intestinal L cell, and selected neurons in the hypothalamus and brain stem[1-3]. Much interest in the pGdp's has been generated over the past several years with the isolation and sequence analysis of both the gene and the cDNA for proglucagon[4-6]. Proglucagon was found not only to contain the sequences for glucagon and several large glucagon-containing peptides (glicentin and oxyntomodulin), but was also predicted to give rise by proteolytic cleavage to two glucagon-like peptides (GLP-1 and -2) (Fig. 1). Synthesis and secretion of the proglucagon-derived peptides (pGdp's) are now known to be subject to tissue-specific regulation, at the levels of gene expression, post-translational processing and peptide secretion. For example, plasma and tissue levels of the intestinal pGdp's rise 2- to 3-fold in streptozotocin-diabetic rats, while concentrations in the pancreas and brain are not markedly altered[7]. Furthermore, changes in proglucagon post-translational processing occur in the brain of the developing rat, but not in the pancreas or intestine[8]. In this review, I will discuss some of the work we have done to elucidate the mechanisms underlying tissue-specific differences in the production of the pGdp's.

1. IN VITRO MODELS

To study the tissue-specific production of pGdp's, we have developed several in vitro cell culture models, including fetal rat intestinal cells (FRIC), fetal rat hypothalamic cells (FRHC), and pancreatic islets, as described below.

In all experiments, pGdp's are detected by radioimmunoassay for glucagon-like immunoreactive (GLI) peptides and immunoreactive glucagon (IRG). GLI peptides are detected using an antiserum (K4023; Novo Alle, Bagsvaerd, Denmark) that recognizes the mid-sequence of glucagon and all N- and/or C-

Fuel Homeostasis and the Nervous System, Edited by M. Vranic *et al.*
Plenum Press, New York, 1991

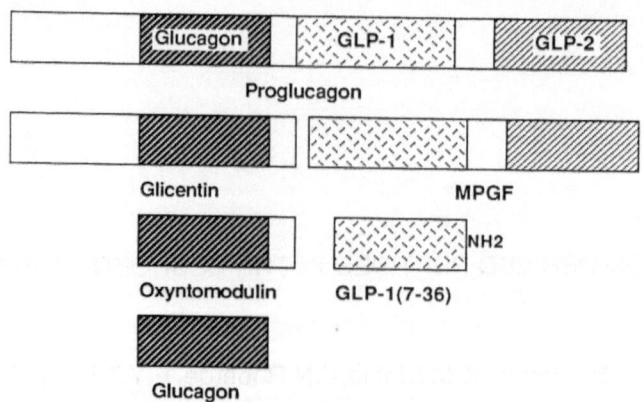

Fig. 1. Proglucagon and known proglucagon-derived peptides (GLP: glucagon-like peptide; MPGF: major proglucagon fragment).

terminally extended forms of glucagon. IRG peptides are detected using antiserum 04A (Dr. R.H. Unger, Dallas, TX) that recognizes primarily glucagon.

1.1 Fetal Rat Intestinal Cells in Culture

Proglucagon-producing L cells are dispersed diffusely in the epithelial layer of the intestinal mucosa, with highest concentrations in the crypts of the terminal ileum[9]. Few cell culture models of intestinal endocrine cells exist[10-12], in part due to difficulties encountered with bacterial contamination. To develop a cell culture model for the intestine, we therefore utilized the sterile tissue of the fetal rat intestine[13]. Fetal rat intestinal cells are enzymatically dispersed and placed into culture for 24 hr. The cells are heterogeneous, with a variety of cell types notable by light miscroscopy (Fig. 2A). Numerous epithelial cells can be found in the cultures by immunofluorescent localization of low molecular weight cytokeratins, markers for the gastrointestinal epithelium. Within this population of epithelial-like cells, a smaller group of cells stains specifically for the GLI peptides but not for IRG, suggesting the presence of the intestinal pGdp's and the absence of glucagon in the intestinal cultures (Fig. 2B).

To establish that FRIC cultures can synthesize proglucagon, and thus the pGdp's, total cellular RNA was examined by Northern blot for the presence of proglucagon mRNA transcripts[3]. A single mRNA transcript was detected in FRIC cultures that was identical in size to proglucagon mRNA transcripts detected in pancreas and intestine (Fig. 3A). FRIC cultures were also found to contain mRNA transcripts for other intestinal regulatory peptides, including somatostatin (Fig. 3A), peptide YY and cholecystokinin (not shown), suggesting that the cells are at least partially representative of the normal rat intestinal epithelium with respect to regulatory peptide gene expression.

The ability of FRIC cultures to correctly process proglucagon was established by chromatographic analysis of cellular pGdp's peptides. Cell extracts were found to contain glicentin and oxyntomodulin, but not glucagon, a distribution of peptides identical to that of adult rat ileum (Fig. 4A,B). Thus, FRIC cultures were

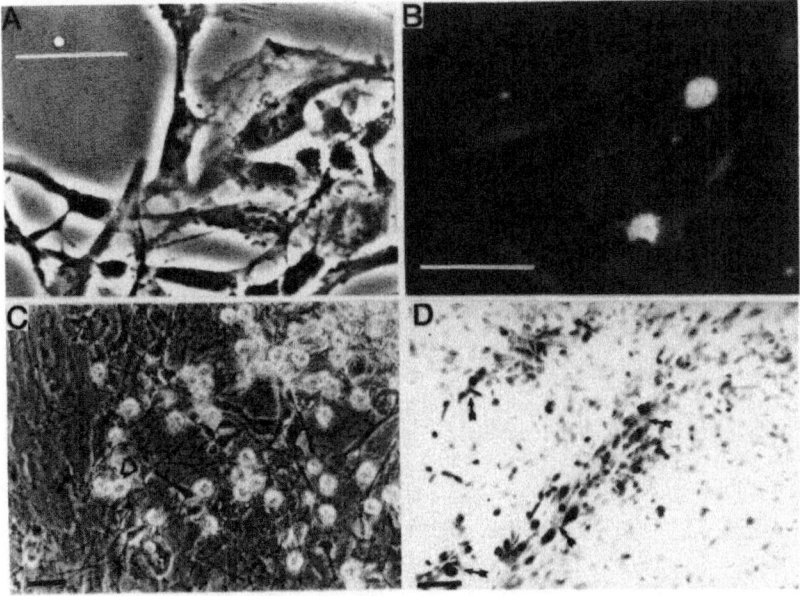

Fig. 2. FRIC (A,B) and FRHC (C,D) cultures. (A and C) Light microscopy. (B) Immunofluorescent staining for FRIC GLI peptides. (D) Immunostaining for FHRC IRG peptides. The scale bar represents 50 um.

demonstrated to be a suitable model for further investigations of the factors regulating intestinal pGdp synthesis and secretion.

1.2 Fetal Rat Hypothalamic Cells in Culture

The presence of pGdp's in the hypothalamus has been clearly demonstrated by a variety of immunological techniques[14,15]; however, whether the hypothalamus is capable of <u>de novo</u> synthesis of these peptides was not established by these studies. Recently, proglucagon mRNA transcripts have been detected in fetal and adult rat brain stem, and in adult rat hypothalamus[2]. The level of expression of proglucagon mRNA was very low in the adult hypothalamus, however, and no transcripts could be detected in fetal hypothalamus. Therefore, we developed a hypothalamic culture system utilizing fetal hypothalami[16], to establish whether the pGdp's are synthesized in this tissue, and to then investigate the regulation of hypothalamic pGdp synthesis and secretion.

Fetal rat hypothalami (19-21 d gestation) are mechanically dispersed by passage through a series of needles of decreasing inner diameter, and the cells are placed into culture for 1 week. FRHC cultures contain a mixed cell population (Fig. 2C), with cells containing neurofilament, glial-fibrillary acidic protein or neuron-specific enolase detectable by immunocytochemistry. Strong immunopositivity for GLI and IRG (Fig. 2D) peptides was also observed, both in clusters of cells and in individual cells, suggesting the presence of glucagon and glicentin and/or oxyntomodulin.

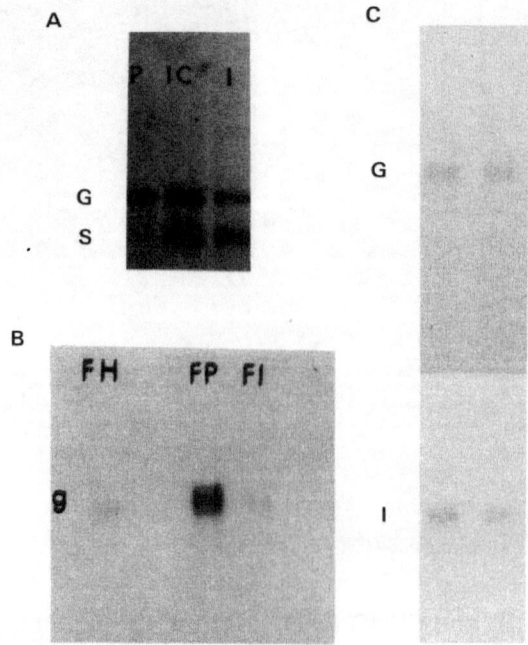

Fig. 3. Northern blot analysis of RNA from (A) FRIC cultures, (B) FHRC cultures, and (C) pancreatic islets. (A) Total cellular RNA from rat pancreas (P), fetal rat intestinal cultures (IC) and rat intestine (I) was hybridized to ^{32}P-labelled cDNA probes for proglucagon (G) and prosomatostatin (S). (B) Poly-A$^+$ RNA from FRHC cultures (FH) and total cellular RNA from fetal rat pancreas (FP) and fetal rat intestine (FI) was hybridized to a cDNA probe for proglucagon (g). (C) Total cellular RNA from rat pancreatic islets was hybridized to cDNA probes for proglucagon (G) and proinsulin (I).

To establish that FRHC cultures, and thus the fetal hypothalamus, actually synthesize the pGdp's contained therein, Northern blot analysis was carried out on total RNA extracted from the cultured cells. A single mRNA species was detected upon hybridization of FRHC poly A+ mRNA to cDNA for proglucagon (Fig. 3B). This transcript was identical in size to proglucagon mRNA transcripts detected in fetal rat pancreas and intestine. RNase protection analysis of the hypothalamic mRNA transcripts further demonstrated identity of the fetal hypothalamic, fetal brain stem and adult rat intestine proglucagon mRNA[16].

Thus, analysis of proglucagon mRNAs indicated that the sequence of proglucagon in the brain is identical to that of the intestine. However, immunologic analyses of pGdp's in the cultured intestinal cells demonstrated the presence of GLI but not IRG peptides, whereas the cultured hypothalamic cells clearly stained for both the GLI and IRG peptides. To establish whether the FRHC staining for IRG was due to the presence of glucagon in the cells, FRHC cultures and fetal rat hypothalamic pGdp's were examined by HPLC (Fig. 4C and E). In both cases, a clear peak of immunoreactive glucagon was detected, in addition to variable amounts of glicentin and oxyntomodulin. Thus, fetal rat hypothalamus and the FRHC cultures produce an identical proglucagon to that of the intestine, yet tissue-

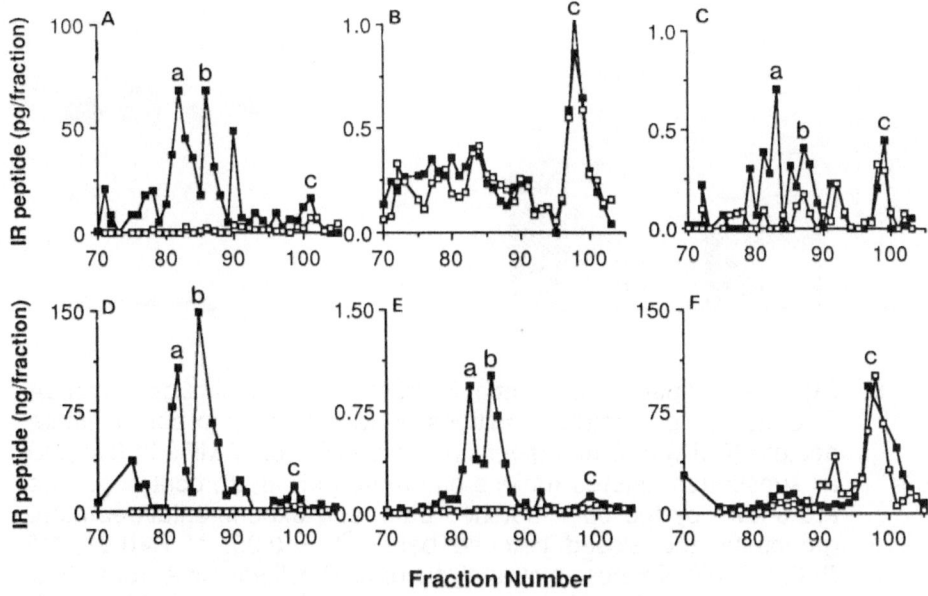

Fig. 4. HPLC analysis of GLI (closed squares) and IRG (open squares) peptides extracted from (A) FRIC cultures, (B) adult rat ileum, (C) FRHC cultures, (D) adult rat hypothalamus, (E) fetal rat hypothalamus, and (F) adult rat pancreas. Peaks a, b and c have been tentatively identified as glicentin, oxyntomodulin and glucagon, respectively.

specific post-translational processing results in differences in the pGdp's generated from this precursor molecule.

1.3 Pancreatic Islets

To compare the results obtained for intestinal and hypothalamic pGdp's with those obtained for the best studied of the pGdp-producing tissues, the pancreas, we have established a method for preparation of adult rat pancreatic islets[17]. The islets are collected by hand picking after collagenase digestion, and are placed into culture for 48 hr. Northern blot analysis of total islet RNA has demonstrated that the cultured islets contain mRNA transcripts for both proglucagon and proinsulin (Fig. 3C).

Thus, we have 3 distinct models to investigate pGdp's in the neuroendocrine system. These models have been used to examine the factors regulating tissue-specific proglucagon gene expression and pGdp synthesis and secretion.

2. SYNTHESIS OF PROGLUCAGON AND DERIVED PEPTIDES

Over the past 5 years, the regulation of proglucagon gene expression has been studied primarily using cell lines derived from pancreatic tumours[18-23]. Evidence from rat RIN1056 cells has suggested that proglucagon gene expression

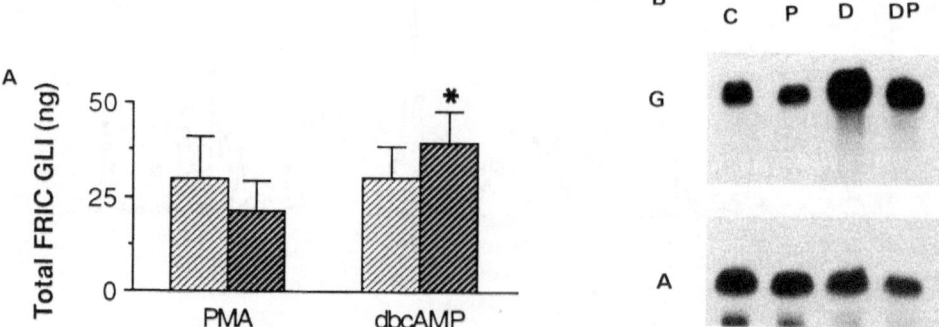

Fig. 5.　(A) Total GLI peptide content of FRIC cultures after incubation for 24 hr under either control conditions or with 1 uM phorbol myristate acetate (PMA) or 5 mM dibutyryl cyclic AMP (dbcAMP). In this and all subsequent figures (unless otherwise indicated), control groups are shown by the open hatched bars, and experimental/treatment groups by the closed hatched bars.　* P<0.05, ** P<0.01, *** P<0.001. (B) Northern blot analysis of total cellular RNA from FRIC cultures treated for 24 hr under either control conditions (C), or with 1 uM PMA (P), 5 mM dbcAMP (D) or both agents (DP). The blot was hybridized to cDNA probes for proglucagon (G) and actin (A).

is under the control of a protein kinase (PK) C-, but not a PKA-dependent pathway[18], whereas hamster InR1-G9 cells lack responsiveness to both PKA and PKC[23]. The ability to induce gene transcription by activation of either PKA or PKC is now known to be dependent upon the presence of a cyclic AMP response element (CRE) or a tumour-promoter (phorbol ester) response element (TRE), respectively, in the 5' flanking region of the gene[24]. Gene activation through a CRE is dependent upon the catalytic subunit of protein kinase A (PKA) for phosphorylation and activation of the nuclear transcription factor CREB, or CRE-binding protein[25]. The proglucagon gene has been demonstrated to contain a CRE between bp 5'-298 and -291[21]. The contextual sequences surrounding a CRE appear to be important to it's ability to modulate gene transcription, however, and the proglucagon CRE has been reported to be inactive in the presence of it's flanking sequences when transfected into a pancreatic tumour-derived cell line[26].

We hypothesized that the proglucagon CRE might be active in tissues other than the pancreas, and that such tissue-specific activity might account for differences in proglucagon gene expression between pancreas, intestine and brain. FRIC cultures were therefore treated with agents that increase the activity of PKA, including dibutyryl cyclic AMP (dbcAMP), forskolin and cholera toxin, or with phorbol myristate acetate (PMA) to activate PKC[27]. Interestingly, in 2 hr experiments, dbcAMP treatment of FRIC cultures increased the total pGdp content of the cultures by 10% (P<0.05), suggesting an effect on pGdp synthesis. In contrast, treatment of the cultures with the control agent, sodium butyrate, or with PMA did not significantly alter FRIC culture pGdp content. Therefore, to further establish whether PKA activation has an effect on intestinal pGdp synthesis, 24 hr incubations of FRIC cultures with dbcAMP or PMA were undertaken (Fig. 5A). As in the short-term experiments, activation of PKC had no effect on the total content of pGdp's. Stimulation of the PKA pathway for 24 hr increased the total culture pGdp

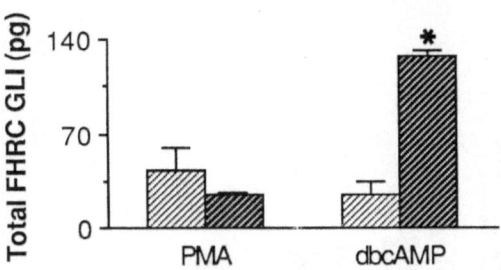

Fig. 6. Total GLI peptide content of FRHC cultures after incubation for 24 hr under either control conditions or with 1 uM PMA or 5 mM dbcAMP. Other details as in Fig. 5A.

content, however, consistent with an increase in pGdp synthesis[3]. Northern blot analysis of total cellular RNA from FRIC cultures confirmed that activation of PKA, but not of PKC, was associated with increases in proglucagon mRNA transcript levels (Fig. 5B), in a time-dependent fashion[3]. Thus, these data implicated PKA and the proglucagon CRE in pGdp synthesis in the intestinal L cell.

To establish whether the pGdp responses to PKA and PKC are tissue-specific, we have also examined the role of these signal transducers in pGdp synthesis using the hypothalamic culture model. DbcAMP, but not PMA treatment of FRHC cultures was found to increase the total cell content of pGdp's (Fig. 6), consistent with an effect through PKA on the proglucagon CRE[28].

Thus, our data has demonstrated that pGdp synthesis in the intestine and hypothalamus is increased by PKA activation, suggesting that the proglucagon CRE is active in the two different cell types. Molecular analysis of the proglucagon CRE and data from pancreatic tumour cell lines, however, had suggested that the proglucagon CRE is not active. To reconcile these opposing findings, we examined the effects of PKA activation in pancreatic islets[17]. Activation of PKA in pancreatic islets significantly increased cellular levels of both immunoreactive glucagon and proglucagon mRNA transcripts, although this occurred in a glucose-dependent fashion (Fig. 7). Thus, the proglucagon CRE appears to be active in pancreatic islets but not in pancreatic tumour cells. That this is not due to an inactivity of the proglucagon CRE or to an absence of CREB in the tumour cells, was demonstrated in co-transfection studies using the hamster InR1-G9 cells[17]. In these studies, the 5' flanking region of the proglucagon gene (GLU) was linked to the coding region of chloramphenicol acetyltransferase (CAT), and gene activation was monitored by enzymatic assay for CAT activity. Transfection of InR1-G9 cells with GLU-CAT alone or co-transfection of GLU-CAT with the gene for CREB did not induce transcriptional activity. In contrast, co-transfection of GLU-CAT with the gene for the catalytic subunit of PKA significantly induced gene transcription. Partial deletion of the proglucagon CRE prevented this activation. Thus, in the presence of active PKA, the proglucagon CRE does induce transcriptional activity in hamster InR1-G9 cells[17].

In summary, all of the data we have obtained using primary cultures of intestinal, hypothalamic and pancreatic islet cells is consistent with our hypothesis

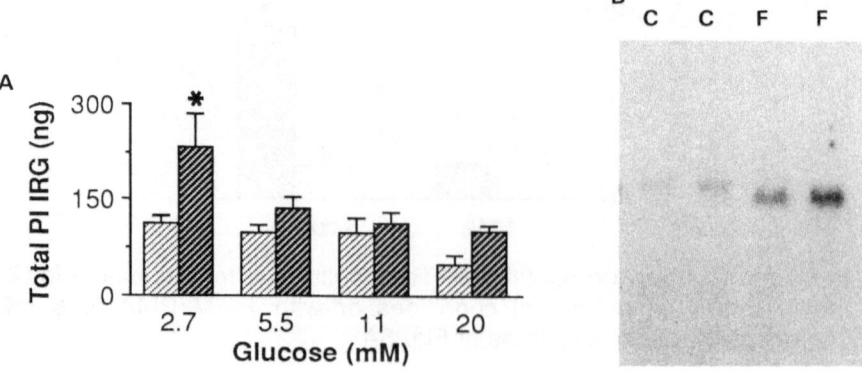

Fig. 7. (A) Total IRG peptide content of cultured pancreatic islets after incubation for 24 hr with varying concentrations of glucose and under either control conditions or with 10 uM forskolin plus 10 uM isobutylmethylxanthine (IBMX). Other details as in Fig. 5A. (B) Northern blot analysis of total cellular RNA from pancreatic islets treated for 24 hr under either control conditions (C), or with forskolin plus IBMX (F). The blot was hybridized to a cDNA probe for proglucagon.

that the proglucagon CRE is active. A recent report further supports this hypothesis in that a specific 'glucagon' CRE-binding protein has now been detected in InR1-G9 cells that exhibits a binding specificity for the proglucagon CRE which is different from that of the cloned placental CREB[29]. As pGdp synthesis was found to be increased by PKA activation in intestine, hypothalamus and pancreas, our data does not support the notion that the proglucagon CRE plays a role in the tissue-specific regulation of proglucagon gene expression. Finally, the results of our studies suggest caution in the interpretation of data obtained using immortalized cell lines, as such cells may have inherent limitations due to abnormalities in some component(s) of the signal transducing systems.

3. POST-TRANSLATIONAL PROCESSING OF PROGLUCAGON

As for many prohormones, the post-translational processing of proglucagon is tissue-specific. HPLC analysis of the pGdp's stored in the adult rat pancreas, intestine and hypothalamus shows that each of the pGdp-producing tissues has a distinct pattern of GLI and IRG peptides (Fig. 4B,D,F)[7]. In pancreas, processing of the N-terminal region of proglucagon is extensive, with 29 amino acid glucagon predominanting. In the intestine, processing of the N-terminal region of proglucagon is less complete, such that glicentin and oxyntomodulin but no glucagon is produced. In contrast to pancreas and intestine, the pattern of proglucagon processing in the adult rat hypothalamus is similar to that of intestine, with glicentin and oxyntomodulin predominating, but detectable amounts of glucagon also being produced.

The C-terminal end of proglucagon has only recently been studied with respect to post-translational processing[30-32]. In contrast to the N-terminal end of proglucagon, however, it appears that processing is incomplete in the pancreas,

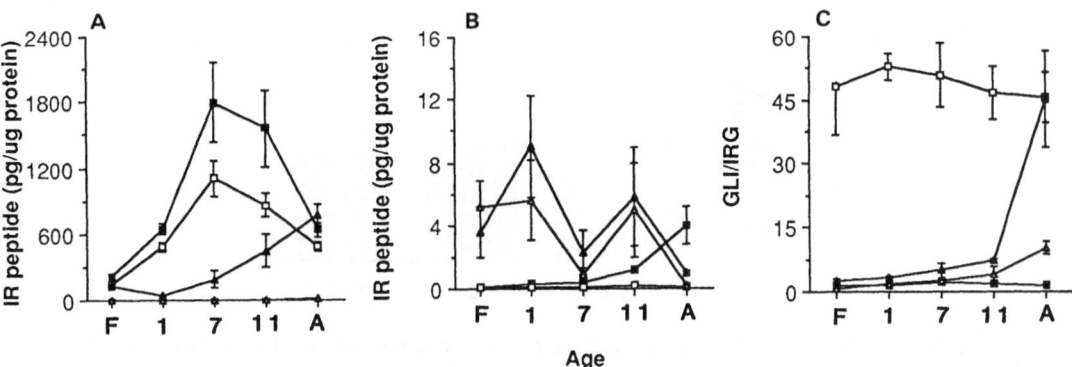

Fig. 8. Levels of GLI (closed symbols) and IRG (open symbols) in (A) pancreas (squares) and intestine (triangles), and (B) hypothalamus (squares) and medulla oblongata (triangles), in 20-21 d fetal (F), 1-2, 7, and 11-21 d neonatal, and adult (A) rat. (C) The ratio of GLI to IRG in developing rat pancreas (closed squares), intestine (open squares), hypothalamus (closed triangles) and medulla oblongata (open triangles).

with the predominant peptide synthesized being the major proglucagon fragment containing GLP-1 and GLP-2 (Fig. 1). These peptides are separated in the intestine, resulting primarily in a truncated form of GLP-1 [GLP-1(7-36NH$_2$)] and GLP-2. Processing in the adult rat hypothalamus appears to be intermediate between that of pancreas and intestine, with both MPGF and GLP-1(1-37) detectable[33].

Immunocytochemical studies have suggested that post-translational processing of proglucagon in the pancreas may undergo developmental changes. Only immunoreactive glicentin was detected in 1st trimester human fetal islets, whereas the mature human A cell stains for both glucagon and glicentin[34]. We have therefore examined proglucagon post-translational processing during development in the rat[8,16,35]. Adult patterns of post-translational processing were detected in fetal and neonatal rat pancreas and intestine, with glucagon predominating in pancreas, and glicentin and oxyntomodulin in intestine, at all ages studied. Distinct profiles of tissue concentrations of these peptides were found, however, during development. GLI concentrations in the intestine were low in the fetus but increased with development, whereas immunoreactive glucagon levels peaked in the neonatal pancreas and then decreased towards adult levels (Fig. 8A). The ratio of GLI to IRG peptides did not change in either tissue during development (Fig. 8C). Thus, although the patterns of pGdp accumulation differed between pancreas and intestine, no evidence for altered processing of proglucagon in either the developing pancreas or intestine could be detected.

As found in the developing intestine, hypothalamic GLI peptides increased during development in the rat (Fig. 8B). In marked contrast to the fetal intestine, however, significant amounts of glucagon were detected in the fetal hypothalamus (Fig. 4E vs 4A). Interestingly, the levels of glicentin and oxyntomodulin relative to glucagon increased progressively towards adult levels in the post-natal hypothalamus. Thus, the ratio of GLI to IRG peptides increased in the hypothalamus with development, from 2.6 $\pm$ 0.5 in the fetus to 46 $\pm$ 11 in the adult

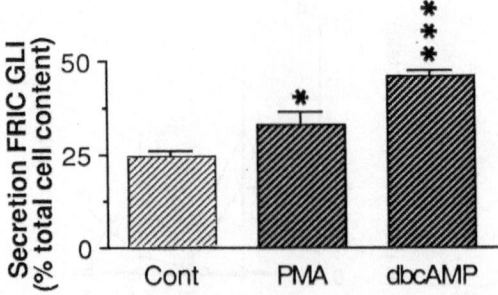

Fig. 9.   Secretion of GLI peptides by FRIC cultures during 2 hr of incubation under either control conditions or with 1 uM PMA or 5 mM dbcAMP. Other details as in Fig. 5A.

(P<0.05). A similar but smaller increase in the GLI to IRG ratio was also noted in the developing medulla oblongata (Fig. 8C). The brain stem differed from the other pGdp-producing tissues, however, in that peptide levels were highest in the fetus and decreased during development.

Thus, increased production of glucagon relative to the other GLI peptides has been demonstrated in the fetal hypothalamus and brain stem, suggesting a developmental role for glucagon in the immature brain. Furthermore, distinct patterns of peptide accumulation have been noted for all of the pGdp-producing tissues, consistent with diverse roles for these peptides in development. The factors that regulate these changes are currently not known.

4. SECRETION OF THE PROGLUCAGON-DERIVED PEPTIDES

Pancreatic glucagon secretion is well known to be regulated by glycemia, insulin, somatostatin and epinephrine, all factors involved in carbohydrate homeostasis[36]. In contrast, the regulation of intestinal pGdp secretion has been only poorly characterized, due to the complexity of this endocrine organ, and to a lack of suitable in vitro models. Similarly, the regulation of hypothalamic pGdp secretion has not been examined to date. We have therefore utilized the FRIC and FRHC culture models to investigate the control of intestinal and hypothalamic pGdp secretion.

4.1 Intracellular Signals

To establish which of the 2nd messenger pathways are involved in the transduction of extracellular signals in the intestinal L cell, FRIC cultures were treated for 2 hr with agents to activate PKA (dbcAMP, forskolin, theophylline or isobutylmethylxanthine), calcium fluxes (ionomycin, A23187) or PKC (PMA)[13] (Fig. 9). A significant and additive 2- to 3-fold stimulation of pGdp secretion was observed in response to each of the signal activators. Furthermore, stimulation of pGdp secretion did not alter proglucagon post-translational processing in the cells, such that glicentin and oxyntomodulin were the predominant secretory products of the intestinal L cell under both basal and stimulated conditions. The ability to

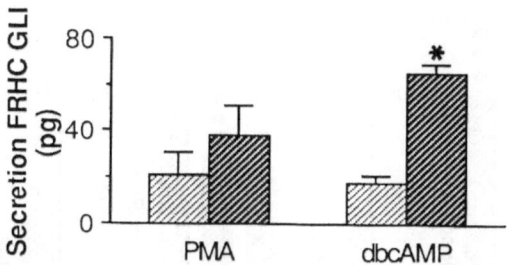

Fig. 10.   Secretion of GLI peptides by FRHC cultures during 24 hr of incubation under either control conditions or with 1 uM PMA or 5 mM dbcAMP. Other details as in Fig. 5A.

modulate pGdp secretion through three distinct 2nd messenger pathways may provide the L cell with a means by which an integrated response to a diversity of extracellular signals can be provided.

Hypothalamic cultures were also examined for the effects of PKA and PKC on pGdp secretion, as for the intestinal cells. Interestingly, although dbcAMP treatment increased peptide release 3-fold, no significant effect of PKC activation could be demonstrated (Fig. 10)[28]. Thus, the intracellular regulation of hypothalamic pGdp secretion appears to differ from that of the intestinal L cell at the level of PKC.

Finally, in a similar examination of the effects of PKA activation on the pancreatic islet, glucagon secretion was stimulated above basal levels only under conditions of low glucose concentrations (2.5 mM) (Fig. 11). Thus, the response of the A cell to PLA appears to be modulated by the prevailing levels of glucose.

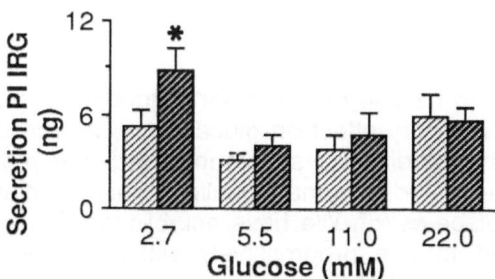

Fig. 11.   Secretion of IRG peptides by cultured pancreatic islets during 24 hr incubation with varying concentrations of glucose under either control conditions or with 10 uM forskolin plus 10 uM IBMX. Other details as in Fig. 5A.

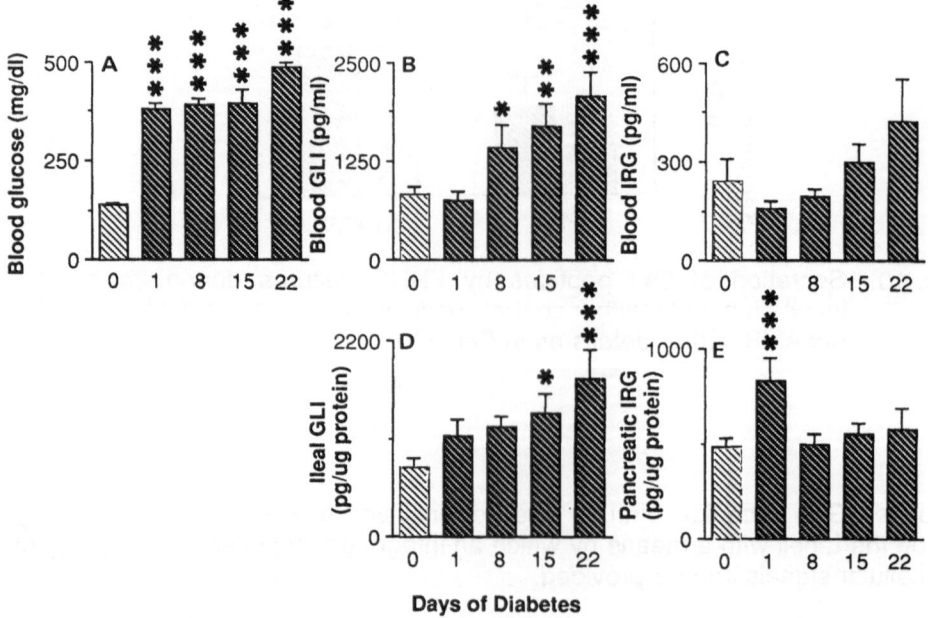

Fig. 12. Plasma levels of glucose (A), GLI (B) and IRG (C) peptides, and concentrations of ileal GLI (D) and pancreatic IRG (E) peptides, during the first 3 weeks of streptozotocin-induced diabetes.

Our results indicate that some of the tissue-specific differences between the pGdp-secreting tissues are related to the intracellular signals that regulate secretion of the peptides. For example, PKA modulates intestinal, hypothalamic and pancreatic pGdp secretion, whereas intestinal but not hypothalamic cells are responsive to PKC. A further degree of complexity is added when the effects of glycemia on the responses of the pancreatic A cell are considered.

4.2 Extracellular Signals

Insulin appears to be one of the major regulators of pGdp synthesis and secretion, having an inhibitory effect on glucagon gene transcription in hamster InR1-G9 cells[22]. In vivo studies have demonstrated that plasma levels of the intestinal pGdp's are elevated in human insulin-dependent diabetes mellitus and in animal models of diabetes[37,38]. We have established that circulating levels of intestinal pGdp's start to rise progressively within 1 week of the onset of streptozotocin-diabetes, and that these changes occur in parallel with increases in the intestinal concentrations of pGdp's (Fig. 12)[7]. In contrast to the gut pGdp's, the pancreatic levels of glucagon rise only transiently in diabetes, and brain levels of pGdp's fluctuate slightly. Thus, diabetes is associated with tissue-specific increases in both the synthesis and the secretion of the intestinal pGdp's. The intracellular mechanisms underlying these changes remain to be elucidated.

Table 1. Secretion of GLI peptides by FRIC cultures in response to treatment for 2 hr with various peptides. Data is expressed as a percent of control secretion.

Peptide	n	Minimal Effective Dose (M)	Secretion	P
CGRP	5	10^{-8}	167±37	0.05
GIP	5	10^{-10}	230±82	0.05
GRP	5	10^{-10}	160±20	0.01
Galanin	8	10^{-12}	82± 9	0.05
S-14	6	10^{-8}	81±11	0.0001
S-28	6	10^{-10}	57± 6	0.0001
CCK-8	6	10^{-6}	92±15	NS
NT	5	10^{-6}	90± 9	NS
PYY	5	10^{-6}	100±17	NS
VIP	6	10^{-6}	145±28	NS

We have also recently examined various intestinal regulatory peptides for effects on intestinal pGdp secretion[39]. Gut regulatory peptides were found to fall into 3 groups, based upon their effects on intestinal pGdp secretion: stimulatory (calcitonin gene-related peptide, CGRP; gastric-inhibitory peptide, GIP; and gastrin-releasing peptide, GRP), inhibitory (galanin; somatostatin-14, S-14; and S-28), and ineffective (cholecystokinin, CCK; neurotensin, NT; peptide YY, PYY; and vasoactive intestinal peptide, VIP) (Table 1). Interestingly, each of the inhibitory peptides was an enteric neuropeptide that exerts effects through PKA (as determined by pertussis toxin sensitivity of the pGdp response). In contrast, stimulatory peptides were endocrine (GIP) or neural (CGRP, GRP) in origin, and acted through either PKA (CGRP, GIP) or PKC (GRP).

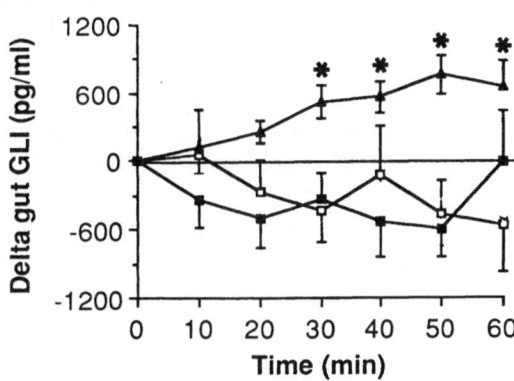

Fig. 13. Changes in plasma gut GLI peptides in response to perfusion of the rat ileum with 0.9% saline (open squares), 200 mM glucose (closed squares) or an emulsified fat:bile salt solution (closed triangles). Other details as in Fig. 5A.

The effects of nutrients on secretion of the intestinal pGdp's have also been examined, as the intestinal L cell is an 'open' type epithelial cell and is therefore exposed to the luminal contents. In FRIC cultures, glucose concentrations of up to 11 mM did not signficantly alter pGdp secretion. In contrast, sodium oleate (0.1 mM) induced a dose-dependent 2-fold increment in pGdp secretion (P<.05). To confirm that nutrients exert their effect on the intestinal L cell through the luminal membrane, we have perfused the ileal lumen with glucose (200 mM) or emulsified corn oil emulsified with bile salts and have examined the effects on intestinal pGdp secretion[40]. Glucose did not alter gut pGdp secretion, whereas emulsified fats increased circulating intestinal pGdp levels signficantly (Fig. 13), confirming our in vitro findings.

A model derived from the results of our studies on the intra- and extracellular regulation of the intestinal L cell is shown in Fig. 14.

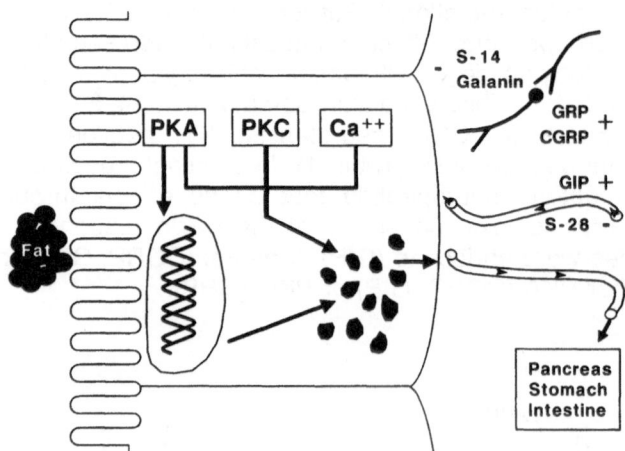

Fig. 14. A model for the regulation of proglucagon-derived peptide synthesis and secretion by the intestinal L cell. All peptides exert effects through PKA except GRP which acts through PKC.

5. SUMMARY

Using several novel in vitro culture systems, we have examined the tissue-specific regulation of the proglucagon-derived peptides, at the levels of proglucagon gene expression and pGdp synthesis and secretion. Our studies indicate that proglucagon gene expression in intestine, hypothalamus and pancreas is under the regulatory control of protein kinase A- but not a protein kinase C-dependent pathway. PKA and PKC stimulate secretion of the intestinal pGdp's, whereas only PKA stimulates secretion of the hypothalamic peptides. Pancreatic glucagon secretion in response to PKA is subject to further modulation by prevailing glucose concentrations. This diversity in intracellular regulation of the pGdp's may account for some of the tissue-specific differences in synthesis and secretion of the pGdp's that we have observed in diabetes and during development.

6. REFERENCES

1. U. Novak, A. Wilks, G. Buell and S. McEwen, Identical mRNA for preproglucagon in pancreas and intestine, Eur J Biochem. 164:553 (1987).

2. D.J. Drucker and S. Asa, Glucagon gene expression in vertebrate brain, J Biol Chem. 263:13475 (1988).

3. D.J Drucker and P.L. Brubaker, Proglucagon gene expression is regulated by a cyclic AMP-dependent pathway in rat intestine, Proc Natl Acad Sci USA. 86:3953 (1989).

4. G.I. Bell, R.F. Santerre and G.T. Mullenbach, Hamster preproglucagon contains the sequence of glucagon and two related peptides, Nature. 302:716 (1983).

5. L.C. Lopez, M.L. Frazier, C-J. Su, A. Kumar and G.F. Saunders, Mammalian pancreatic preproglucagon contains three glucagon-related peptides, Proc Natl Acad Sci USA. 80:5485 (1983).

6. G. Heinrich, P. Gros and J.F. Habener, Glucagon gene sequence: four of six exons encode separate functional domains of rat pre-proglucagon, J Biol Chem. 259:14082 (1984).

7. P.L. Brubaker, D.C.Y. So and D.J. Drucker, Tissue- specific differences in the levels of proglucagon- derived peptides in streptozotocin-induced diabetes, Endocrinol. 124:3003 (1989).

8. Y.C. Lee, P.L. Brubaker and D.J. Drucker, Developmental and tissue-specific regulation of proglucagon gene expression, Endocrinol. 127:In Press (1990).

9. L-I. Larsson, J. Holst, R. Hakanson and F. Sundler, Distribution and properties of glucagon immunoreactivity in the digestive tract of various mammals: an immunohistochemical and immunochemical study, Histochem. 44:281 (1975).

10. A.M.J. Buchan, D.L. Barber, M. Gregor and A.H. Soll, Morphologic and physiologic studies of canine ileal enteroglucagon-containing cells in short-term culture, Gastroenterol. 93:791 (1987).

11. D.L. Barber, A.M.J. Buchan, J.H. Walsh and A.H. Soll, Isolated canine ileal mucosal cells in short-term culture: a model for study of neurotensin release, Am J Physiol. 250:G374 (1986).

12. A.H. Soll, T. Yamada, J. Park and L.P. Thomas, Release of somatostatinlike immunoreactivity from canine fundic mucosal cells in primary culture, Am J Physiol. 247:G558 (1984).

13. P.L. Brubaker and M. Vranic, Fetal rat intestinal cells in monolayer culture: a new in vitro system to study the glucagon-related peptides, Endocrinol. 120:1976 (1987).

14. H. Tager, M. Hohenboken, J. Markese and R.J. Dinerstein, Identification and localization of glucagon-related peptides in rat brain, Proc Natl Acad Sci USA. 77:6229 (1980).

15. S-L.C. Jin, V.K.M. Han, J.G. Simmons, A.C. Towle, J.M. Lauder and P.K.

Lund, Distribution of glucagonlike peptide 1 (GLP-1), glucagon, and glicentin in the rat brain: an immunocytochemical study, J Comp Neurol. 271:519 (1988).

16. E.Y.T. Lui, S.L. Asa, D.J. Drucker, Y.C. Lee and P.L. Brubaker, Glucagon and related peptides in fetal rat hypothalamus in vivo and in vitro, Endocrinol. 126:110 (1990).

17. D.J. Drucker, R. Reynolds, K. Stobie, R. Campos and P.L. Brubaker, The rat glucagon gene is regulated by a cyclic AMP-dependent pathway in pancreatic islet cells, Endocrinol. 128:in press (1991).

18. J. Philippe, D.J Drucker and J.F. Habener, Glucagon gene transcription in an islet cell line is regulated via a protein kinase C-activated pathway, J Biol Chem. 262:1823 (1987).

19. D.J. Drucker, J. Philippe, L. Jepeal and J.F. Habener, Glucagon gene 5'-flanking sequences promote islet cell-specific gene transcription, J Biol Chem. 262:15659 (1987).

20. J. Philippe, D.J Drucker, W.L. Chick and J.F. Habener, Transcriptional regulation of genes encoding insulin, glucagon, and angiotensinogen by sodium butyrate in a rat islet cell line, Mol Cell Biol. 7:560 (1987).

21. J. Philippe, D.J. Drucker, W. Knepel, L. Jepeal, Z. Misulovin and J.F. Habener, Alpha-cell-specific expression of the glucagon gene in conferred to the glucagon promoter element by the interactions of DNA-binding proteins, Mol Cell Biol. 8:4877 (1988).

22. J. Philippe, Glucagon gene transcription is negatively regulated by insulin in a hamster islet cell line, J Clin Invest. 84:672 (1989).

23. D.J Drucker, J. Philippe and S. Mojsov, Proglucagon gene expression and posttranslational processing in a hamster islet cell line, Endocrinol. 123:1861 (1988).

24. M.R. Montminy, K.A. Sevarino, J.A. Wagner, G. Mandel and R.H. Goodman, Identification of a cyclic AMP-responsive element within the rat somatostatin gene, Proc Natl Acad Sci USA. 83:6682 (1986).

25. K.K. Yamamoto, G.A. Gonzalez, W.H. Biggs III and M.R. Montminy, Phosphorylation-induced binding and transcriptional efficacy of nuclear factor CREB, Nature. 334:494 (1988).

26. P.J. Deutsch, J.P. Hoeffler, J.L. Jameson and J.F. Habener, Cyclic AMP and phorbol ester-stimulated transcription mediated by similar DNA elements that bind distinct proteins, Proc Natl Acad Sci USA. 85:7922 (1988).

27. P.L. Brubaker, Control of glucagon-like immunoreactive peptide secretion from fetal rat intestinal cultures, Endocrinol. 123:220 (1988).

28. K.M. Stobie and P.L. Brubaker, Regulation of the production of glucagon-like peptides in the fetal hypothalamus, in: The Nervous System and Fuel Homeostasis: 1st Toronto-Stockholm Symposium on Perspectives in Diabetes Research, M. Vranic and S. Effendic, ed., Plenum Press, Toronto, 21A (1990).

29. W. Knepel, J. Chafitz and J.F. Habener, Activation of glucagon gene expression by cAMP-dependent protein kinase A in a pancreatic cell line, 72nd Annual Meeting of the Endocrine Society. 533A (1990).

30. S.K. George, L.O. Uttenthal, M. Ghiglione and S.R. Bloom, Molecular forms of glucagon-like peptides in man, FEBS Lett. 192:275 (1985).

31. C. Orskov, J.J. Holst, S. Knuhtsen, F.G.A. Baldissera, S.S. Poulsen and O.V. Nielsen, Glucagon-like peptides GLP-1 and GLP-2, predicted products of the glucagon gene, are secreted separately from pig small intestine but not pancreas, Endocrinol. 119:1467 (1986).

32.   S. Mojsov, M.G. Kopczynski and J.F. Habener, Both amidated and nonamidated forms of glucagon-like peptide I are synthesized in the rat intesine and the pancreas, J Biol Chem. 265:8001 (1990).

33.   K. Suda, H. Takahashi, N. Fukase, H. Manaka, M. Tominaga and H. Sasaki, Distribution and molecular forms of glucagon-like peptide in the dog, Life Sci. 45:1793 (1989)

34.   Y. Stefan, S. Grasso, A. Perrelet and L. Orci, A quantitative immunofluorescent study of the endocrine cell populations in the developing human pancreas, Diabetes. 32:293 (1983).

35.   P.L. Brubaker, Ontogeny of the glucagon-like immunoreactive peptides in rat intestine, Reg Pep. 17:319 (1987).

36.   R.H. Unger and L. Orci, Glucagon and the A cell: physiology and pathophysiology (first of two parts), New Engl J Med. 304:1518 (1981).

37.   M.D. Wider, T. Matsuyama, J.C. Dunbar and P.P. Foa, Elevated gut glucagon-like immunoreactive material in human and experimental diabetes and its suppression by somatostatin, Metabol. 25 (Suppl. 1):1487 (1976).

38.   B. Kreymann, Y. Yiangou, S. Kanse, G. Williams, M.A. Ghatei and S.R. Bloom, Isolation and characterisation of GLP-1 7-36 amide from rat intestine. Elevated levels in diabetic rats, FEBS Lett. 242:167 (1988).

39.   P.L. Brubaker and Q.S. Fong, Regulation of intestinal proglucagon-derived peptide secretion by gut regulatory peptides, 72nd Annual Meeting of the Endocrine Society. 591A (1990).

40.   J.N. Roberge and P.L. Brubaker, Luminal nutrient effects on ileal glucagon-like immunoreactive peptide secretion in diabetes, Diabetes. 38(Suppl. 2):894A (1989).

EFFECT OF STRESS ON GLUCOREGULATION

IN PHYSIOLOGY AND DIABETES

M. Vranic[+], P. Miles, K. Rastogi, K. Yamatani[++], Z. Shi, L. Lickley[+],
G. Hetenyi, Jr.[#]

Departments of Physiology, Medicine and Surgery[+]
University of Toronto, and Women's College Hospital
Toronto, Canada,   Department of Physiology[#], University of Ottawa
Canada,   Yamagata University School of Medicine[++], Japan

ABSTRACT

To examine the glucoregulatory responses to stress and their impact on
diabetes, we used the following models of stress: A) Hypoglycemia; B)
Epinephrine infusion; C) Intracerebroventricular (ICV) injection of carbachol, an
analog of acetylcholine.

A) Hypoglycemia induces release of all counterregulatory hormones. During
acute hypoglycemia, glucose production increases initially mainly due to
glucagon release but eventually also due to a very large increment in
catecholamines. In newborn dogs, neither epinephrine nor glucagon respond to
a decrease in plasma glucose. This lack of a safeguard against hypoglycemia
may indicate that the brain in pups is less dependent on a normal supply of
glucose as a fuel, than in adult dogs. Counterregulation is enhanced when the
effects of endogenous opiates are blocked by naloxone, indicating that
endogenous opiates play a regulatory role during hypoglycemia. However, beta-
endorphins which can be released with epinephrine during various stress
situations, potentiate the peripheral effect of epinephrine. Glucoregulatory
responses, even to slight changes in plasma glucose, are greatly enhanced
during glucocorticoid treatment. This apparently reflects the greater sensitivity of
the liver to glucagon. In diabetic dogs, similar to human diabetics, the glucagon
response is abolished and the response of the catecholamines is partially
decreased. On the basis of histological studies, we proposed that the deficient
glucagon response in diabetes could be related to an increase in the
somatostatin-glucagon ratio in the diabetic pancreas. This ratio is further
augmented when normoglycemia is maintained with insulin. In response to a
decrease in plasma glucose, there is a biphasic increment in glucose production
in normal dogs, which is missing in diabetes. When normoglycemia is restored
in diabetic dogs with phlorizin treatment, the second but not the first increment
in glucose production is restored. We postulated, therefore, that the toxic effect
of hyperglycemia, in addition to the lack of glucagon response, is the main
reason why in diabetes, glucose production cannot respond promptly to a
decrease in plasma glucose. The low rate of metabolic clearance of glucose
seen in diabetes in the post-absorptive state, also reflects, at least in part, the
toxic effect of glucose, because with acute normalization of glucose with

phlorizin, metabolic glucose clearance substantially improves. Hyperglycemia is the main reason for the decreased number of glucose transporters in diabetic muscle.

B) Epinephrine infusion in normal dogs mimics some effects of stress, in that it increases glucose production, inhibits metabolic glucose clearance and increases lipolysis. These metabolic effects of epinephrine are independent of glucagon release. In diabetes, however, epinephrine-induced hyperglycemia is exaggerated which is mainly due to glucagon. This occurs both because of excessive glucagon release, and increased hepatic sensitivity to the effects of glucagon related to hypoinsulinemia.

C) ICV injection of a small amount of carbachol induces a release of all counterregulatory hormones. Interestingly, insulin secretion is not affected, possibly because the alpha- and beta-adrenergic pancreatic inputs are balanced. Surprisingly, this release of counterregulatory hormones induces only a marginal change in plasma glucose, since increased glucose production is matched by a similar increase in glucose uptake. In contrast, in hyperglycemic diabetic dogs, the same carbachol injection induces a much larger increment in plasma glucose. This occurred because the metabolic clearance rate of glucose did not increase. We therefore postulated a neural mechanism which controls peripheral glucose uptake and requires a permissive effect of insulin. Somatostatin, injected ICV before carbachol, abolishes most counterregulatory responses as well as the increment in glucose turnover. Lipolysis is also decreased but FFA re-esterification is abolished, reflecting a decrease in glucose uptake in the adipocyte.

INTRODUCTION

This chapter summarizes our work related to glucoregulation in stress and is not meant to be a comprehensive review of the literature. We have examined the following models of stress: A) Hypoglycemia induced by insulin or phlorizin, in both normal and diabetic dogs. We examined the role of counterregulatory hormones and in addition, the role of endogenous opiates and the effect of chronic treatment with methyl-prednisolone are discussed. Histological examination of the pancreas in diabetic dogs under hyperglycemic and normoglycemic conditions resulted in a new hypothesis regarding the mechanism of glucagon responsiveness to hypoglycemia in diabetic dogs. B) Epinephrine infusions were used as a partial model of stress and the mechanism of exaggerated hyperglycemic response in diabetes was examined. C) Intracerebroventricular injections of an analog of acetylcholine were used to induce mild stress in conscious dogs. This led to a new hypothesis which postulates an adrenergic neural mechanism which controls peripheral glucose uptake. This putative neural mechanism requires a permissive effect of insulin.

HYPOGLYCEMIA

Hormonal and metabolic responses

Hypoglycemia provides a model for moderate to severe stress. Initially, glucagon and the catecholamines are the hormones that safeguard against hypoglycemia and which play a role in recovery from hypoglycemia. The release of cortisol and growth hormones also contribute to glucoregulation in prolonged hypoglycemia[1-7]. The release of glucagon is modulated by the autonomic nervous system and by the direct action of glucose and insulin on the alpha cells. The release of the other regulatory hormones, however, is mainly by neural mechanisms activated by insufficient supply of glucose to the brain (neuro-glucopenia). The relative importance of glucagon and epinephrine depends on the degree and duration of hypoglycemia and on both the concentration and rate of change of plasma insulin.

If hypoglycemia is due to excess insulin, then insulin itself is likely to interfere with the counterregulatory response, primarily by its effect on the release of glucagon and insulin from the islets and possibly also of catecholamines from the adrenals and the sympathetic nerve endings. In addition, hyperinsulinemia decreases the liver's sensitivity to the counterregulatory hormones.

To distinguish between the effects of insulin and of hypoglycemia proper on glucoregulatory responses, phlorizin was infused to non-anaesthetized dogs[5]. The ability of phlorizin to lower plasma glucose independently of insulin also became a valuable tool in assessing the effects of hyperglycemia in diabetes[8-11]. Phlorizin inhibits the absorption of glucose in the renal tubuli and thus increases the rate of glucose removal from the circulation. In the dose applied, phlorizin does not interfere with either the release of hormones or metabolic processes in the liver[5]. During the phlorizin infusion, the concentration of glucose in plasma settled within 15-20 minutes at a new steady level of approximately only 15% below its pre-infusion concentration, since a compensatory increase in hepatic glucose production matched the increased removal of glucose at the new equilibrium level. This small decrease in glucose concentration was accompanied by a small decrease in plasma insulin and a threefold increase in plasma glucagon concentration. When the latter was abolished by the simultaneous infusion of somatostatin, the increase in hepatic glucose production was also abolished and the concentration of plasma glucose dropped to 60 mg/dl, i.e. to overt hypoglycemia. These experiments revealed glucagon as the physiological regulator of plasma glucose in the range near the euglycemic level.

When the concentration of glucose in the plasma was reduced to about 60 mg/dl by an infusion of insulin, and plasma IRI was raised to 300 µU/ml, the concentration of plasma glucagon nearly doubled. However, while somatostatin abolished the rise in plasma glucagon, it failed to decrease the rate of hepatic glucose production. Although in these experiments plasma catecholamines were not monitored, the observed rise in plasma FFA concentration was interpreted as an index of an increase in plasma epinephrine. The interpretation of the results was that contrary to the regulation near euglycemia, in overt hypoglycemia catecholamines are the main regulatory hormones. Their effect would be to prevent the fall in plasma glucose to life-threatening levels. In similar experiments a 40-fold increase in plasma epinephrine and a 3-fold increase in plasma norepinephrine were observed in insulin-induced hypoglycemia[12].

To assess the role of counterregulatory hormones in regulating gluconeogenesis in response to an increased rate of glucose removal from the plasma, phlorizin was infused in 4-day fasted dogs, in which hepatic glucose release (Ra) mainly represents the rate of gluconeogenesis. Ra quickly doubled simultaneously with the initial fall in plasma glucose. Gluconeogenesis from alanine[13] and from glycerol[14] increased. Isotopic studies have demonstrated that the contribution of glycerol to gluconeogenesis depends primarily on the amount of glycerol made available by lipolysis, whereas the contribution of alanine to gluconeogenesis is mainly due to the increased fraction of alanine converted to glucose in the liver. When phlorizin was injected, plasma glucose declined by about 20%. This small decrease in glucose was accompanied by an increase in plasma glucagon, ACTH and cortisol, but not in epinephrine and norepinephrine concentration. The combined effect of a decrease in glucose and insulin concentration might have been responsible for the increase in plasma glucagon. It has been shown that corticotropin-releasing factor (CRF) can stimulate both ACTH secretion and also activate sympathetic nervous outflow[15]. Since in our studies with phlorizin during mild hypoglycemia, the activation of the ACTH-cortisol axis occurred without the release of catecholamines[14], it appears that the activation of the sympathetic nervous system under a variety of stress stimuli may require more CRF than the activation of the ACTH-cortisol axis. During exercise, in contrast to the resting state, even a marginal drop in plasma glucose (5-10 mg%) can trigger a marked epinephrine response[16]. It would be interesting to know whether this reflects a synergistic effect of stimuli related to exercise and to hypoglycemia.

Hypoglycemia in the newborn

The range of concentration of plasma glucose varies widely in the neonate and no correlation between Ra and plasma glucose levels has been found in newborn dogs[17]. In 1-11 day old pups, neither glucagon nor adrenaline were mobilized in insulin-induced hypoglycemia, although the concentration of ACTH did increase in the plasma[18,19]. This finding indicated that the mechanisms responsible for the release of CRF and adrenaline are not identical: one being more sensitive to the lack of glucose than the other. Whereas the lesser dependence of the newborn brain on glucose offers a plausible hypothesis of the absence of adrenaline release, the lack of glucagon release[18] also indicates a reduced sensitivity of the alpha cells to changes in the concentration of glucose in the newborn than in the adult dog. This is even more evident in pups fasted for 1-2 days. Contrary to grown dogs, fasting plasma glucose in pups falls to hypoglycemic levels (39 ± 4 mg/dl in pups less than 4 days old) and the concentration of plasma IRG decreases in parallel. The likely biological significance of this response is in a reduction of energy expenditure (including that for gluconeogenesis) in an organism in which food supply may not be easily forthcoming at all times and in which maintaining a relatively high plasma glucose level is not critical for the functioning of the brain. It is noteworthy that in newborn pups the alpha cells do not respond to somatostatin either[18]. It is also conceivable that glucagon's unresponsiveness to hypoglycemia may also result because some step in glucagon biosynthesis is not yet fully developed.

Effect of blockade of endogenous opiates by naloxone

During both insulin-induced hypoglycemia[20], and strenuous exercise[21], there is a substantial rise in plasma β-endorphin levels. It has also been reported that β-endorphin can stimulate glucagon release from the pancreas[22]. We have also observed that infusions of β-endorphins, both intracarotid and intrajugular, stimulate glucagon secretion independently of circulating catecholamines. They also increased cortisol release[23], presumably by activating the pituitary-adrenocortical axis. In contrast, when β-endorphin was administered intracerebroventricularly (ICV), there was a 3-fold increase in plasma epinephrine and a 6-fold increase in norepinephrine. This resulted in hyperglycemia, which was due to both an increase in glucose production and inhibition of glucose clearance. There was also an increase in plasma cortisol and insulin but no change in plasma glucagon[24]. This would suggest that β-endorphins affect glucagon release directly rather than through a central mechanism, while the stimulation of the sympathetic nervous system is largely central. Thus, the release of endorphins during hypoglycemia probably participates in the control of the release of some counterregulatory hormones. We have shown that during insulin-induced hypoglycemia, plasma β-endorphins increased 5-fold. Interestingly, with opiate-receptor blockade by naloxone during hypoglycemia, there was an earlier release of epinephrine, glucagon, β-endorphin, ACTH and cortisol as well as a greater release of glucagon and cortisol[12]. Enhanced release of counterregulatory hormones was observed at all levels of hypoglycemia (Figure 1). This resulted in a greater increase of glucose production, thus lessening the insulin-induced hypoglycemic excursion. This has been confirmed in man[25]. In the resting state, naloxone treatment increased glucagon, epinephrine and cortisol, and inhibited insulin release, but did not effect glucose kinetics and norepinephrine. The observation that some effects of naloxone, an inhibitor of opiates, are similar to the effects of the central infusion of β-endorphins[24] suggests that the effects of naloxone are exerted

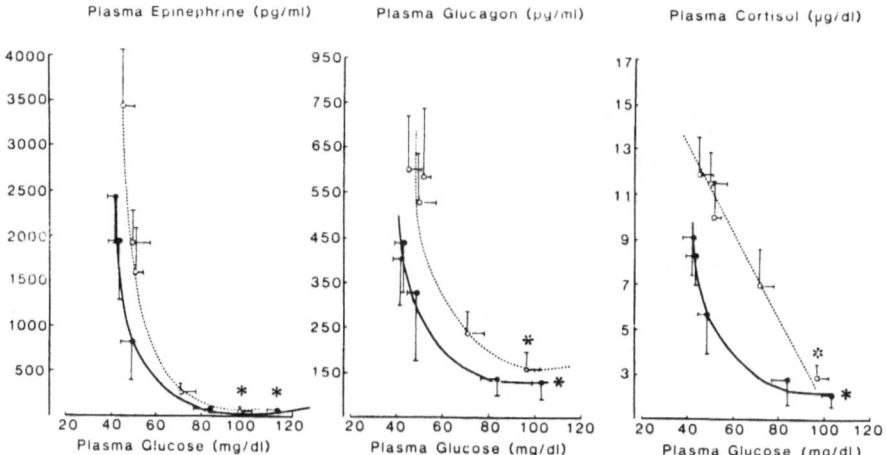

Figure 1. Relationships between plasma glucose and plasma epinephrine, glucagon, and cortisol in insulin-induced hypoglycemia with (O——O) and without naloxone (●——●). Asterisks represent average basal values taken before insulin infusion. Other points are taken at 10, 30, 50 and 70 min after the start of insulin infusion. (Reproduced with permission from reference 12).

through the blockade of peripheral rather than central opiate receptors, and/or that an opiate blockade that affects a large variety of opiates does not necessarily reflect a predominant suppression of β-endorphins.

Although β-endorphins infused peripherally do not affect epinephrine release, they can, nevertheless, potentiate the peripheral effect of epinephrine as illustrated in Figure 2. β-endorphins (0.3 or 0.06 µg/kg/min) were infused concurrently with two repeated epinephrine infusions (0.1 µg/kg/min), given to mimic a repeated stress stimulus. In the dog, in contrast to man, the early response to epinephrine is stimulation rather than inhibition of insulin release[27]. Interestingly, with the second infusion of epinephrine there was diminished insulin response (β-cell is apparently desensitized), resulting in a greater hyperglycemic response due to an exaggerated suppression of MCR. β-endorphin diminished both the insulin and glucagon responses to epinephrine. This resulted in a decrease in glucose production and greatly enhanced inhibition of glucose metabolic clearance. During the first epinephrine infusion, hyperglycemia was also exaggerated, but during the second epinephrine infusion β-endorphin suppressed glucose production, but this was balanced by a suppression of glucose clearance and glycemia was unaffected. The interaction between β-endorphins and epinephrine on both the liver and in the periphery seems to be due to modulation of epinephrine-induced insulin and glucagon release by β-endorphins rather than to a direct effect on the target organs. This was supported by similar studies in alloxan diabetic dogs, in which endorphins did not suppress epinephrine-induced insulin and glucagon release, and consequently, did not affect glucose kinetics[28]. The concept that the actions of β-endorphin on glucose kinetics are indirect is also supported by experiments demonstrating that β-endorphins did not affect basal or glucagon-stimulated glucose production from isolated liver cells in vitro. It is apparent that the release of endorphins during stress may have a variety of effects and it is particularly interesting to speculate on the physiological impact of their combined regulatory effect on the perception of pain and glucoregulation.

Effect of methyl-prednisolone

Injection of methyl-prednisolone (MP) at a dose of 4 mg/kg-day over one week increases the rate of hepatic production (Ra) significantly, with a marginal

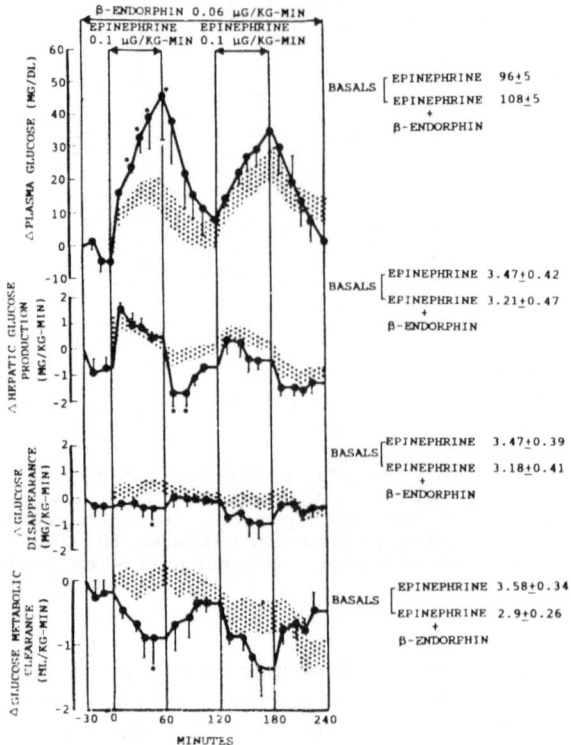

Figure 2. Effects of repeated 0.1 μg/kg-min epinephrine infusion alone (▓) or together with 0.6 μg/kg-min beta-endorphin (●——●), in normal dogs. Plasma glucose concentrations and rates of production, utilization, and metabolic clearance of glucose are shown as deviations (Δ) from the average control values indicated. Glucose production, utilization and metabolic clearance are plotted at midpoints of the time intervals for which they were calculated. Values are expressed as mean±SEM. N=9 for epinephrine infusion alone; N=4 for epinephrine and beta-endorphin infusions. Significant differences between the responses in the two groups are indicated (*). (Reproduced with permission from reference 26).

7% increase in the level of plasma glucose[29] . The elevated level of Ra has been interpreted to be the result of an increased rate of gluconeogenesis, compensated by an increased secretion of insulin[30]. Later experiments revealed that a similar effect can be elicited with 4 daily injections of 3 mg/kg-day[31], and that the increase in Ra is indeed due to gluconeogenesis, largely from an increased supply of precursors to the liver[32]. Thus MP-treated dogs offer the possibility of investigating non-hypoglycemic glucoregulation (by the infusion of phlorizin) in the presence of increased gluconeogenesis and hyperinsulinemia. In contrast to untreated dogs, in animals treated with MP (4 days, 3 mg/kg-day), during the infusion of phlorizin the decrease in plasma glucose was found to be small and transient, lasting less than 10 minutes. The increase in the disappearance of glucose from the circulation was almost instantaneously compensated by a rapid and appropriate increase in Ra[33]. The latter seemed to be triggered by only a 26% increase in the concentration of plasma IRG, compared to the increment of 150% observed in untreated animals[5]. This smaller increase in plasma IRG is presumably a reflection of the lack of the stimulus caused by the larger and longer decrease in plasma glucose observed in control animals. It also indicates a heightened sensitivity of the liver to glucagon (i.e. a response to a smaller ambient concentration) despite a steady supraphysiological level of insulin due to the glucocorticoid treatment.

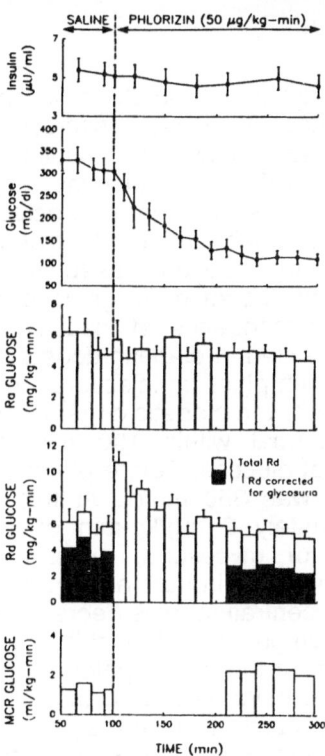

Figure 3. Effect of an infusion of phlorizin (50 µg/kg-min) on plasma concentration of insulin (IRI), glucose, rate of appearance (Ra) of glucose, and the rate of disappearance (Rd) in insulin-deprived alloxan diabetic dogs. Standard errors of means are shown as vertical bars. Correction for glycosuria was made only during steady state both before and at the end of phlorizin infusion. (Modified from Reference 9).

The rise in the concentration of glucagon in the plasma in response to insulin induced hypoglycemia was also less in MP-treated dogs than in untreated animals but the difference might be accounted for by a greater nadir of plasma glucose which did not drop below 60 mg/dl during the infusion, compared to 40 mg/dl in untreated dogs[5,33]. There was no significant difference in the release of catecholamines. The recovery to the basal level of plasma glucose after the cessation of the infusion of insulin in the MP-treated animals, however was faster, possibly a reflection of the high rate of gluconeogenesis that, aided by the increased sensitivity of the liver to an elevated level of plasma IRG, seemed to have overridden the expected contrary effect of hyperinsulinemia.

Effect of diabetes on hormonal and metabolic responses

It is well known that, in some diabetics, the release of glucagon in response to hypoglycemia is impaired, and as time progresses, the rise in plasma epinephrine compared with the response observed in normal subjects also becomes smaller[34-39]. Accordingly, the increase in the rate of glucose release is markedly reduced or absent. This creates a major problem in treatment of diabetic patients. On one hand, intensive insulin therapy is desirable because it minimizes daily fluctuations of glucose but on the other hand it increases the threat of hypoglycemia. The reason for this defect in counterregulation is not known, and the question was raised whether this was a specific defect of human diabetes. However, we have shown that a rather similar defect in glucoregulation occurs also in insulin-deprived alloxan diabetic dogs and this defect is observed as early as 1-2 weeks following alloxan administration[9,10]. Thus the impaired response in glucagon secretion is, at least in our animal model, a defect that appears soon after insulin deprivation. We could also observe some attenuation of the catecholamine response, however to a lesser extent than reported in patients in whom protracted diabetes may have led to a greater deterioration, presumably related to diabetic neuropathy.

In order to find out whether the decreased metabolic glucose clearance in diabetes and the deficient increase in glucose production in response to hypoglycemia were the consequence of insulin deficiency or hyperglycemia, plasma glucose was normalized in diabetic dogs either acutely with an infusion of phlorizin or chronically with phlorizin injections. In the first set of experiments, phlorizin was infused in 7 alloxan diabetic dogs[9] (Figure 3). Interestingly, plasma glucose became near normal within 100 minutes, as a consequence of increased renal excretion of glucose. Before phlorizin infusion, Rd was 6 mg/kg-min and plasma glucose was 350 mg/dl. At this level of glycemia we have determined that approximately 2 mg/kg-min of glucose is excreted by the kidney[39]. Rd, corrected for glycosuria, amounted to 4 mg/kg-min, which corresponds to glucose utilization in normal dogs[10]. However, metabolic glucose clearance (Rd/glucose concentration) was decreased by 60% (1.3 ml/kg-min, as compared to 3.3 ml/kg-min in normal dogs[10]), indicating a diabetes-induced defect in glucose transport. At the end of phlorizin infusion, the average Rd was also 5.4 mg/kg-min and we have measured that phlorizin induces glycosuria of 2.7 mg/kg-min[9] in normoglycemic alloxan diabetic dogs. Thus, with lowering of plasma glucose, glucose uptake by the tissues decreased only slightly. This is reflected in the MCR which equals 2.3 ml/kg-min which, when compared to normal dogs, is decreased by only 30%. This indicates that with normoglycemia, despite marked hypoinsulinemia (5.5 µU/ml), glucose transport improved. These in vivo data are similar to data related to measurements of the number of glucose transporters in the muscle of streptozotocin diabetic rats. We have shown[40] that in diabetes there is a decrease of glucose transporters both in plasma and in internal membranes which reflects decreased glucose clearance. When plasma glucose was normalized with phlorizin in diabetic rats, the number of glucose transporters measured with cytochalasin-B was normalized both in plasma and internal membranes. This corresponds to normal metabolic glucose clearance[11]. It should be noted that diabetic rats were less diabetic than alloxan diabetic dogs. Their insulin in the post-absorptive state was normal, while in diabetic dogs it was 70% reduced. These data indicate the possibility that, in the fasting state, glucose concentration, even in the presence of low plasma insulin, is an important regulator of glucose transport. With hyperglycemia, glucose transport becomes deficient but the mass effect of elevated plasma glucose can, at least compensate in part, for this deficiency. Hyperglycemia, in the post-absorptive state, could be explained in the following way: hypoinsulinemia causes increased glucose production which causes hyperglycemia. The excess glucose is removed by glycosuria. Glucose uptake by peripheral tissues remains unchanged because the mass action of glucose compensates for the lower number of glucose transporters.

In the second set of experiments, diabetic dogs were treated with intramuscular injections of phlorizin for 2 days prior to the experiment. Their glucoregulatory responses were compared to normal dogs and diabetic dogs which were not treated with phlorizin. In normal dogs, during the infusion of 10 mU/kg-min insulin for 150 minutes, the concentration of glucose in plasma stabilized at about 50 mg/dl whereas the concentration of glucagon in plasma increased by a factor of 3, the concentration of norepinephrine and dopamine by a factor of 4, and epinephrine was increased 15-fold. Ra increased in two phases, an early peak (350% basal) was followed by a plateau at about 2 times basal (Figure 4). In diabetic dogs it was necessary to increase the dose of

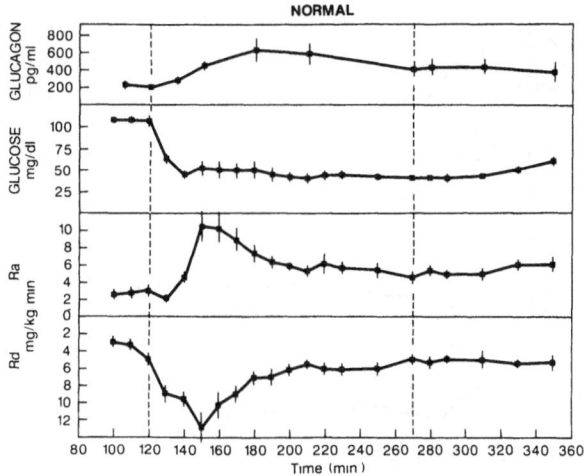

Figure 4. Effect of insulin-induced hypoglycemia on concentration of plasma glucagon (IRG) and glucose as well as on rates of hepatic glucose production (Ra) and disappearance (Rd) in normal dogs. Ordinates are as indicated. Abscissa: time in minutes from injection of tracer (t=0). Insulin (10 mU/kg-min) was infused between t=120 and 270 min. Average values ± SE of 7 experiments are shown. (Reproduced with permission from reference 10).

insulin up to 10-fold in order to achieve hypoglycemia comparable to normal dogs. In diabetes, irrespective of its duration, plasma IRG was decreased during hypoglycemia and the increase in epinephrine was significantly reduced (it increased by a factor of 2- to 10-fold). Ra remained unchanged (Figure 5). In phlorizin treated dogs the level of plasma glucose was reduced to normal (97±3 mg/dl). With insulin, Rd increased less than in normal dogs, indicating insulin resistance. With the normalization of plasma glucose, the glucagon levels in

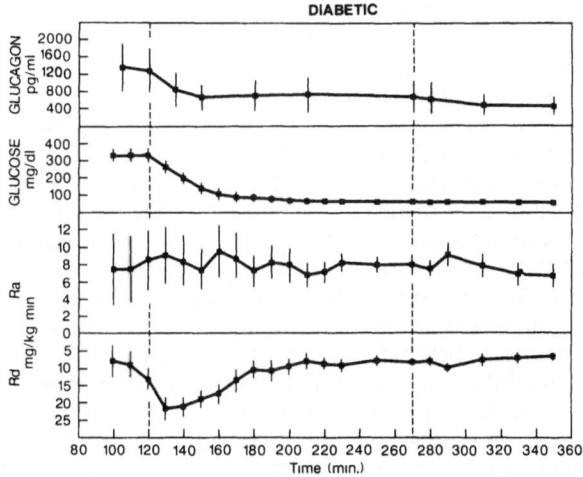

Figure 5. Effect of insulin-induced hypoglycemia on concentration of plasma glucagon (IRG) and glucose as well as on the rate of hepatic glucose production (Ra) and disappearance (Rd) in insulin-deprived diabetic dogs 14 days after induction of diabetes with alloxan. Insulin was infused between t=120 and 270 min. Ordinates and abscissa as on Figure 4. Average values ± SE of 4 experiments are shown. (Reproduced with permission from reference 10).

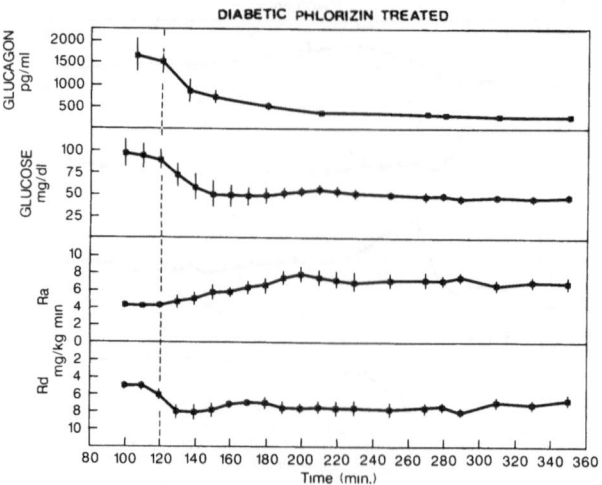

Figure 6. Effect on insulin-induced hypoglycemia on concentration of plasma glucagon (IRG) and glucose as well as on rate of hepatic glucose production (Ra) and disappearance (Rd) in insulin-deprived diabetic dogs treated with phlorizin. Insulin was infused between t=120 and 350 min. Ordiantes and abscissa as on Figure 4. Average values $\pm$ SE of 5 experiments are shown. (Reproduced with permission from Reference 10).

plasma remained high and were reduced by insulin infusion. The responses of epinephrine and norepinephrine remained blunted. There was no early rise in Ra but eventually the same elevated plateau was reached at the same time as in normal animals (Figure 6).

We concluded that in diabetic dogs, plasma glucagon concentrations decreased because the sensitivity of the α-cell to insulin is maintained but that to hypoglycemia is lost. The increase in plasma catecholamines is less than in normals, and with no parallel increase in plasma glucagon, is not sufficient to increase Ra. It seems that the lack of the first (transient) increase in Ra reflects the lack of an increase in plasma IRG, because we have previously found that a biphasic response in glucose production occurs when glucagon and epinephrine are infused simultaneously[41]. Euglycemia, however, partially restores the response of the liver to catecholamines, perhaps by eliminating the inhibitory effect of glucose itself on gluconeogenesis. High ambient glucose concentration may inhibit gluconeogensis by decreasing the activity of pyruvate carboxylase[42] and/or by a putative increase in the level of D-fructose-2-6-diphosphate, especially when the synthesis of glycogen is inhibited[43], as it is in diabetes. The effect of the catecholamines on gluconeogenesis is not impaired in glycogen depleted liver cells taken from alloxan diabetic animals[44].

As shown in Figure 3 infusion of phlorizin in diabetic dogs decreased plasma glucose to near normal levels within 100 minutes. During this phlorizin infusion over 300 minutes, glucose production did not increase. In contrast, however, when the animals were intensely treated with insulin for one week, and plasma glucose was restored to normal, the infusion of phlorizin did not affect plasma glucose concentrations, because glucose production increased simultaneously with an increase in glucose removal. There was also a significant increase in plasma glucagon. It seems, therefore, that in intensive insulin treatment, glucoregulatory responses can be largely restored[9].

A new hypothesis regarding the mechanism of glucagon irresponsiveness to hypoglycemia in diabetic dogs - histochemical study

We have argued that in diabetic dogs the impaired glucagon release in

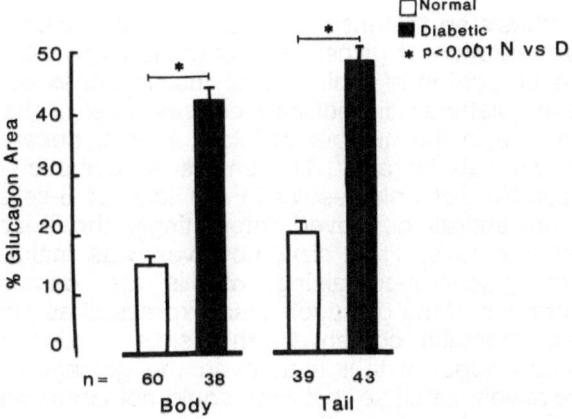

Figure 7. Morphometric analysis of glucagon area as percent of pancreatic islet area in 4 normal and 4 hyperglycemic alloxan diabetic dogs. Values are presented as mean±SEM. (Reproduced with permission from reference 45).

response to hypoglycemia could either result from the observed decreased adrenergic response or that some change in responsiveness in the diabetic islets. We therefore examined the insulin and glucagon content and morphology of the pancreatic islets in normal dogs, hyperglycemic alloxan diabetic dogs and in alloxan diabetic dogs where normoglycemia was maintained by insulin[45]. In diabetic dogs the insulin content decreased to 1% of normal (from 5519±958 µg/gm to 55.2±23.1 µg/gm). Restoration of euglycemia did not change insulin content. We hypothesised that restoration of euglycemia with insulin, might have decreased endogenous insulin release and that the concentration of insulin in the pancreas would increase. On the other hand, normalization of glucose concentration and the increase of insulin concentration in the plasma may

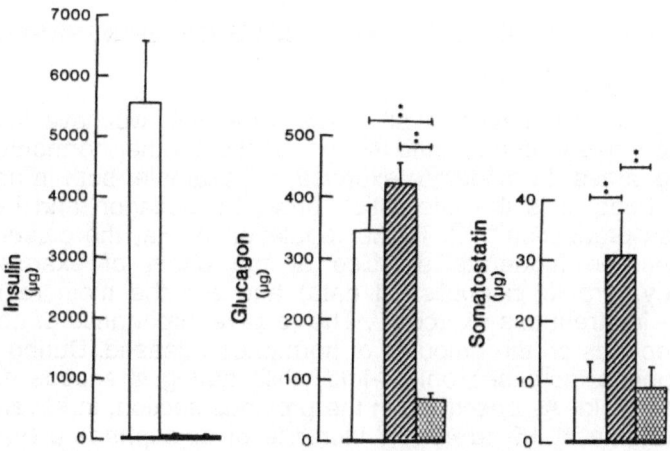

Figure 8. Total amount of pancreatic insulin, glucagon and somatostatin in normal (□), and alloxan diabetic dogs, hyperglycemic-hypoinsulinemic (▨, 48 hrs after insulin injection) and normoglycemic-normoinsulinemic (▨ , insulin treated), p<0.03. (Modified from Reference 45).

decrease insulin biosynthesis so that insulin storage would decrease. The latter proved to be true. Normalization of glycemia in diabetic dogs did not increase insulin content and the proportion of proinsulin to insulin (index of biosynthetic activity) did not increase. Diabetes did not only destroy most of the β-cells, but there was a 70% decrease in the number of islets. Similar observations were made in the streptozotocin diabetic rat[46,47]. This shows the diabetic state induced by alloxan or streptozotocin not only results in the loss of β-cells but that a large number of islets are entirely destroyed. Interestingly, the mean size of the surviving islets did not decrease. Their size, however, was maintained by an increased number of glucagon-containing α-cells as demonstrated by morphometric measurements of the glucagon area expressed as percent of islet area. Although the somatostatin content of the islets is also increased in diabetes, and to a much larger extent than even the glucagon content, the number of D-cells is relatively small so that they could not contribute in a major way to the maintenance of mean islet size. The glucagon area, expressed as percent of the islet area, increased 3-fold (Figure 7). Despite a 70% loss of islet tissue, total glucagon content did not decrease (Figure 8), indicating a 3-fold increase in glucagon content in each residual islet. This also indicates that the augmented amount of glucagon in each islet is due to hyperplasia of α-cells and not to an increased content of glucagon in each α-cell. Somatostatin content increased 3-fold, or by a factor of 10 in an average islet. Thus the somatostatin/glucagon ratio increased 4-fold. Normalization of glycemia with insulin significantly decreased the glucagon content in the pancreas to 20% of its normal level. It appears that normoglycemia, due to insulin administration, essentially normalized glucagon content per diabetic islet. In addition, the somatostatin content normalized in the pancreas indicating that in normoglycemia, somatostatin content in the islet was still increased 3-fold. Therefore, the somatostatin/glucagon ratio increased further so that in normoglycemic diabetic dogs, the somatostatin/glucagon ratio was about 6 times higher than in normal dogs. Thus, a relative abundance of somatostatin over glucagon (less marked in hyperglycemia) combined with extremely low total glucagon content in normoglycemia could explain, at least in part, the impaired glucagon response to hypoglycemia in diabetes. If the same change of somatostatin/glucagon ratio occurs in human diabetes, that could explain the low glucagon response to hypoglycemia not only in diabetic dogs[9,10], but also in IDDM[35].

STRESS AND THE GLUCOREGULATORY ROLE OF EPINEPHRINE

Studies in normal dogs

Since epinephrine is released in all stress situations, we have investigated the glucoregulatory role of epinephrine independently of other hormonal changes that occur during stress. In moderate exercise, for example, both in normal and alloxan diabetic dogs, it is the interaction between glucagon and insulin that regulates glucose production[16,48,49]. In the regulation of Ra, the catecholamines are of importance in moderate exercise at the onset of exercise (Miles, Finegood, Lickley, Vranic, unpublished data) and are the main regulators of glucose turnover in strenuous exercise[50]. The relative importance of epinephrine and glucagon depends on the amounts of hormones released. During moderate exercise, epinephrine increases only 4-fold while during strenuous exercise is increased up to 18-fold. As described in the previous section, in severe, insulin-induced hypoglycemia, a 15- to 40-fold increase of epinephrine plays a pivotal role in increasing glucose production independently of glucagon[5].

To establish whether or not glucagon is indispensable in the metabolic responses to epinephrine, we examined the response in normal, conscious dogs to the infusion of epinephrine, with or without concomitant glucagon suppression by somatostatin[26]. This model allowed us to examine the selective effect of epinephrine in the following way: in the dog, epinephrine induces a transient increase in both plasma insulin and glucagon. A small dose of somatostatin suppressed glucagon well below its basal level, but it did not affect the release of insulin. Epinephrine increased glucose production transiently and this response was not affected by glucagon suppression. Thus, epinephrine can exert its full hyperglycemic effect independently of glucagon in normal dogs, and that has also been demonstrated in normal humans[51]. Hyperglycemia induced by epinephrine is due not only to the transient increase in glucose production, but also to a sustained decrease in metabolic glucose clearance. In man at least, epinephrine's stimulatory effect on the liver and inhibitory effect on glucose clearance reflects β-adrenergic mechanisms[34]. Interestingly, the effect of an epinephrine infusion is transient, while during hypoglycemia, the increase in glucose production is sustained. The difference could be due to the fact that hypoglycemia by itself stimulates hepatic glucose production, whereas hyperglycemia inhibits glucose production by the liver[52-54]. In a recent extensive study, various factors responsible for control of glucose production during hypoglycemia were analyzed[7]. It is of particular interest that under marked hyperinsulinemic conditions, the brain is the primary director of glucagon release and it is responsible for 75% of the life-sustaining glucose production.

Studies in diabetic dogs

It was suggested more than 50 years ago that glucagon contributes to the diabetic state[55]. Conclusive evidence for the diabetogenic role of glucagon was provided during the last decade[56,57]. Although glucagon does not appear to be involved to any major degree in the action of epinephrine on the liver in normal dogs or humans, in diabetes, glucagon may be a major mediator of the hepatic effect of epinephrine for two reasons: first, when insulin concentration in the pancreas is diminished, the glucagon response to secretagogues is increased, and second, hypoinsulinemia can greatly increase liver sensitivity to glucagon[58-61]. Indeed, in insulin-dependent diabetic subjects, epinephrine-stimulated glucagon release accounted for 50% of epinephrine's effect on hepatic glucose production[62] and the responsiveness to epinephrine and glucagon was augmented in insulin-infused juvenile-onset diabetics[63].

To determine the extent to which glucagon release could contribute to the diabetic instability of stress, we infused epinephrine into alloxan diabetic dogs[41]. In these dogs, the basal rate of glucose production was elevated. This is due, at least in part, to glucagon, because hypophysectomy in alloxan diabetic dogs restores basal values for plasma glucagon and Ra[64], and glucagon replacement in these dogs increases Ra and induces hyperglycemia. Epinephrine was given alone or together with somatostatin to alloxan diabetic dogs. A direct effect of somatostatin was excluded by experiments in which glucagon was infused in addition to somatostatin and epinephrine. In diabetic dogs, epinephrine infusion resulted in greatly exaggerated hyperglycemia. This was mainly due to exaggerated glucagon release and a resulting larger and more sustained increase in glucose production which was markedly attenuated by somatostatin.

The question was then raised of whether increased liver sensitivity to glucagon plays an important role in addition to exaggerated glucagon release. Therefore, we infused epinephrine in depancreatized dogs[65]. In these dogs, despite the absence of pancreatic tissue, there is a source of glucagon (3500MW) in the fundus of the stomach which has the same biological activity as pancreatic glucagon[66,67]. However, the response of this extrapancreatic glucagon to epinephrine infusion is small. Epinephrine was infused in depancreatized dogs in both normoglycemic (80 mg/dl) and hyperglycemic state (200 mg/dl) which was maintained by subbasal and basal insulin infusion respectively. Under both conditions, epinephrine induced a 2-fold increase in plasma extrapancreatic glucagon. This rise was abolished when somatostatin was added to the epinephrine infusion. Under normoglycemic and

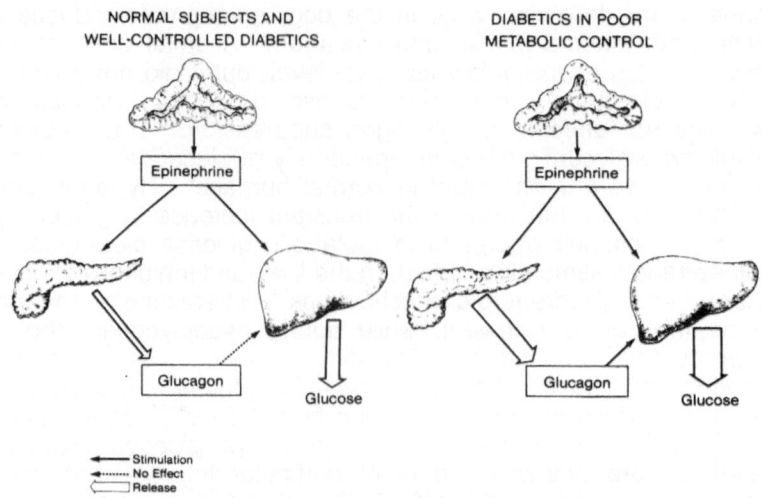

normoinsulinemic conditions, as in normal dogs, epinephrine's effect on glucose production was independent of glucagon release. Under hyperglycemic, hypoinsulinemic conditions, however, the effect of epinephrine on glucose production was excessive and this was due to the epinephrine-induced glucagon release. This shows the pivotal importance of even a small glucagon release when the liver is sensitized by hypoinsulinemia[60,61]. The importance of these epinephrine/glucagon interactions in diabetes is summarized in Figure 9.

THE MECHANISMS OF THE GLUCOREGULATORY RESPONSES TO STRESS AND THEIR DEFICIENCY IN DIABETES

The obligatory requirement of the brain for glucose necessitates a precise mechanism for glucose homeostasis. The hormonal and metabolic responses to stress have been examined under a variety of stress conditions such as severe injury[68,69], major surgery[70,71], myocardial infarction[71,72] and emotional stress[73], however, the glucoregulatory mechanisms are not fully understood. In moderate exercise, another form of stress, the peripheral energy requirements necessitate changes in glucose metabolism whereby the augmented rate of glucose utilization (Rd) is precisely matched with an elevated rate of hepatic glucose production (Ra), thus averting the threat of hypoglycemia. There is also a safeguard against hyperglycemia which is dependent upon insulin[74]. With insulin deficiency or resistance, the exercise-induced increment in the rate of metabolic clearance of glucose (MCR) is diminished, whereas depending on the diabetic state, the increment in Ra is either normal or increased, hence leading to further deterioration of glycemic control and resultant hyperglycemia[74,75] (Figure 10).

The intracerebroventricular (ICV) injection of carbachol (an analog of the neurotransmitter acetylcholine) in anesthetized rats, elicits hormonal responses similar to those seen in clinical stress and exercise[76]. In these studies, glucose fluxes were not examined and, therefore, the mechanism of the ensuing hyperglycemia could not be determined. In addition, it is well known that anaesthesia can affect both hormonal responses and their peripheral effects[77]. In a recent study[78], we injected carbachol into the third cerebral ventricle to compare the physiological glucoregulatory responses to those in diabetes. We wished to find out whether the increase in glucose production induced by neural

RESPONSE OF GLUCOSE PRODUCTION
AND CLEARANCE TO EXERCISE

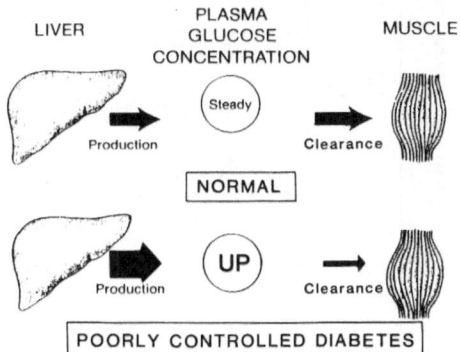

Figure 10. In normal dogs during moderate exercise, glucose production and clearance are well matched and therefore plasma glucose concentration remains unchanged. In poorly controlled diabetic dogs, insulin deficiency and resistance leads to increased glucose production by the liver and to increased clearance by the muscle resulting in exaggerated hyperglycemia.

inputs and release of counterregulatory hormones would also be accompanied by suppression of MCR, as during epinephrine infusions, or if the increased energy demands would lead to an increase in MCR, as observed in exercise. Specifically, we wanted to know whether the expected excessive hyperglycemia in the diabetic dogs is due to excessive glucose production and/or suppressed glucose clearance.   We injected a small dose of carbachol (17 nmol/50µl) into the third ventricle of the brain via a chronic indwelling ICV cannula in conscious dogs, before and after the induction of alloxan diabetes. Figure 11 indicates the various sites that will be rapidly exposed to carbachol following the injection. In 5 control dogs, injection of 100 µl of water did not affect concentrations of metabolites, hormones or glucose kinetics. The injection of carbachol in normal

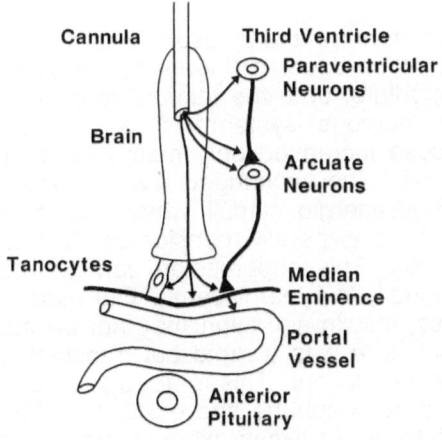

Figure 11. Flow of neuropeptides injected into third ventricle of the dog.

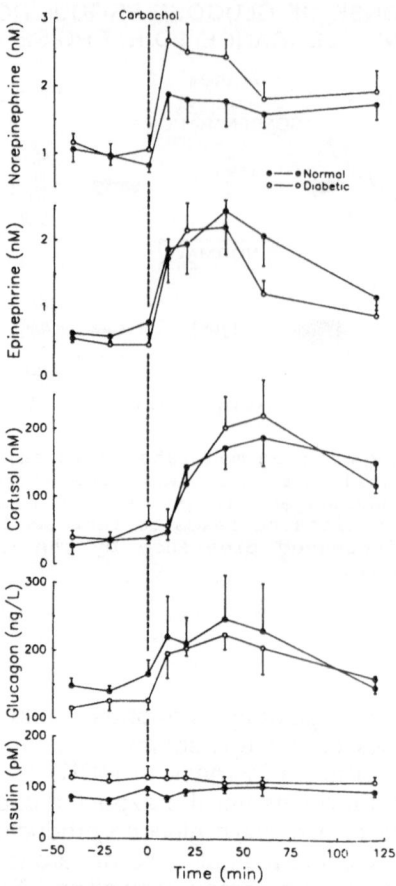

Figure 12. The effect of carbachol (27 nmol/50 µl water) injected into the third ventricle (t=0 min) in normal (●——●) and moderately controlled diabetic (O——O) dogs (n=7). In diabetic dogs, plasma glucose was maintained at 9 mM with a constant infusion of insulin (138 µU/kg-min). Despite subbasal insulin infusion, plasma insulin levels were increased by 40% because of residual insulin secretion. Mean ± SEM. (Reproduced with permission from Reference 78).

dogs increased plasma epinephrine and cortisol 4-5 fold, while the concentration of norepinephrine and glucagon doubled (Figure 12). It was suggested that the activation of CRF could mediate, both the responses of the ACTH-cortisol axis and of the sympathetic nervous system[15,79,80] as indicated in Figure 13. Interestingly, plasma glucose increased only marginally (5 mg/dl) and plasma insulin remained unchanged. The unchanged levels of insulin could indicate a balance of the α and β-adrenergic stimuli which inhibit or stimulate insulin secretion, respectively. It is generally considered that insulin secretion is suppressed in severe stress. This suppression can be relative when insulin increases, but not proportionally to hyperglycemia. Our model demonstrates that, at least in moderate stress, insulin secretion may not be suppressed.

As shown in Figure 14, there was a rapid but transient 2.5-fold increase in tracer-determined glucose production. This is not unexpected, as it reflects the increases in epinephrine, norepinephrine and glucagon. The magnitude of the increase corresponds well to our previous experiments in depancreatized dogs, which were kept normoglycemic by constant basal insulin infusion[65]. What was unexpected however, was that MCR also increased transiently and

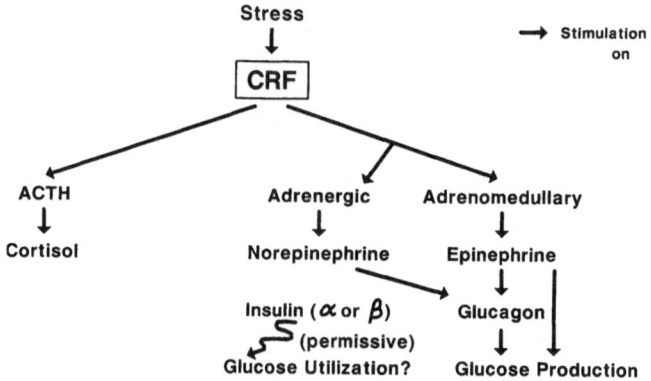

Figure 13. Scheme indicating that, during stress, the release of CRF can concurrently activate the ACTH-cortisol axis and the outflow of the autonomic nervous system. In our experiments, insulin concentration did not change, possibly because the alpha- and beta-adrenergic pancreatic inputs were balanced. It is postulated that the permissive effect of insulin is needed for increments in glucose utilization.

proportionately to glucose production, and this is the reason that glucose concentration increased only marginally. This is surprising because epinephrine infusion decreases metabolic glucose clearance by 50% when insulin is kept constant. As a consequence, there is a large increase in plasma glucose during epinephrine infusion[85], but only a marginal increment during carbachol injection[78]. The increase in glucose metabolic clearance following carbachol is similar to that observed during exercise in normal dogs, where insulin exerts only a permissive effect. The rapid increment in MCR following carbachol implies the possibility of a neural mechanism which could increase glucose uptake independently of an increment in plasma insulin. Carbachol injection not only mobilized glucose but also increased lipolysis.

Alloxan diabetic dogs were given ICV carbachol, when plasma glucose was maintained at moderate hyperglycemia with a subbasal insulin infusion of 138 μU/kg/min. Plasma insulin concentration, however, was up to 40% higher than in normal dogs (Figure 12), reflecting the fact that alloxan diabetic dogs still have residual insulin secretion. There was clear evidence of insulin resistance, because despite higher insulin levels, MCR was decreased by 50% (Figure 14) and lipolysis was exaggerated. As expected, there was no change in plasma insulin following carbachol. Increments in the counterregulatory hormones were similar to those seen in normal dogs, except for a 50% higher norepinephrine release. The metabolic effect of ICV carbachol was greatly exaggerated in the diabetic dogs. The increment in plasma glucose was six times and that of glycerol 2 1/2 times larger than in normal dogs (Figure 14). The initial rise in plasma glucose could have been due, in part, to a more prolonged increase in hepatic glucose production. However, the main reason for hyperglycemia is that the increment in Rd was delayed and much smaller than in normal dogs, and MCR remained suppressed. Therefore, the glucoregulatory effects of ICV carbachol were comparable to those of epinephrine infusion in diabetic, hyperglycemic dogs[65]. Thus, normal increases in glucose uptake during both stress and exercise, appear to require some insulin. Our experimental design does not permit us to delineate the site of increased glucose uptake in normal dogs. One can surmise that, in addition to muscle, there is also increased uptake of glucose in adipose tissue, because, with carbachol injection, the reesterification of FFA and lipolysis are both increased. There was a positive

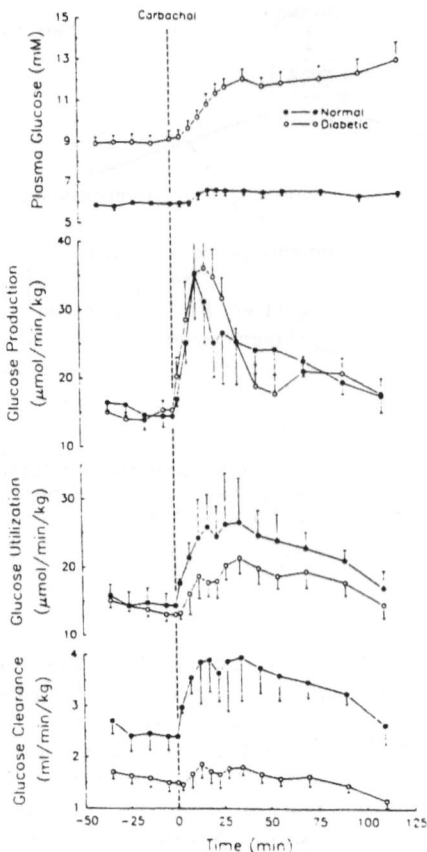

Figure 14. The effect of carbachol (27 nmol/50 uL water) injected into the third ventricle (t=0 min) of normal (●——●) and moderately controlled diabetic (○——○) dogs (n=7). Mean ± SEM. (Reproduced with permission from reference 78).

correlation between total glucose uptake and the rate of FFA reesterification both in normal and diabetic dogs. It could be that during this type of stress there is a change in blood flow in the muscle. This factor by itself however, cannot explain the increased glucose uptake, because in mildly diabetic dogs with the same stress stimulus metabolic glucose clearance remained low and did not increase.

Among the probes which can be used to dissect hormonal responses during stress are analogs of somatostatin[15,79]. When we injected ICV somatostatin octapeptide (20 nmol), 20 min before ICV carbachol[81] in both normal and diabetic dogs, the epinephrine and cortisol responses were abolished, while those of glucagon and norepinephrine were diminished. As a consequence, the rise in glucose production and clearance was prevented in normal dogs. It is not surprising that the control of glucose production was related to adrenergic pathways and glucagon release. What was surprising, however, was that these data indicate the possibility that an adrenergic neural pathway, perhaps through α-receptors, can increase glucose clearance. In diabetic dogs, a small rise in glucose production persisted, possibly related to the excessive release of norepinephrine that was not completely abolished by somatostatin. The already low glucose clearance was not further reduced. The ICV injection of the somatostatin analog alone did not induce any hormonal or metabolic changes.

Somatostatin decreased the carbachol-induced glycerol release in both normal and diabetic dogs. However, the release of FFA was augmented 7-fold in normal and 3-fold in diabetic dogs. This presumably indicates an inhibition of FFA reesterification in adipocytes. We speculate that this could be related to a decrease in glucose uptake consistent with ongoing lipolysis in the adipocytes. We speculate, therefore, that during mild stress there are neural pathways that can augment glucose uptake in both muscle and adipocytes. These inputs require a permissive effect of insulin, but are independent of insulin increments. This hypothesis is consistent with increased energy requirements which occur in a variety of tissues during stress.

Acknowledgements

The work reported in this chapter was supported by grants from Medical Research Council of Canada, Canadian Diabetes Association and Juvenile Diabetes Foundation International. Mladen Vranic is a Killam Scholar of Canada Council. Drs Giacca, Shi and Yamatani are Postdoctoral Fellows of Canadian Diabetes Foundation and P. Miles was supported by Ontario Graduate Studentship. The authors would like to thank D. Bilinski, L. Cook, L. Lam and M. VanDelangeryt for excellent assistance and L. Vranic for editorial work.

REFERENCES

1. A. D. Cherrington, J. E. Liljenquist, G. I. Shulman, Importance of hypoglycemia-induced glucose production during isolated glucagon deficiency, Am J Physiol, 236:E263-E271 (1979).
2. N. J. Christensen, K. G. M. M. Alberti, and O. Brandsborg, Plasma catecholamines and blood substrate concentrations: studies in insulin-induced hypoglycemia and after adrenaline infusions, Eur J Clin Invest, 5:415-423 (1975).
3. Sacca, L., Perez, G., Carteni, G., and F. Rengo, Evaluation of the role of the sympathetic nervous system in the glucoregulatory response to insulin-induced hypoglycemia in the rat. Endocrin, 101:1016-1022 (1977).
4. R. A. Rizza, P. E. Cryer, and J. E. Gerich, Role of glucagon, catecholamines and growth hormone in human glucose counterregulation. Effects of somatostatin and combined alpha- and beta-adrenergic blockade in plasma glucose recovery and glucose flux rate after insulin-induced hypoglycemia. J Clin Invest, 64:62-70 (1979).
5. C. Gauthier, M. Vranic, and G. Hetenyi, Jr., Importance of glucagon in regulatory rather than emergency responses to hypoglycemia, Am J Physiol, 238:E131-E140 (1980).
6. A. J. Garber, I. E. Karl, and D. M. Kipnis, Alanine and glutanine synthesis and release from skeletal muscle. IV. Beta-adrenergic inhibition of amino acid release, J Biol Chem, 251:1851-1857 (1976).
7. D. W. Biggers, S. R. Myers, D. Neal, R. Stinson, N.B. Cooper, J. B. Jaspen, P. E. Williams, A. D. Cherrington, and R. T. Frizzell, Role of brain in counterregulation of insulin-induced hypoglycemia in dogs, Diabetes, 37:7-16 (1989).
8. L. Rosetti, D. Smith, G. L. Shulman, D. Papachristou, and R. A. DeFronzo, Correlation of hyperglycemia with phlorizin normalizes tissue sensitivity to insulin in diabetic rats, J Clin Invest 79:1510-1515 (1987).
9. B. Lussier, M. Vranic, N. Kovacevic, and G. Hetenyi, Jr., Glucoregulation in alloxan diabetic dogs, Metab, 35:18-24 (1986).
10. G. Hetenyi, Jr., C. Gauthier, M. Byers, and M. Vranic, Phlorizin induced normoglycemia partially restores glucoregulation in diabetic dogs, Am J Physiol, 256:E277-E283 (1989).
11. A. Klip, T. Ramlal, D. Dimitrakoudis, P. J. Bilan, G. Cartee, E. Gulve, J. O.

Holloszy, and M. Vranic, The subcellular distribution of glucose transporters (GTs) in normal and diabetic rat skeletal muscle is regulated by hyperglycemia and insulin, Endocrine Society, p. 47, (Abstract 89), (1990).

12. K. M. A. El-Tayeb, P. L. Brubaker, H. L. A. Lickley, E. Cook, and M. Vranic, Effect of opiate receptor blockade on normoglycemic and hypoglycemic glucoregulation, Am J Physiol, 250:E236-E242 (1986).

13. C. Gauthier, and G. Hetenyi, Jr., Origin of glucose released in the regulatory response against hypoglycemia, Metab, 31:147-153 (1982).

14. C. Gauthier, M. Vranic, and G. Hetenyi, Jr., Nonhypoglycemic glucoregulation: role of glycerol and glucoregulatory hormones, Am J Physiol, 244:E373-E379 (1983).

15. M. R. Brown, and L. A. Fisher, Central nervous systems effects of corticotropin releasing factor in the dog, Brain Research, 280:75-79 (1983).

16. D. H. Wasserman, H. L. A. Lickley, and M. Vranic, Interactions between glucagon and other counterregulatory hormones during normoglycemic and hypoglycemic exercise, J Clin Invest, 74:1404-1413 (1984).

17. G. Hetenyi Jr., S. Varma and J. S. Cowan, Relations between blood glucose and hepatic glucose production in newborn dogs, Brit Med J (2)625-627, #5814 (1972).

18. G. Hetenyi Jr., N. Kovacevic, S. E. H. Hall, and M. Vranic, Plasma glucagon in pups, decreased by fasting, unaffected by somatostatin or hypoglycemia, Am J Physiol 231:1377-1382 (1976).

19. J. S. Cowan, and G. Hetenyi Jr., Hypoglycemia in newborn dogs provokes substantial ACTH but probably not adrenaline secretion, Can J Physiol Pharm 57:476-484 (1979).

20. K. Nakao, Y. Nakai, H. Jingami, S. Oki, J. Fukata, and H. Imura, Substantial rise of plasma beta-endorphin levels after insulin-induced hypoglycemia in human subjects. J Clin Endocrinol Metab, 49:838-841 (1979).

21. F. Fraioli, C. Moretti, D. Paolucci, E. Alicicco, F. Crescenzi, and G. Fortunio, Physical exercise stimulates marked concomitant release of beta-endorphin and adrenocorticotropic hormone (ACTH) in peripheral blood in man, Experimientia, 36:987-989 (1980).

22. M. Feldman, R. S. Kiser, R. H. Unger, and C. H. Li, Beta-endorphin and the endocrine pancreas, studies in healthy and diabetic human beings, N Engl J Med, 300:349-353 (1983).

23. K. M. A. El-Tayeb, C. J. T. Gauthier, P. L. Brubaker, H. L. A. Lickley, and M. Vranic, Hormonal and metabolic responses to intracarotid and intrajugular infusion of beta-endorphin in normal dogs, Can J Physiol Pharmacol, 64:306-310 (1986).

24. J. A. Nash, P. M. Radosewich, B. Lacy, N. Rizk, H. Hourani, P. E. Williams and N. Abumrad, Effects of naloxone on glucose homeostasis during insulin-induced hypoglycemia, Am J Physiol, 257:E367-E373 (1989).

25. S. Caprio, G. Gerety, M. Diamond, W. V. Tamborlane, and R. S. Sherwin, Naloxone enhances the hepatic response to hypoglycemia in diabetics with defective counterregulation, Diabetes, 38(suppl.2):4A, (Abstract) (1989).

26. K. M. A. El-Tayeb, P. L. Brubaker, M. Vranic, and H. L. A. Lickley, Beta-endorphin modulation of the glucoregulatory effects of repeated epinephrine infusion in normal dogs. Diabetes 34:1293-1300 (1985).

27. D. E. Gray, H. L. A. Lickley, and M. Vranic, Physiologic effects of epinephrine on glucose turnover and plasma free fatty acid concentrations mediated independently of glucagon, Diabetes, 29:600-609 (1980).

28. K. M. A. El-Tayeb, M. Vranic, P. L. Brubaker, and H. L. A. Lickley, Beta-endorphin modulation of the glucoregulatory effects of epinephrine infusion in alloxan diabetic and normal dogs, Diabetologia, 30:745-754 (1987).

29. R. Ninomiya, N. F. Forbath and G. Hetenyi Jr., Effect of adrenal steroids

on glucose kinetics in normal and diabetic dogs, Diabetes 14:729-739 (1965).

30. J. Campbell, and S.K. Rastogi, Elevation of serum insulin, albumin and FFA with gains of liver lipid and protein induced by glucocorticoid treatment in dogs, Can J Physiol Pharm 46:421-429 (1968).

31. B. Issekutz Jr. and M. Allen, Effect of catecholamines and methylprednisolone on carbohudrate metabolism of dogs, Metabolism 21:48-59 (1972).

32. G. Hetenyi Jr., B. Pagurek, E.A. Dittmar, C. Ferrarotto, The effects of methylprednisolone on the turnover of alanine and on the transfer of carbon atoms from alanine to pyruvate and glucose, Can J Physiol Pharm 58:787-796 (1980).

33. B. Lussier, M. Vranic, C. Gauthier, and G. Hetenyi, Jr., Glucoregulation in dogs treated with methylprednisolone, Metabolism 34:906-911 (1985).

34. P. E. Cryer, and J. G. Gerich, Relevance of glucose counterregulatory system to patients with diabetes, Diabetes Care, 6:95-99 (1983).

35. G. Bolli, P. De Feo, P. Campagnucci, M. G. Cartechini, G. Angeletti, F. Santeusanio, P. Brunetti, and J. E. Gerich, Abnormal glucose counterregulation in insulin dependant diabetes mellitus: interaction of anti-insulin antibodies and impaired glucagon and epinephrine secretion, Diabetes, 32:134-141 (1983).

36. P. E. Cryer, Hypoglycemic glucose counterregulation in patients with insulin dependant diabetes mellitus, J Clin Lab Med, 99:451-456 (1982).

37. H. Drost, D. Gruneklee, K. Kley, W. Wiegelman, H. L. Kruskemper, and F. A. Gries, Untersuchungen zur gluckagon-, STK-, cortisolsekretion bei insulininduzierter hypoglykamie bie insulabhangigen diabetikern ohne neuropathie, Klin Wochenschr, 58:1197-1202 (1980).

38. J. E. Gerich, M. Langlois, C. Noacco, J. Karam, and P. H. Forsham, Lack of glucagon response to hypoglycemia in diabetes: evidence for an intrinsic pancreatic alpha-cell defect, Science, 182:171-173 (1973).

39. O. Bjorkman, P. Miles, D. Wasserman, L. Lickley and M. Vranic, Regulation of glucose turnover during exercise in pancreatectomized totally insulin deficient dogs: Effects of β-adrenergic blockade, J. Clin. Invest 81:1759-1767 (1988).

40. T. Ramlal, S. Rastogi, M. Vranic, and A. Klip, Decrease in glucose tranporter number in skeletal muscle of mild diabetic (streptozotocin-treated) rats, Endocrinology 125:890-897 (1989).

41. G. Perez, F. W. Kemmer, H. L. A. Lickley, and M. Vranic, Importance of glucagon in mediating epinephrine-induced hyperglycemia in alloxan-diabetic dogs, Am J Physiol, 241(4):E328-E335 (1981).

42. F. G. McDaniel, Acute suppression of hepatic gluconeogenesis by glucose in the intact animal Am J Physiol, 229:E569-E575 (1975).

43. L. Hue, Gluconeogenesis and its regulation, Diabetes/Metabolism Reviews, 3:111-126 (1987).

44. M. E. Wernette-Haymond, and H. A. Landy, Regulation of gluconeogenesis in hepatocytes in fasted alloxan diabetic rats, Diabetes, 34:767-773 (1985).

45. K. S. Rastogi, L. Lickley, M. Jokay, S. Efendic, and M. Vranic, Paradoxical reduction in pancreatic glucagon with normalization of somatostatin and decrease in insulin in normoglycemic alloxan diabetic dogs: A putative mechanism of glucagon irresponsiveness to hypoglycemia, Endocrin, 126:1096-1104 (1990).

46. E. Chen, I. Komiya, L. Inman, K. McCorkle, T. Alam, and R. H. Unger, Metabolic and cellular responses of islet during pertubations of glucose homeostasis determined by in situ hybridization histochemistry, Proc Natl Acad Sci, U.S.A., 86:1367 (1989).

47. M. Vranic, H. L. A. Lickley and J. K. Davidson, Exercise and stress in diabetes mellitus. In: Clinical Diabetes Mellitus: A problem oriented

approach. (ed J. K. Davidson). Thieme-Stratton Inc., New York, pp. 172-205 (1986).

48. D. H. Wasserman, H. L. A. Lickley, and M. Vranic, Important role of glucagon during exercise and diabetes, J Appl Physiol, 59(4):1272-1281 (1985).

49. D. H. Wasserman, H. L. A. Lickley, and M. Vranic, Role of beta-adrenergic mechanisms during exercise in poorly-controlled insulin deficient diabetes, J Appl Physiol, 59:1282-1289 (1985).

50. E. Simatirakis, P. D. G. Miles, M. Vranic, R. Hunt, R. Gougen-Rayburn, C. J. Field and E. B. Marliss, Glucoregulation during single and repeated bouts of intense exercise and recovery in man Clin Invest Med 13(4), Abstract 134 (1990).

51. R. A. Rizza, M. Haymond, P. Cryer, and J. Gerich, Differential effects of epinephrine on glucose production and disposal in man Am J Physiol, 237:E356-E362 (1979).

52. R. N. Bergman, Integrated control of hepatic glucose metabolism in the dog Ann NY Acad Sci, 148:441-468 (1977).

53. B. Issekutz, Role of beta-adrenergic receptors in mobilization of energy sources in exercising dogs J Appl Physiol, 44:869-876 (1978).

54. G. I. Shulman, J. E. Liljenquist, P. E. Williams, W. W. Lacy, and A. D. Cherrington, Glucose disposal during insulinopenia in somatostatin-treated dogs J Clin Invest, 62:478-491 (1978).

55. L. Kepinov, and S. Petit-Dutaillis, Action hyperglycemiate du sang du chien diabetique, Arch Int Physiol, 34:48-100 (1931).

56. R. H. Unger, Role of glucagon in the pathogenesis of diabetes: The status of the controversy, Metab, 27:1691-1709 (1978).

57. R. H. Unger, and L. Orci, Hypothesis: The essential role of glucagon in the pathogenesis of diabetes mellitus, Lancet, 1:14-16 (1975).

58. N. Altszuler, B. Gottlieb, and J. Hamshire, Interaction of somatostatin, glucagon and insulin on hepatic glucose output in the normal dog, Diabetes, 25:116-121 (1976).

59. A. D. Cherrington, J. L. Chiasson, A. S. Jennings, U. Keller, and W. Lacy, The role of glucagon and insulin in the regulation of basal glucose production in the postabsorptive dog, J Clin Invest, 58:1407-1418 (1976).

60. A. D. Cherrington, W. W. Lacy, and J. L. Chiasson, The effects of glucagon on glucose production during insulin deficiency in the conscious dog, J Clin Invest, 62:664-677 (1978).

61. H. L. A. Lickley, G. G. Ross, and M. Vranic, Effects of selective insulin or glucagon deficiency on glucose turnover, Am J Physiol, 236:E255-E262 (1979).

62. J. E. Gerich, M. Lorenzi, E. Tsalikian, and J. H. Karam, Studies on the mechanism of epinephrine-induced hyperglycemia in man, Diabetes, 25:67-71 (1976).

63. H. Shamoon, R. Hendler, and R. Sherwin, Altered responsiveness to cortisol, epinephrine and glucagon in insulin-infused juvenile onset diabetics: a mechanism for diabetic instability, Diabetes, 29:284-291 (1980).

64. F. W. Kemmer, A. Sirek, O. V. Sirek, G. Perez, and M. Vranic, Glucoregulatory mechanisms following hypophysectomy in diabetic dogs with residual insulin secretion, Diabetes, 23:26-34 (1983).

65. F. W. Kemmer, H. L. A. Lickley, D. E. Gray, G. Perez, and M. Vranic, The state of metabolic control determines the role of epinephrine-glucagon interactions of glucoregulation in diabetes, Am J Physiol, 242(4):E428-E436 (1982).

66. K. Doi, M. Prentiki, C. Yip, W. Muller, B. Jeanrenaud, and M. Vranic, Identical biological effects of pancreatic glucagon and a purified moiety of canine gastric glucagon, J Clin Invest, 63:525-531 (1979).

67. T. W. Hatton, C. C. Yip, and M. Vranic, Biosynthesis of glucagon (IRG3500) in canine gastric mucosa, Diabetes, 34:38-46 (1985).
68. R. R. Wolfe, M. J. Durkot, J. R. Allsop, and J. F. Burke, Glucose metabolism in severly burned patients, Metab, 28:1031-1039 (1979).
69. G. L. Clifton, M. G. Ziegler, and R. G. Grossman, Circulating catacholamines and sympathetic activity after head injury Neurosurgery, 8:10-14 (1981).
70. J. B. Halter, A. E. Pflug, and D. Porte, Jr., Mechanism of plasma catecholamine increases during surgical stress in man J Clin Endocrinol Metab, 45:936-944 (1977).
71. P. E. Cryer, Physiology and pathophysiology of the human sympathoadrenal neuroendocrine system, N Engl J Med, 303:436-444 (1980).
72. N. J. Christensen, and J. Videbaek, Plasma catecholamines and carbohydrate metabolism in patients with acute myocardial infarction, J Clin Invest, 54:278-286 (1974).
73. F. W. Kemmer, R. Bisping, H. J. Steingruber, H. Baar, F. Hartman, R. Schlagheche, and M. Berger, Psychological stress and metabolic control in patients with type I diabetes mellitus, N Engl J Med, 314:1078-1084 (1986).
74. M. Vranic, R. Kawamori, S. Pek, N. Kovacevic and G. A. Wrenshall, The essentiality of insulin and the role of glucagon in regulating glucose utilization and production during strenuous exercise in dogs, J Clin Invest, 57:245-256 (1976).
75. M. Vranic, and G. A. Wrenshall, Exercise, insulin and glucose turnover in dogs, Endocrin, 85:165-171 (1969).
76. M. R. Brown, J. Rivier, and W. Vale, Somatostatin: Central nervous systems action on glucoregulation, Endocrin, 140:1709-1715 (1979).
77. P. S. Sebel, in: Hazards and complications of anaesthesia, T. H. Taylor and E. Major, eds. New York, NY, Churchill Livingston (1987).
78. P. Miles, K. Yamatani, L. Lickley, and M. Vranic, Mechanism of glucoregulatory responses to stress and their deficiency in diabetes. Proc. Natl. Acad. Sci, U.S.A., (1990) (in press).
79. M. R. Brown, and L. A. Fisher, Brain peptide regulation of adrenal epinephrine secretion, Am J Physiol, 247:E41-E46 (1984).
80. M. R. Brown, Neuropeptides: Central nervous systems effects on nutrient metabolism, Diabetologia, 20:299-304 (1981).
81. P. D. G. Miles, K. Yamatani, H. L. A. Lickley, and M. Vranic, The intracerebroventricular injection of a somatostatin analog (ODT8-SS) suppresses the stress response in normal and diabetic dogs, Program of International Symposium on Somatostatin, Montreal, Canada. Abstract 68, (1989).

SEARCH FOR THE HYPOGLYCEMIA RECEPTOR

USING THE LOCAL IRRIGATION APPROACH

Casey M. Donovan, Patricia Cane and Richard N. Bergman

Dept. Physiology and Biophysics
University of Southern California
Los Angeles, CA

INTRODUCTION

The central nervous system (CNS) has been implicated in glucose homeostasis since Claude Bernard's [1] observation in the mid-19th century that puncture of the fourth ventricle of the brain resulted in glycosuria. Observations in the early part of this century that specific lesions in the hypothalamus were associated with hyperglycemia and glycosuria, provided added support for a role of the CNS in glucose metabolism. [2,3] Extensive research has subsequently revealed that the ventromedial hypothalamus (VMH) and lateral hypothalamus (LH) are capable of exerting reciprocal control over glucose metabolism. [4-6] Electrical stimulation of the VMH leads to hyperglycemia, hepatic glycogenolysis, hyperglucagonemia, and hypoinsulinemia. [7,8] Conversely, lesions to the VMH are characterized by hyperinsulinemia. [4,5] Chemical stimulation of the LH with epinephrine or norepinephrine leads to elevated secretion of insulin. [9] A similar insulin response has been observed with electrical stimulation of the dorsal motor nucleus of the vagus, with which the LH interacts. It has been further shown that the VMH and LH are capable of direct effects upon hepatic glycogenolysis, gluconeogenesis and glycogenesis. Stimulation of the VMH activates glycogen phosphorylase and phosphoenolpyruvate carboxykinase (PEPCK), while reducing pyruvate kinase activity. [10,11] Alternatively, stimulation of the LH leads to activation of glycogen synthetase and suppression of PEPCK activity. [10,11] Thus the VMH and LH have the potential to exert a strong control over hepatic glycogen metabolism via direct hepatic innervation and innervation of endocrine glands.

While the efferent limb for neural control of glucose metabolism has been well characterized, less is known about the afferent limb which senses the changes in glucose concentration. Early observations that hypoglycemia elicited a sympathetic response, [12] taken in conjunction with the above mentioned observations related to CNS lesioning, suggested that the afferent limb resides in the CNS. Additional support for this view came from studies involving direct application of 2-deoxyglucose (2-DG), a potent glucopenic agent, into the CNS. Miselis and Epstein [13] demonstrated increased feeding following the injection of 2-DG into the lateral cerebral ventricles. More recently, the existence of glucose sensitive neurons within the VMH and LH has been confirmed. [14]

BRAIN CLAMP

While studies employing nonphysiological stimuli (eg. electrical stimulation, pharmacological intervention) may elucidate potential afferents, they do not guarantee that those cells sensing these nonphysiological stimuli also respond to variations in blood glucose concentration. We proposed to ascertain the role of the putative CNS glucoreceptors in monitoring blood glycemia, by comparing the counterregulatory response to systemic hypoglycemia in the presence and absence of forebrain euglycemia. To accomplish this we introduced the 'brain clamp' technique, by which euglycemia is established across the brain via localized glucose infusion, during insulin induced systemic hypoglycemia. [15] Under these conditions any suppression of the counterregulatory responses (ie. elevations in glucagon, epinephrine, norepinephrine, or HGO) could be attributed to putative CNS glucoreceptors.

Moderate Hypoglycemia

Dogs (21-31 kg) were chronically cannulated in the carotid arteries (cannulas advanced 2cm rostrally) for glucose infusion, and femoral artery (cannula advanced 7cm into the aorta) for arterial blood sampling. Following one week of recovery the animals were exposed to the first of three protocols. Prior to initiating any experiments acute catheters were placed in the cephalic vein for tracer (^{3}H-glucose) and insulin infusion, the saphenous vein for venous sampling, and the jugular vein (advanced rostrally) for sampling cerebral venous effluent. A 30 minute basal sampling period preceeded the insulin infusion (150 mU/min) which was initiated at time zero and maintained for the experimental period. Serial blood samples were drawn for measurement of glucose, insulin, glucagon, epinephrine, norepinephrine and ^{3}H-glucose. Protocol 1 consisted of insulin infusion with no glucose replacement (Fig. 1). Owing to the substantial hypoglycemia incurred under this protocol, experiments were terminated at 60 minutes to minimize discomfort to the animal. In protocol 2 glucose was infused via the carotid catheters at a rate that 'clamped' jugular venous effluent at basal glycemia. Owing to substantial glucose escape from the carotid bed, ~82% of carotid glucose infusion, the prevailing systemic glycemia in protocol 2 was substantially higher than protocol 1. Protocol 3 was effected in which glucose was infused peripherally, so as to clamp the systemic glycemia at the level achieved in protocol 2.

In protocol 1, subsequent to insulin infusion, glucose was observed to fall from a basal value of 100 ± 2 mg/dl to a nadir of 37 ± 2 mg/dl. This severe hypoglycemia led to dramatic increases in glucagon, epinephrine, and norepinephrine, from basal values of 256 ± 35, 63 ± 8, and 209 ± 33 pg/ml, to peak values of 467 ± 35, 1762 ± 582, and 650 ± 133 pg/ml, respectively (Fig. 2). Elevating arterial glucose concentration, via peripheral glucose infusion, to 61 ± 5 mg/dl during insulin infusion (ie. protocol 3), substantially suppressed the counterregulatory response. Despite this, both epinephrine and norepinephrine demonstrated significantly elevated levels, 523 ± 104 and 445 ± 71 pg/ml, during this moderate hypoglycemia. However, clamping the carotid perfused

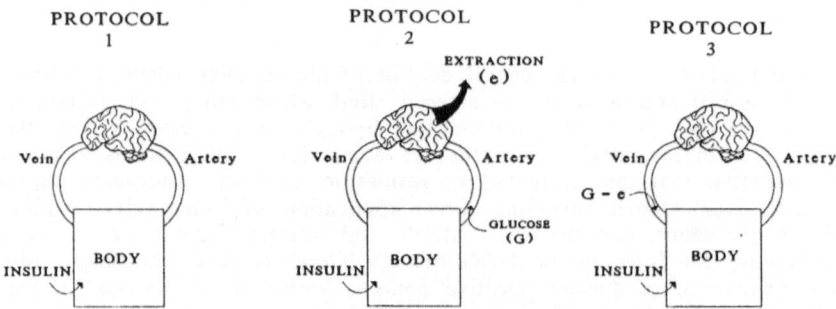

Fig. 1. Schematic representation of the three protocols employed for the "brain clamp" study. Protocol 1 involved insulin infusion w/o glucose infusion. Protocol 2, the "brain clamp", involved insulin infusion with carotid glucose infusion (G) to establish forebrain euglycemia. Extraction (e) represents the carotid perfused tissue glucose uptake. Protocol 3, the "matched infusion", involved insulin infusion with peripheral glucose infusion (G-e).

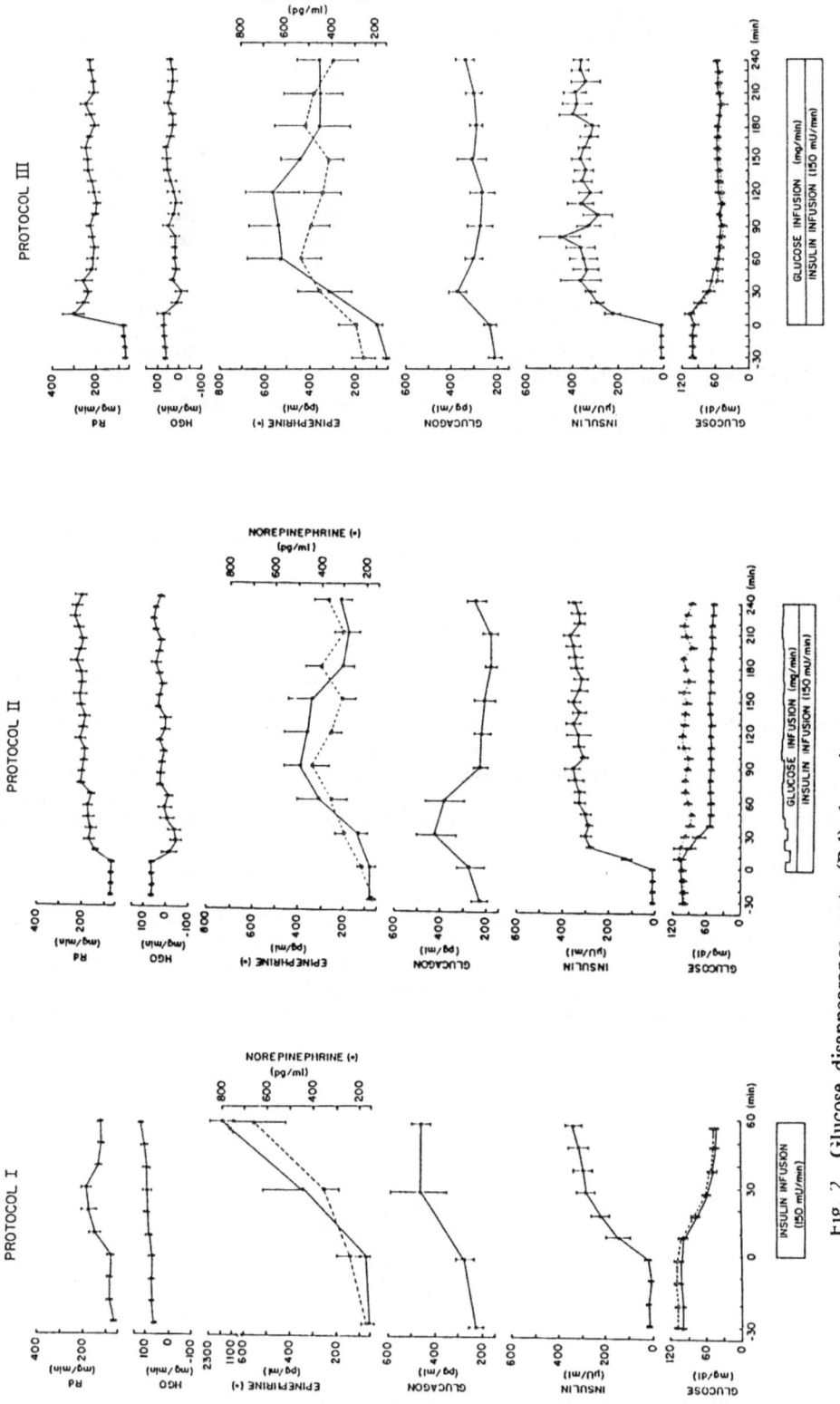

Fig. 2. Glucose disappearance rate (Rd), hepatic glucose output (HGO), epinephrine (solid line), norepinephrine (dashed line), glucagon, insulin, peripheral vein glucose (solid line) and jugular vein glucose (dashed line) as a function of time during protocols 1,2, and 3. Basal period is –30 to 0 minutes. Experimental period is 0 to 240 minutes. (From Cane, P., et al, Diabetes, 35:268, 1986)

187

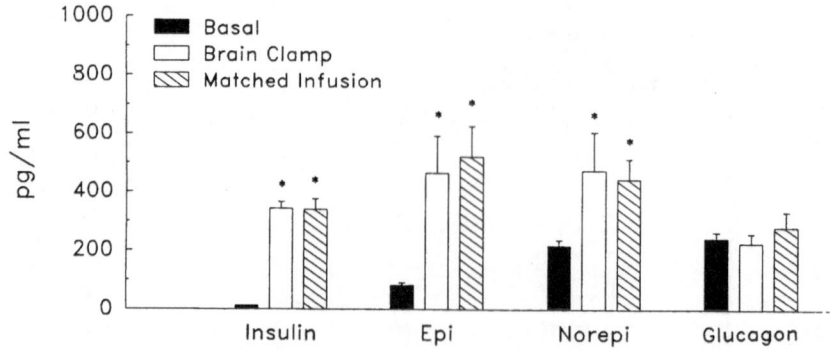

Fig. 3. Basal and steady-state (90-240 minutes) values for insulin, epinephrine (Epi), norepinephrine (Norepi), and glucagon during the brain clamp (protocol 2) and matched infusion (protocol 3). * Significantly different from basal (P<0.05). No significant differences were observed between the brain clamp and matched infusion experiments. (Adapted from Cane, P., et al, <u>Diabetes</u>, 35:268, 1986)

tissues at euglycemia did not significantly impact upon any of the counterregulatory responses (Fig. 3).

The rationale for this study suggested that if there were essential forebrain glucoreceptors, then intracarotid glucose replacement would be expected to suppress the counterregulatory response. As demonstrated above (Fig. 3) when the brain was clamped at euglycemia (protocol 2) the counterregulatory responses were not significantly different from those observed under similar systemic glycemia with no carotid glucose replacement (protocol 3). These results indicate the lack of any essential forebrain glucoreceptors governing the response to moderate hypoglycemia. However, it is apparent that a full-blown counterregulatory response was not elicited during moderate hypoglycemia (Fig. 4). The counterregulatory responses to protocol 1, where systemic glucose fell to 37 mg/dl, were 168%-388% of those observed at ~62 mg/dl (protocols 2 & 3).

Deep Hypoglycemia

Was the failure to prevent the counterregulatory responses due to insufficient hypoglycemia during intracarotid glucose infusion? To address this question we conducted a second series of experiments to ascertain if there were essential forebrain glucoreceptors for counterregulation to 'deep' hypoglycemia.[16] The surgical preparation, experimental procedures and sampling protocols were identical to those described

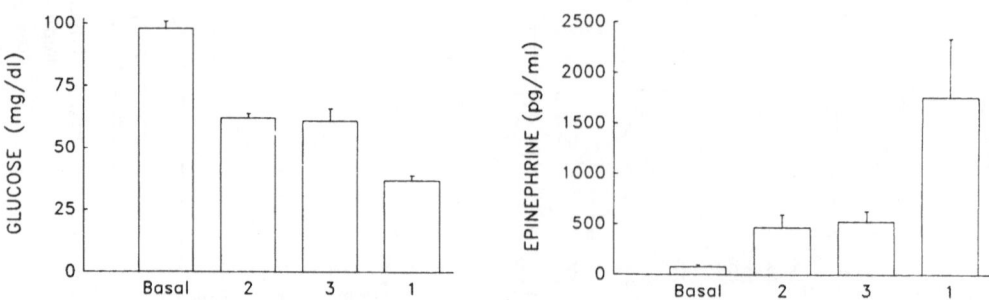

Fig. 4. Basal and steady-state values for glucose and epinephrine during protocols 1, 2, and 3.

above. Principal differences were the elimination of protocol 1, ie. insulin infusion without glucose infusion (it should be noted that protocols 1 and 2 for this study were analogous to protocols 2 and 3 for the moderate hypoglycemia studies), and the implementation of a low rate of cerebral glucose replacement. The primary limitation in the level of established glycemia during our initial study was the substantial 'escape' of glucose from the carotid bed. Thus to effect a lower level of systemic glycemia during protocol 1 the carotid artery infusion was set so as to establish jugular glycemia at ~55 mg/dl. Assuming a similar fractional extraction by the carotid perfused tissues under current conditions, the carotid arterial glucose was estimated to be ~66 mg/dl. For protocol 2 glucose was infused peripherally at a rate determined to clamp systemic glucose levels at those observed for protocol 1.

The lower rate of central glucose replacement resulted in a substantially lower level of systemic hypoglycemia in protocol 1, 39 ± 2 mg/dl, which was essentially matched in protocol 2, 40 ± 2 mg/dl. As a result the elevations in plasma epinephrine and norepinephrine concentration (Fig. 5) were substantially greater than previously observed during moderate hypoglycemia. However, when the carotid perfused tissues were clamped at a level of 55-66 mg/dl, previously shown to significantly suppress the counterregulatory response (see protocol 3 above), this failed to have any effect upon the current sympathoadrenal response. Thus, the previous observation that forebrain glucoreceptors were not essential for counterregulation could not be attributed to the level of hypoglycemia.

Distribution of Flow

To determine the distribution within the brain of the infused glucose several animals (n=4) were employed in acute experiments. Under anesthesia, the carotid arteries were catheterized as described above, then radioactive microspheres (^{125}I, 10 mCi/ml) were infused simultaneously into both carotid arteries. Following the microsphere infusion animals were sacrificed (O.D. pentobarbital), the brain excised, and sectioned samples taken from both hemispheres. The carotid was observed to largely perfuse the anterior and posterior cortex, and anterior hypothalamus (Fig. 6). In similar experiments four dogs were catheterized in the vertebral arteries and subsequently infused with microspheres. The vertebrals perfused primarily the hindbrain, ie. cerebellum, medulla and pons, and posterior hypothalamus, with a significant contribution to the posterior cortex. This distribution of flow for both the carotid and vertebrals was similar to previous reports.[17,18]

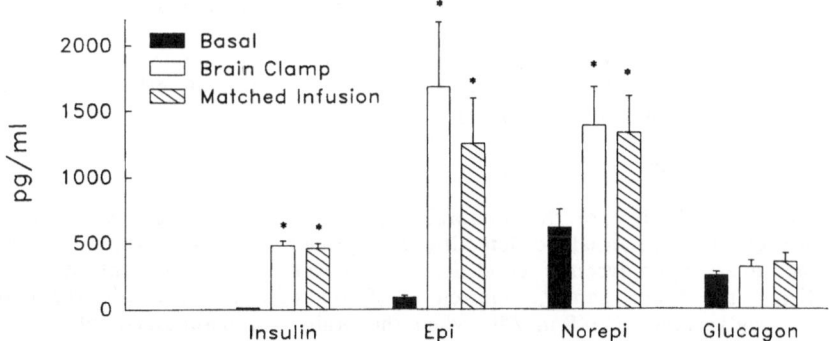

Fig. 5. Basal and steady-state (90-160 minutes) values for insulin, epinephrine (Epi), norepinephrine (Norepi), and glucagon during the brain clamp (protocol 1) and matched infusion (protocol 2) at 'deep hypoglycemia'. * Significantly different from basal (P<0.05). No significant differences were observed between the brain clamp and matched infusion experiments. (Adapted from Cane, P., et al, Am J Physiol, 255:E680, 1988)

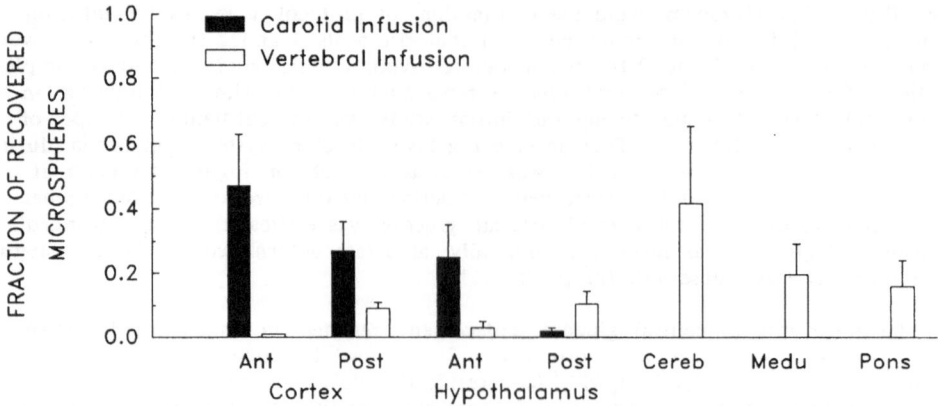

Fig. 6. Fractional distribution of microspheres infused via the carotid or vertebral arteries. Carotid infused microspheres were recovered primarily in the anterior (Ant.) and posterior (Post.) cortex, and anterior hypothalamus. Recovery of the vertebral infused microspheres was primarily in the hindbrain, ie. cerebellum (Cereb.), medulla (Medu.), and pons, with significant recovery in posterior cortex and posterior hypothalamus. (From Cane, P., et al, Am J Physiol, 255:E680, 1988)

Vertebral Infusion

It remained a distinct possibility that the essential glucoreceptors reside in the hindbrain. While Miselis and Epstein[13] demonstrated increased feeding following 2-DG injection into the lateral cerebral ventricles, they were unable to demonstrate the same when 2-DG was directly injected into either the VMH or LH. DiRocco and Grill[19] reported increased feeding and hyperglycemia following administration of 2-DG in decerebrated rats (ie. forebrain removed) indicating a nonessential role for the hypothalamus in sensing glucoprivation. Subsequent findings by Ritter, et al[20] suggested the hindbrain as the possible location for the CNS glucosensors responding to 2-DG.

To ascertain a potential role for the hindbrain glucoreceptors, we repeated the experimental protocols above using animals chronically cannulated in the vertebral arteries. Protocol 1 consisted of insulin infusion with vertebral glucose replacement, while protocol 2 consisted of insulin infusion with peripheral glucose infused so as to clamp the systemic glycemia at the level observed in protocol 1. As a result of hypoglycemia, both epinephrine and norepinephrine demonstrated significant elevations above basal. However, vertebral glucose replacement had no significant effect upon the response for either catecholamine (Fig. 7).

Brain Clamp: Conclusions

Neither the forebrain nor the hindbrain appears to be the exclusive locus for the putative glucoreceptors essential to detecting hypoglycemia. This is true for both moderate and severe hypoglycemic conditions. However, in a subsequent study, Biggers, et al[21], utilizing the 'brain clamp', reported a 75% decrease in the counterregulatory response to hypoglycemia (45-48 mg/dl) when the brain was maintained at an estimated glycemia of 85 mg/dl. The primary difference was the simultaneous perfusion of both the carotid and vertebral arteries (ie. 'clamping' the entire brain) by Biggers and coworkers. These data, taken together with the results discussed above suggest the existence of a redundant glucoreceptor system within the brain. Such a system would ostensibly provide for a general sympathoadrenal response to hypoglycemia within any region of the brain. However, the limited number of reports and discrepant findings cannot be interpreted as confirming the existence of any essential central glucoreceptors.

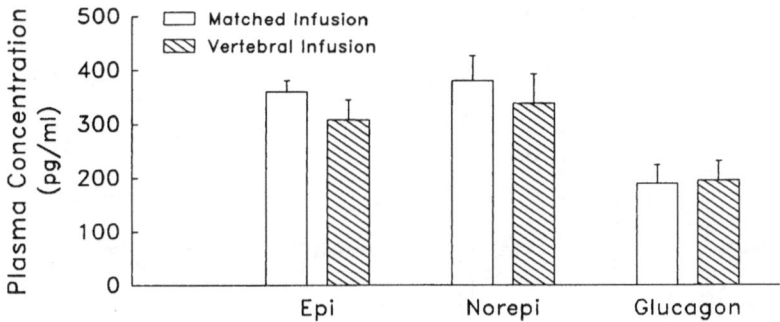

Fig. 7. Basal and hypoglycemic steady-state (90-160 minutes) values for insulin, epinephrine (Epi), norepinephrine (Norepi), and glucagon during vertebral glucose replacement (protocol 1) and matched infusion (protocol 2). No significant differences were observed between the vertebral glucose replacement and matched infusion experiments.

LIVER CLAMP

Equivocal results regarding the existence of essential glucoreceptors within the CNS offered the possibility that essential glucoreceptors reside in the periphery. Glucoreceptors have been identified at a number of loci other than the CNS, eg. oral cavity, duodenum, portal-hepatic region. Among these alternative glucoreceptors, those of the portal-hepatic region have received the greatest attention. Russek[22] originally proposed the existence of the portal-hepatic glucoreceptor based on temporal relationships between anorexic behavior and arterial-portal glucose differences. The portal-hepatic region is richly innervated by afferent fibers, some of which are sensitive specifically to glucose. Niijima[23,24] reported afferent discharges in the hepatic branch of the vagus, which were inversely related to the portal glucose concentration. Neurons of the hypothalamus, which are sensitive to elevations in portal glucose concentration have been reported.[25,26] Shimizu, et al[25] observed that the majority of identifiable glucose sensitive neurons in the LH are also responsive to portal glucose infusions. These observations along with the established efferent connections to the liver and endocrine glands have led to the proposal of a glucoregulatory reflex.[5,27] However, the potential role for these putative portal-hepatic glucoreceptors in the regulation of blood glucose concentration has remained controversial.[28]

While the appropriate neural network for a glucoregulatory reflex appeared to exist, a physiological role for the putative portal-hepatic glucoreceptors in glucoregulation remained to be elucidated. To address this issue we effected a 'liver clamp' (LC) preparation (Fig. 8), analogous to the 'brain clamp', to yield hepatic euglycemia during systemic hypoglycemia.[29] As with the 'brain clamp', it was reasoned that if essential glucoreceptors reside in the portal-hepatic region, then establishing euglycemia across the liver during systemic hypoglycemia should suppress the sympathoadrenal response.

Dogs were chronically cannulated under anesthesia one week prior to initiating the experiments. Cannulas were placed in the portal vein (glucose infusion for LC), the carotid artery (serial blood sampling), and jugular vein (insulin infusion). The femoral vein was cannulated with the tip of the catheter advanced to the inferior vena cava, rostral to the hepatic vein(s). An inflatable cuff was surgically implanted around the inferior vena cava just caudal to the hepatic vein (inflating the cuff temporarily occludes flow inferior to the cuff, and hepatic venous blood can then be sampled from the femoral catheter). Acute catheters were placed in the cephalic vein(s) as needed for indocyanine green dye (hepatic plasma flow measurement) and peripheral glucose infusion during the matched systemic infusion. A 30 minute basal collection period was followed by 150 minutes of insulin infusion (3.5 mU $\cdot$ kg^{-1} $\cdot$ min^{-1}).

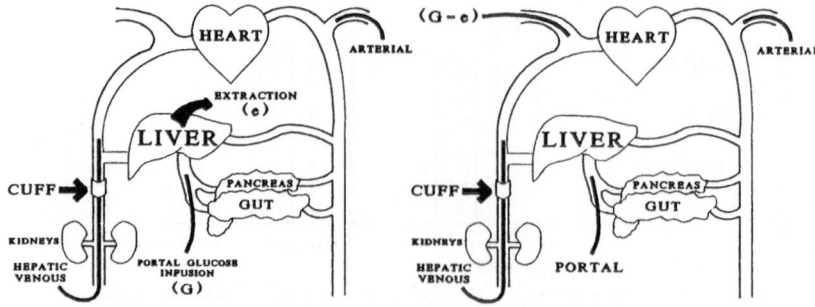

Fig. 8. Schematic representation of the two protocols for the "liver
clamp" study. The "liver clamp", involved insulin infusion with
portal glucose infusion (G) to establish hepatic euglycemia.
Extraction (e) represents hepatic glucose uptake. The "matched
infusion", involved insulin infusion with peripheral glucose
infused (G-e) to match the systemic glycemia observed during the
liver clamp.

For LC the portal glucose infusion rate was adjusted every 10 minutes to maintain the hepatic glycemia at basal arterial values, ~100 mg/dl. In control matched
infusion experiments (MI) glucose was infused peripherally at a rate calculated to
match arterial glycemia between LC and MI. Thus the principal difference between the
two treatments was the level of hepatic glycemia, ie. hypoglycemia for MI and euglycemia for LC.

Insulin infusion led to a rapid rise in plasma insulin concentration followed by a
sustained plateau (Fig. 9). In response to the rise in insulin, arterial glucose

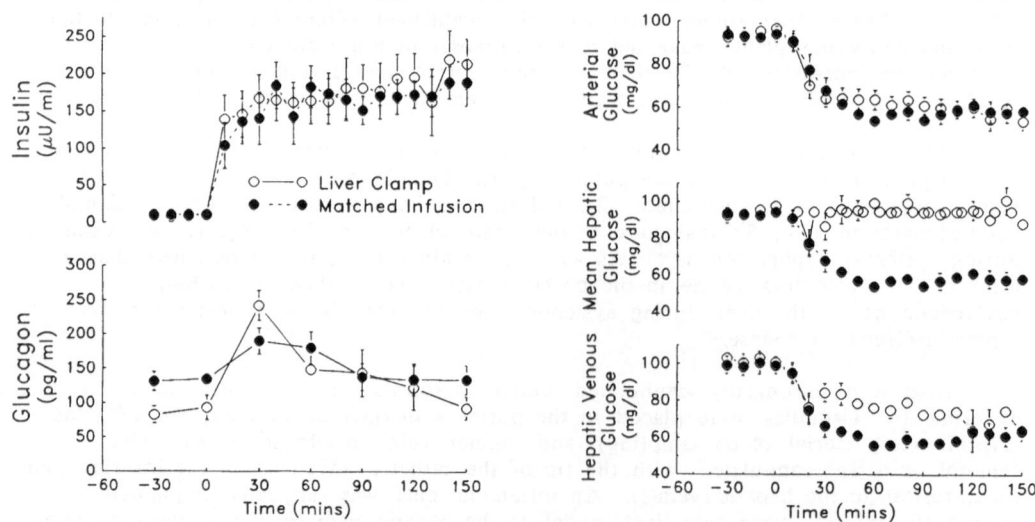

Fig. 9. Arterial glucose, mean hepatic glucose (concentration calculated to be reach-
ing the liver), hepatic venous glucose, insulin, and glucagon concentrations
as a function of time during the liver clamp and matched infusion protocols.
Basal period is -30 to 0 minutes. Experimental period is 0 to 150 minutes.
(Adapted from Donovan, C.M., et al, <u>Diabetes</u>, In press)

declined from 94 ± 4 mg/dl during the basal period to 58 ± 4 mg/dl during the liver clamp. The matched infusion yielded arterial blood glucose concentrations that were not significantly different from those obtained during LC. The calculated mean hepatic glucose concentration for MI, 59 ± 4 mg/dl, was significantly lower than that for LC, 95 ± 4 mg/dl. Hepatic venous glucose levels were significantly elevated between 30 and 110 minutes of LC when compared with values for MI. Hepatic venous glucose values during LC were lower than the calculated hepatic glucose concentrations. This was reflected in an average arterial-venous difference of 22 ± 5 mg/dl between 90 and 150 minutes. Despite systemic hypoglycemia, there was an apparent hepatic glucose uptake equivalent to 60% of the portal glucose infusion rate during the 'liver clamp'.

Hypoglycemia resulted in elevated catecholamine concentrations that peaked between 60-90 minutes and were sustained during the experiment. Basal values for epinephrine and norepinephrine were 103 ± 16 and 201 ± 32 pg/ml, respectively. During the matched infusion experiments, epinephrine and norepinephrine rose to plateau values of 676 ± 201 and 396 ± 58 pg/ml, respectively. Establishing euglycemia across the liver during systemic hypoglycemia resulted in a suppression of the epinephrine response (42 ± 7%, P=0.015) for all seven animals (Fig. 10). Six of seven animals demonstrated a similar response in norepinephrine values, ie. a 41 ± 10 % suppression during the liver clamp. One animal failed to demonstrate a similar response for norepinephrine, therefore the mean response for all animals was 32 ± 13%. The parallel increases for epinephrine and norepinephrine (r=0.97) suggest a common hepato-sympathoadrenal reflex governing the catecholamine response in hypoglycemia. This may reflect a common source, the adrenal medulla, for both catecholamines or a coordinated sympathetic activation of the adrenals and postganglionic neurons.

Clamping the liver at euglycemia failed to significantly impact either peak glucagon values (246 ± 22 and 214 ± 24 pg/ml for LC and MI, respectively) or glucagon values during the final hour of hypoglycemia (Fig. 9). Unlike the catecholamines, the glucagon response was transient peaking between 30-90 minutes of hypoglycemia then declining back towards basal.

Liver Clamp: Conclusions

A hepato-sympathoadrenal reflex, in which input from the liver afferents regarding ambient glucose concentration is integrated at higher centers yielding the appropriate

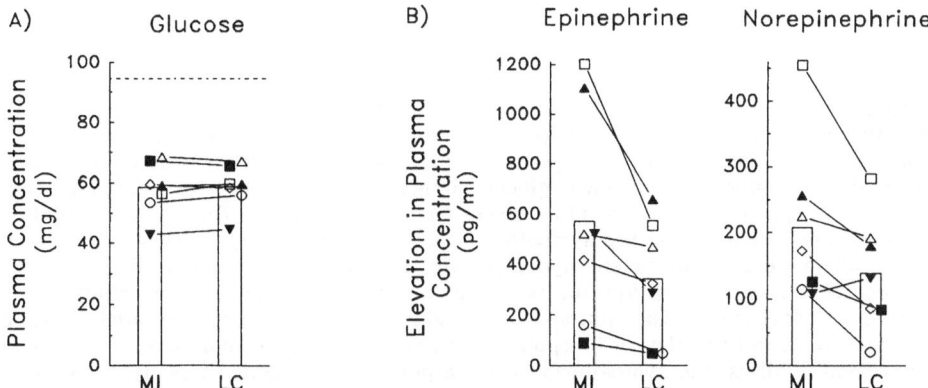

Fig. 10. A) Steady-state (90-150 mins) arterial glucose concentrations for the liver clamp (LC) and matched infusion (MI) experiments. Dashed line represents the mean basal glucose arterial glucose concentration. B) Average elevation in epinephrine and norepinephrine during the liver clamp and matched systemic infusion (90-150 mins). Bars represent mean values at steady-state (90-150 mins) for the two conditions. (From Donovan, C.M., et al, Diabetes, In press)

glucoregulatory response, has previously been hypothesized. [5,27] That the portal-hepatic glucoreceptors serve a functional role in the sympathoadrenal response to hypoglycemia was demonstrated. The relative contribution of the portal-hepatic glucoreceptors to neural regulation of this sympathetic response remains to be fully elucidated. That the liver clamp did not fully suppress the rise in catecholamine levels, offers the possibility of other important glucoreceptors, eg. the putative "redundant" CNS glucoreceptors. Alternatively, there may be significant autoregulation by the adrenal glands in response to local hypoglycemia. As epinephrine levels rise exponentially when the blood glucose concentration decreases below 55 mg/dl, [16] the portal-hepatic glucoreceptors may be of greater importance in the response to deep hypoglycemia.

The existence of essential portal-hepatic glucoreceptors raises the possibility that the diminished counterregulatory response often observed in diabetes may reflect hepatic insulin resistance. The specific nature of the diminished hypoglycemic response for glucagon and epinephrine (ie. responses to other stimuli are normal) suggests a deficient "sensory" mechanism. [30-32] Thus, insulin resistance in the putative portal-hepatic glucoreceptors may lead to reduced responsiveness to prevailing glycemia, resulting in a suppressed counterregulatory response. That many diabetics who demonstrate poor hypoglycemic counterregulation also demonstrate "hypoglycemic unawareness" [32] would tend to support this hypothesis. Hypoglycemic unawareness is associated with an inability to sense prevailing hypoglycemia and thus take appropriate corrective measures, eg. eating. This might be expected with impaired hepatic glucoreceptors as they have been strongly implicated in the control of appetite. [33]

A consistent observation among our brain and liver clamp studies was the failure of the clamp to have an impact upon the glucagon response. These findings contrast with Biggers, et al [21] who observed a marked suppression of the glucagon response with the entire brain 'clamped' at euglycemia. Again this discrepancy may be attributable to a redundancy in CNS glucoreceptors. However, if that is true then the neural control of the glucagon and catecholamine responses would appear to be dissociated from one another, at least partially. That is, the glucagon response to hypoglycemia would result from CNS detection of hypoglycemia, while the sympathoadrenal response would result in large part from portal-hepatic glucose detection. Alternatively, the glucagon response may be non-neurally mediated. Glycemia per se is viewed as a potent regulator of pancreatic secretion, with the alpha cells appearing to be relatively more sensitive to changes in blood glucose concentration. In support of non-neural mediation, it has been found that the glucagon response to hypoglycemia is unimpaired by vagotomy, sympathectomy, or cholinergic blockade. [34-36]

SUMMARY

To elucidate the loci for the putative glucoreceptors responding to hypoglycemia we introduced 'brain' and 'liver' clamps. Systemic hypoglycemia was induced by insulin infusion while the area of interest (ie. forebrain, hindbrain, portal-hepatic region) was maintained euglycemic via local glucose irrigation. Utilizing this approach, there appear to be no glucoreceptors residing exclusively in either the forebrain or hindbrain which are essential for the sympathoadrenal response to hypoglycemia. This is true for both moderate and severe hypoglycemic conditions. The possibility of a redundant glucoreceptor system within the brain, as suggested by a subsequent study, [21] remains to be confirmed. The portal-hepatic glucoreceptors appear essential to engendering the full counterregulatory response. Establishing euglycemia across the portal-hepatic region inhibits the sympathoadrenal response to moderate hypoglycemia by over 40%. Further, despite prevailing hypoglycemia and significant elevations in counterregulatory hormones, the liver demonstrated net glucose extraction during the liver clamp, suggestive of overriding neural input to the liver. Thus, the hepatic afferents appear to be very important for the counterregulatory response to hypoglycemia.

Acknowledgements

We are indebted to D. Banks, G. Hayes, and M. Smith for their technical assistance, and thank R. Watanabe for assistance in the statistical analyses. This work was supported by an NIH grant (DK-27619) to R.N. Bergman, and a grant from the American Diabetes Association to C.M. Donovan.

REFERENCES

1. C. Bernard, Chiens rendus diabetiques, Compt. Rend. Soc. Biol. 1:60 (1849).
2. E. Sachs and M.E. MacDonald, Blood sugar studies in experimental pituitary and hypothalamic lesions, Arch. Neurol. Psychiat. 13:335 (1925).
3. L.O. Morgan and C.A. Johnson, Experimental lesions in the tuber cinereum of the dog, Arch. Neurol. Psychiat. 24:696 (1930).
4. C.A. Benzo, The hypothalamus and blood glucose regulation, Life Sci. 32:2509 (1983).
5. T. Shimazu, Neuronal regulation of hepatic glucose metabolism in mammals, Diabetes Metab. Rev. 3:185 (1987).
6. T. Shimazu, A. Fukuda, and T. Ban, Reciprocal influences of the ventromedial and lateral hypothalamic nuclei on blood glucose level and liver glycogen content, Nature 210:1178 (1966).
7. L.A. Frohman and L.L. Bernardis, Effect of hypothalamic stimulation on plasma glucose, insulin and glucagon levels, Am. J. Physiol. 221:1596 (1971).
8. S.C. Woods and D. Porte, Jr, Neural control of the endocrine pancreas, Physiol. Rev. 54:596 (1974).
9. T. Shimazu and K. Ishikawa, Modulation by the hypothalamus of glucagon and insulin secretion in rabbits: studies with electrical and chemical stimulations, Endocrinol. 108:605 (1981).
10. T. Shimazu, H. Matsushita, and K. Ishikawa, Hypothalamic control of liver glycogen metabolism in adult and aged rats, Brain Res. 144:343 (1978).
11. T. Shimazu and S. Ogasawara, Effect of hypothalamic stimulation on gluconeogenesis and glycolysis in rat liver, Am. J. Physiol. 228:1787 (1975).
12. W.B. Cannon, M.A. McIver, and S.W. Bliss, Studies on the conditions of activity in endocrine glands. XIII. A sympathetic and adrenal mechanism for mobilizing sugar in hypoglycemia, Am. J. Physiol. 69:46 (1924).
13. R.R. Miselis and A.N. Epstein, Feeding induced by intracerebroventricular 2-deoxyglucose in the rat, Am. J. Physiol. 229:1438 (1975).
14. Y. Oomura, Glucose as a regulator of neuronal activity, in: "Advances in Metabolic Disorders," vol. 10, A.J. Szabo, ed., Academic Press, New York (1983).
15. P. Cane, R. Artal, and R.N. Bergman, Putative hypothalamic glucoreceptors play no essential role in the response to moderate hypoglycemia, Diabetes 35:268 (1986).
16. P. Cane, C.K. Haun, J. Evered, J.H. Youn, and R.N. Bergman, Response to deep hypoglycemia does not involve glucoreceptors in carotid perfused tissue, Am. J. Physiol. 255:E680 (1988).
17. R.S. Reneman, D. Wellens, A.H.M. Jageneau, and L. Stynen, Vertebral and carotid blood distribution in the brain of the dog and the cat, Cardiovasc. Res. 8:65 (1974).
18. D.L.F. Wellens, J.M.R. Lucien, R.J.J. Wouters, P.B. DeReese, and R.S. Reneman, The cerebral blood distribution in dogs and cats. An anatomical and functional study, Brain Res. 86:42 (1975).
19. R.J. DiRocco and H.J. Grill, The forebrain is not essential for sympathoadrenal hyperglycemic response to glucoprivation, Science 204:1112 (1979).
20. R.C. Ritter, P.G. Slusser, and S. Stone, Glucoreceptors controlling feeding and blood glucose: Location in the hindbrain, Science 213:451 (1979).
21. D.W. Biggers, S.R. Myers, D. Neal, R. Stinson, N.B. Cooper, J.B. Jaspan, P.E. Williams, A.D. Cherrington, and R.T. Frizzell, Role of brain in counterregulation of insulin-induced hypoglycemia in dogs, Diabetes 38:7 (1989).

22. M. Russek, Participation of hepatic glucoreceptors in the control of intake of food, Nature 197:79 (1963).
23. A. Niijima, Afferent impulse discharges from glucoreceptors in the liver of the guinea pig, Ann. New York Acad. Sci. 157:690 (1969).
24. A. Niijima, Glucose-sensitive afferent nerve fibers in the liver and their role in food intake and blood glucose regulation, J. Autonom. Nerv. Syst. 9:207 (1983).
25. N.Y. Shimizu, Y. Oomura, D. Novin, C.V. Grijalva, and P.H. Cooper, Functional correlation between lateral hypothalamic glucose-sensitive neurons and hepatic portal glucose-sensitive units in rats, Brain Res. 265:49 (1983).
26. M. Schmitt, Influences of hepatic portal receptors on hypothalamic feeding and satiety centers, Am. J. Physiol. 225:1089 (1973).
27. A. Niijima and N. Mei, Glucose sensors in viscera and control of blood glucose level, NIPS 2:164 (1987).
28. E.M. Stricker, N. Rowland, C.F. Saller, and M.I. Friedman, Homeostasis during hypoglycemia: Central control of adrenal secretion and peripheral control of feeding, Science 196:79 (1977).
29. C.M. Donovan, J.B. Halter, and R.N. Bergman, Importance of hepatic glucoreceptors in the sympathoadrenal response to hypoglycemia, Diabetes (In Press).
30. J.E. Gerich and P.J. Campell, Overview of counterregulation and its abnormalities in diabetes mellitus, Diabetes Metab. Rev. 4:93 (1988).
31. P.E. Cryer and J.E. Gerich, Relevance of glucose counterregulatory systems to patients with diabetes: Critical roles of glucagon and epinephrine, Diabetes Care 6:95 (1983).
32. S.A. Amiel, W.V. Tamborlane, L.Sacca, and R.S. Sherwin, Hypoglycemia and glucose counterregulation in normal and insulin-dependent diabetic subjects, Diabetes Metab. Rev. 4:71 (1988).
33. E.M. Stricker and M.J. McCann, Visceral factors in the control of food intake, Brain Res. Bull. 14:687 (1985).
34. J. Palmer, D. Henry, J. Benson, D. Johnson, and J. Ensinck, Glucagon response to hypoglycemia in sympathectomized man, J. Clin. Invest. 57:522 (1976).
35. J. Palmer, P. Werner, P. Hollander, and J. Ensinck, Evaluation of the control of glucagon secretion by the parasympathetic nervous system in man, Metabolism 28:549 (1979).
36. R.J.M. Corrall and B.M. Frier, Acute hypoglycemia in man: central control of the pancreatic islet cell function, Metabolism 30:160 (1981).

HYPOGLYCEMIA, GLUCONEOGENESIS AND THE BRAIN

A.D. Cherrington, R.T. Frizzell, D.W. Biggers and C.C. Connolly

Department of Molecular Physiology and Biophysics
Vanderbilt University Medical School
Nashville, TN 37232-0615

The purpose of this chapter is to review the role which gluconeogenesis plays in the response to insulin-induced hypoglycemia and to assess the influence of the brain in driving that response. The chapter is divided into three parts: the response of gluconeogenesis to hypoglycemia; the role of the brain in driving hormone secretion, and as a result gluconeogenesis, during hypoglycemia; and the role of the brain in stimulating gluconeogenesis directly during glucodeprivation.

I. Role of Gluconeogenesis During Insulin-Induced Hypoglycemia

Garber et al. (1) were the first to assess the role of gluconeogenesis during insulin-induced hypoglycemia. They measured the conversion of ^{14}C-alanine to ^{14}C-glucose following a 0.15 U/kg insulin injection in man. This conversion initially decreased but had increased by 200% two hours following insulin injection. Clark et al. (2) also assessed the conversion of alanine to glucose in man following insulin injection (0.05 U/kg). These investigators noted that gluconeogenic conversion had increased by 268% within an hour of the insulin injection. The mechanism by which gluconeogenesis is augmented following insulin injection and the relative importance of gluconeogenesis and glycogenolysis were not determined in their studies. In addition, since hypoglycemia was induced by insulin injection, it is likely that the increases in gluconeogenesis occurred as the insulin concentration fell. Thus, estimates of gluconeogenesis during hypoglycemia resulting from constant hyperinsulinemia also remained unclear.

Gauthier and Hetenyi (3) investigated gluconeogenesis during hypoglycemia induced by phlorizin infusion in dogs. In response to hypoglycemia gluconeogenesis initially accounted for only a small portion of glucose production, but the conversion of alanine to glucose then increased significantly and by 90 minutes was elevated by 125%. This study, taken collectively with those of Garber et al. (1) and Clark et al. (2), demonstrates that gluconeogenesis is important in increasing glucose production and thereby limiting hypoglycemia.

Frizzell et al. (4) attempted to quantitate the importance of gluconeogenesis during hypoglycemia caused by a continuous insulin infusion (5 mU·kg^{-1}·min^{-1} given intraportally) in the overnight-fasted conscious dog. Gluconeogenesis was assessed by measuring the net hepatic balance of the gluconeogenic substrates, as well as by assessing the conversion of ^{14}C-alanine to ^{14}C-glucose. In response to the insulin infusion, the plasma insulin level increased approximately 20-fold and the plasma glucose concentration fell to a steady state value of approximately 42 mg/dl (Figure 1). The levels of glucagon, epinephrine, norepinephrine,

and cortisol all increased in response to the hypoglycemia, but the canine growth hormone (cGH) levels remained unchanged. Glucose production (R_a) initially decreased but then increased two-fold (from 3.1 to 6.1 mg·
kg^{-1}·min^{-1}) by 30 min and remained at that rate throughout the insulin infusion period (Figure 1).

The net hepatic uptake of the gluconeogenic precursors increased 7-fold 60 min into hypoglycemia (Table I). In contrast, by the third hour of insulin infusion and low glucose, the net hepatic uptake of gluconeogenic precursors had increased about 30-fold. At that time the net hepatic uptake of lactate, glycerol, and the gluconeogenic amino acids each accounted for approximately one-third of the total gluconeogenic precursor uptake. In the case of lactate and glycerol it was the increased substrate load to the liver which was primarily responsible for the increase in lactate and glycerol uptake. On the other hand, the increase in uptake of the gluconeogenic amino acids was partly the result of an increase in their fractional extraction by the liver.

The gluconeogenic conversion of alanine to glucose (which is a reflection of the overall gluconeogenic rate) increased 301±85% by the third hour of insulin infusion whereas the efficiency of conversion of alanine to glucose (which is a measure of the intrahepatic gluconeogenic efficiency) rose from 27%±14 to 55%±18 (Figure 2). Since the specific activity of ^{14}C-alanine is diluted within the hepatocyte at the oxaloacetate level, both of these parameters underestimate the actual increase in gluconeogenesis.

The quantitative contribution of gluconeogenesis to glucose production can be calculated in two ways, one giving a maximum and the other a minimum estimate of the process. The maximal gluconeogenic contribution to glucose production can be calculated by summing the net hepatic uptakes of the various gluconeogenic precursors and dividing this sum by two to account for the incorporation of the 3-carbon precursors into the 6-carbon glucose molecule. This quotient can then be divided by the net hepatic glucose balance (μmol·kg^{-1}·min^{-1}) and multiplied by 100 to obtain maximal % contribution to NHGB. Since pyruvate was not measured its contribution to gluconeogenesis was assumed to be 1/10 the contribution of lactate. The minimal contribution of gluconeogenesis to glucose production can be calculated by multiplying the maximal gluconeogenic contribution by the gluconeogenic efficiency (which represents a minimum value). Using these calculations, gluconeogenesis was estimated to account for 3 to 9% of net hepatic glucose production during the control period, 13 to 31% of glucose output at the end of the first hour of insulin infusion and 48 to 87% of net hepatic glucose production by the third hour of insulin infusion. The time course of the gluconeogenic contribution to glucose production during hypoglycemia is illustrated in Figure 3.

At the end of the experiments liver biopsies were taken for analysis of ^{14}C glycogen. The liver contained an amount of ^{14}C-glucose equal to approximately 15% of the total ^{14}C-glucose produced by the liver during the 3 hour hypoglycemic period. Thus, the total glucose synthesis via gluconeogenesis (glucose + glycogen) during hypoglycemia was even greater than thought. In addition, total liver glycogen content at the end of 3 hours of insulin-induced hypoglycemia or saline infusion as 26±3 and 26±4 grams, respectively. This finding suggests that glycogenolysis, after an initial increase in response to insulin-induced hypoglycemia, actually decreased to a rate below basal. Alternatively, the rate of glycogenolysis may have remained elevated in these studies in the face of ongoing synthesis of new glycogen from gluconeogenic precursors (as evidenced by the presence of ^{14}C in the liver). In either situation there was no greater net utilization of glycogen during 3 hours of hypoglycemia than during 3 hours of saline infusion.

Lecavalier, Bolli and Gerich have recently reported that gluconeogenesis also plays a major role in hypoglycemic counterregulation in man

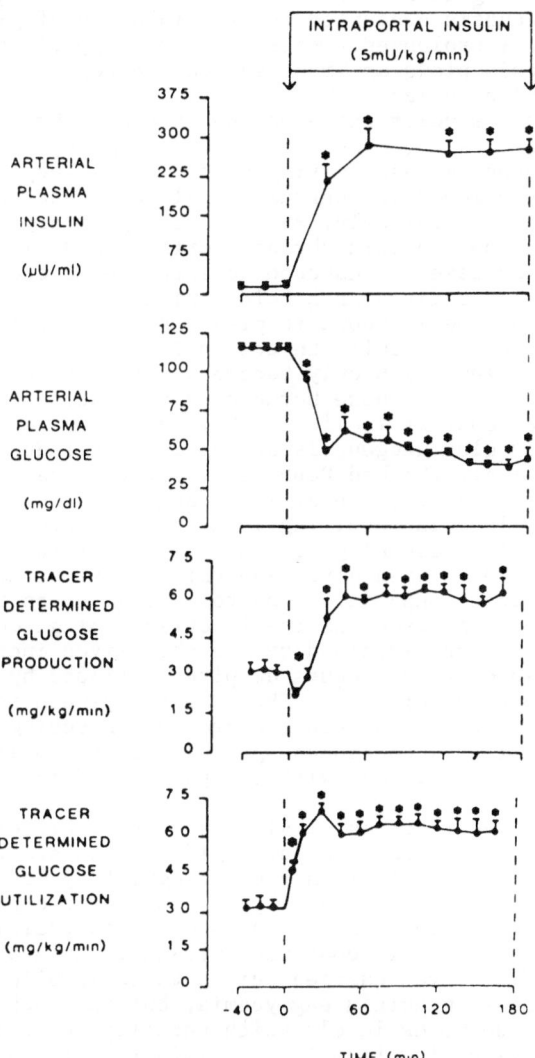

Fig. 1. The effect of intraportal insulin infusion (5.0 mU·kg⁻¹·min⁻¹) on the plasma glucose level and turnover in 5 overnight-fasted, conscious dogs. Data are mean ± SEM. The figure is redrawn from data in reference 4.

(5). These authors assessed glucose production using 6-^{3}H glucose and gluconeogenesis using ^{14}C-lactate. Insulin was infused at 0.35 mU· kg^{-1}·min^{-1} for eight hours, over which time the glucose level fell to 57 mg/dl. During the first three hours of insulin infusion, very little of the glucose produced was derived from ^{14}C- lactate, whereas gluconeo- genesis was estimated to account for 60% of the glucose produced during the last three hours of the study. The authors concluded that the initial increase in glucose production was due to glycogenolysis, whereas gluconeogenesis accounted for the majority of glucose production as hypoglycemia was prolonged. These results support the conclusion that gluconeogenesis plays an important counterregulatory role not only in the dog, but also in man.

While the above studies extended the previous observations by Garber et al. (1), Clark et al. (2) and Gauthier and Hetenyi (3) on the importance of gluconeogenesis during hypoglycemia they did not clarify the mechanisms by which this increase in gluconeogenesis was brought about. As discussed previously, the counterregulatory hormones glucagon, epinephrine, norepinephrine, cortisol and, in man at least, growth hormone, increase in response to insulin-induced hypoglycemia whether the insulin is given as a bolus injection or as a constant infusion. Each hormone is known to play a role in sustaining glucose production following an insulin injection (6-8) although the latter three have minor roles which only become apparent after prolonged or deep hypoglycemia. Since these hormones have gluconeogenic as well as glycogenolytic actions, it is likely that they play an important role in the stimulation of gluconeogenesis during insulin-induced hypoglycemia.

Work by Ivy et al. (9) and Dobbins et al. (10) have confirmed that both epinephrine and glucagon play important roles in the response of the liver to hypoglycemia resulting from insulin infusion. Their work showed that the major role of glucagon is manifest in the first 90 minutes of hypoglycemia when glycogenolysis is the primary process by which glucose is being supplied. The role of glucagon in stimulating glucose production diminishes as the importance of gluconeogenesis is augmented. This may be explained by the shortlived increase in the plasma glucagon level which occurs despite continued hypoglycemia resulting from insulin infusion. The rise in epinephrine, on the other hand, has its major effect on glucose production during more prolonged hypoglycemia at a time when gluconeogenesis was maximally elevated. Again, this role is consistent with the pattern of increase of the hormone in plasma.

The aim of the next series of studies by Frizzell et al. (11) was to assess whether hypoglycemia stimulated gluconeogenesis through mechan- isms in addition to its effect on counterregulatory hormone release. In these studies, a control protocol (hyperinsulinemic-euglycemic clamp) was performed in which insulin was infused intraportally at 5 mU· kg^{-1}·min^{-1} as before, and glucose was infused to maintain euglycemia. An additional protocol was carried out in which insulin was again given, along with glucose to maintain euglycemia, but the increases in the counterregulatory hormones levels which normally occur in response to the hypoglycemia resulting from insulin infusion at 5.0 mU kg^{-1}·min^{-1} were simulated by infusion. The increments of glucagon, epinephrine, norepinephrine, and cortisol in these hormone-replacement experiments were 102±6%, 106±14%, 117±9%, and 124±37%, respectively, of their incre- ments during insulin-induced hypoglycemia.

Glucose production, as assessed by both the tracer and A-V differ- ence methods, was significantly less when hypoglycemia did not accompany the rise in counterregulatory hormone levels. In fact, glucose produc- tion in this setting was only 54±10% (using the tracer method) of that which occurred when hypoglycemia accompanied the rise in the counter- regulatory hormone level (Figure 4). Measurements made using the A-V difference method yielded similar results. Since the difference between the studies was the glycemic level, hypoglycemia per se appeared to play

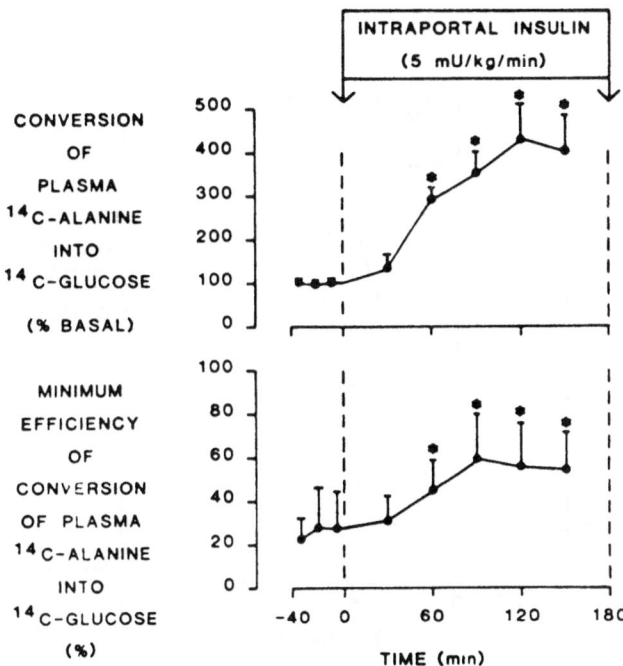

Fig. 2. The effect of intraportal insulin infusion (5.0 mU·kg⁻¹·min⁻¹) on the conversion rate of ^{14}C-alanine to ^{14}C-glucose and on the efficiency of that conversion in 5 overnight-fasted, conscious dogs. Data are mean ± SEM. The figure is taken from reference 4.

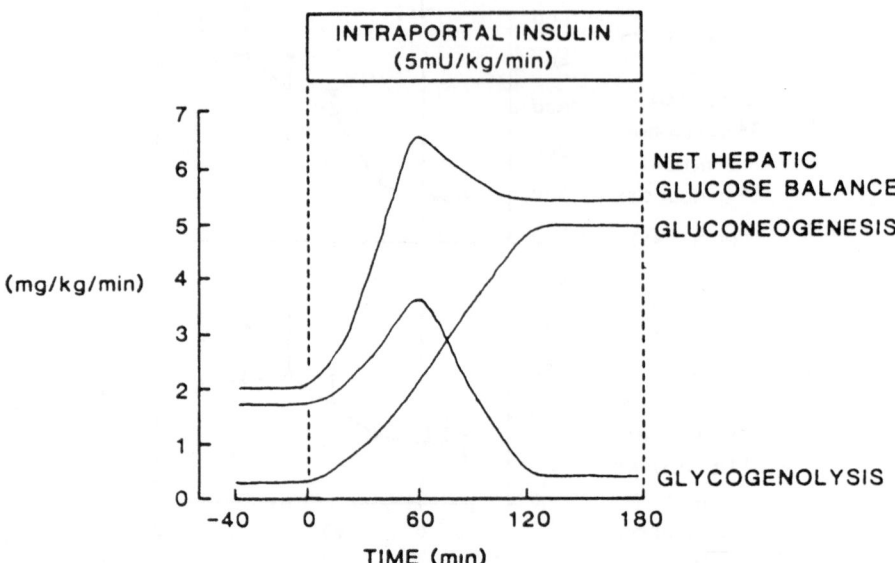

Fig. 3. Estimates of the maximal gluconeogenic and minimal glycogenolytic contributions to net hepatic glucose balance during insulin-induced hypoglycemia. Taken from reference 15 and based on the data shown in Figures 1 and 2 and Table I.

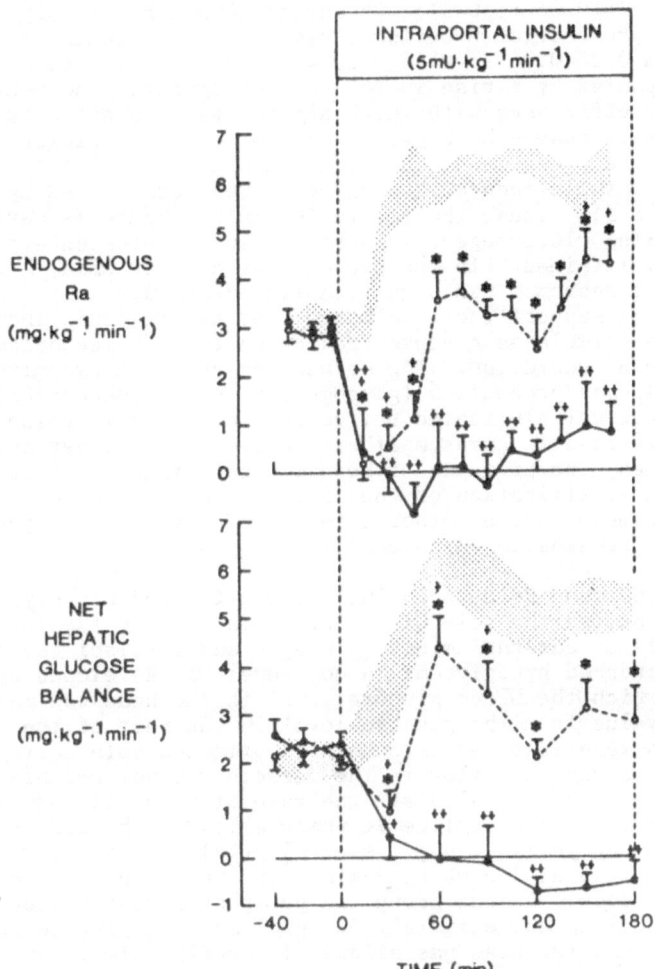

Fig. 4. Glucose production measured with tracer (^{3}H-3 glucose; endogenous R_a) or A-V difference (NHGB) techniques in dogs infused with insulin intraportally at 5.0 mU·kg^{-1}·min^{-1}. The shaded area represents mean ± SEM of the data taken from Figure 1. Data are from animals in which euglycemia was maintained by peripheral glucose infusion and saline (solid line; n=5) or replacement hormone infusions (broken line; n=6) were given. Data are mean ± SEM and are taken from reference 11.

an important role in sustaining glucose production and accounted for approximately 50% of glucose production.

The gluconeogenic substrate load to the liver increased six-fold during counterregulatory hormone infusion in the presence of hyper-insulinemia and euglycemia, an amount equal to approximately 75% of that present when hypoglycemia accompanied the rise in the counterregulatory hormone levels. The fractional extraction of alanine by the liver increased from 0.34 ± 0.06 to 0.51 ± 0.05 during counterregulatory hormone infusion in the presence of hyperinsulinemia and euglycemia, compared to an increase from 0.27 ± 0.06 to 0.64 ± 0.10 when insulin-induced hypogly-cemia was accompanied by a rise in the counterregulatory hormone levels. In addition, the efficiency with which alanine was converted to glucose within the liver increased by $87 \pm 48\%$ when the counterregulatory hormones were infused in the presence of euglycemia and hyperinsulinemia compared to $174 \pm 53\%$ when insulin-induced hypoglycemia was accompanied by hormone secretion (Figure 5). Thus, the counterregulatory hormones were effec-tive in stimulating gluconeogenesis during insulin infusion even if euglycemia was maintained. In the presence of hypoglycemia, however, the gluconeogenic response was significantly enhanced.

These findings support the conclusion that hypoglycemia induced by insulin infusion stimulates glucose production through mechanisms in addition to hormone secretion. The mechanisms by which hypoglycemia per se stimulated glucose production (glycogenolysis, gluconeogenesis) in these studies were not elucidated but could include an increase in sym-pathetic or a decrease in parasympathetic drive to the liver and/or peripheral tissues, the presence of hepatic or peripheral autoregulatory responses, and a sensitization of the liver to the actions of the coun-terregulatory hormones and/or other stimuli. The relative importance of these mechanisms remains to be assessed.

II. Control of Hormone Release During Insulin-Induced Hypoglycemia

In order to assess the role of the brain in driving the increases in the plasma levels of catecholamines, glucagon and cortisol which occur during insulin-induced hypoglycemia a dog model was developed by Biggers et al. (12) in which the blood glucose level in the head was maintained at a different value from the glucose level in the rest of the body. In brief, catheters were inserted into both carotid and both vertebral arteries in such a way that flow in the vessels was not occluded. An algorithm (12) was used to calculate the rate at which to infuse glucose in order to keep the brain euglycemic while allowing the rest of the body to become hypoglycemic. Biggers et al. (12) then infused insulin via a peripheral vein at 3.5 mU$\cdot$kg$^{-1}\cdot$min^{-1} in two groups of conscious, overnight-fasted dogs. In one group the peripheral plasma glucose level was allowed to fall to approximately 50 mg/dl and the glucose level in the plasma perfusing the head was allowed to decline similarly. In the other group the head glucose level was maintained at approximately 85-90 mg/dl while the peripheral glucose level was allowed to fall as in the other group (50 mg/dl). In the absence of marked head hypoglycemia the glucagon level fell, rather than rose (Figure 6), the epinephrine, cortisol and pancreatic polypeptide levels rose only slightly (Figure 6), and the norepinephrine levels did not rise appreciably (Figure 6). Under these circumstances glucose production was markedly reduced (Figure 7). Thus, when insulin was infused at 3.5 mU kg$^{-1}\cdot$min^{-1}, lowering plasma glucose to 50 mg/dl, the primary drive for glucagon secretion came from the CNS. As expected, the secretion of epinephrine from the adrenal medulla, as well as cortisol from the adrenal cortex, was also regulated by neural input. In accord with the previously noted role of each hormone in stimulating hepatic glucose production, ablation of their increases resulted in a marked reduction in the normal increase in glucose production. Thus the CNS, by virtue of its action on the adrenal and pancreas, is responsible for approximately 50% of the increase in glucose production resulting from the infusion of 5.0 mU·kg^{-1}·min^{-1} insulin and the associated hypoglycemia.

III. Autoregulation and Direct Neural Input to the Liver

If the counterregulatory hormones are responsible for the stimulation of half of the normal response to insulin-induced (5.0 mU·kg^{-1}·min^{-1}) hypoglycemia the question arises as to what is responsible for the rest. The two most likely explanations are direct neural drive to the liver and peripheral tissues or autoregulatory responses at these sites. Connolly et al. (13) attempted to dissect these two alternatives by using the "head" clamp technique to control the brain glucose level in dogs which were deprived of all of the classic counterregulatory hormones. Epinephrine and cortisol were eliminated by adrenalectomy while glucagon and canine growth hormone were eliminated by the administration of somatostatin (Table II). Insulin was infused at 5.0 mU·kg^{-1}·min^{-1} intraportally and the peripheral plasma glucose level was allowed to fall to 42 mg/dl. In one protocol the glucose level in the head was also at 42 mg/dl while in the other glucose was infused into the carotid and vertebral arteries to sustain head euglycemia ($\approx$100 mg/dl). Net hepatic glucose production was $\approx$ -1.4 mg·kg^{-1}·min^{-1} in the control period of both protocols in the presence of $\approx$250 µU/ml arterial plasma insulin, a plasma glucose level of 100 mg/dl, and the absence of any measurable amount of the counterregulatory hormones. When the glucose infusion rate was reduced and hypoglycemia was allowed to develop net hepatic glucose production became significant in both groups (Table III). Somewhat less glucose was produced by the liver when the brain was kept euglycemic but this could be accounted for by repression of lipolysis and subsequent reduction in the glycerol and FFA loads to the liver. The gluconeogenic responses in each protocol (Table IV) were similar with the exception of glycerol. These studies thus led to the conclusion that the CNS has little, if any, direct influence on the liver. The non-hormonally stimulated response of the liver to insulin-induced hypoglycemia appears instead to result from autoregulation, a phenomenon also evident in perfused liver preparations (13,14). The CNS does appear to exert an indirect effect on glucose production by the liver in addition to that which it brings about by virtue of its effect on hormone secretion. This indirect action, however, results from a sensitive control of lipolysis mediated presumably through norepinephrine release form nerve endings. Thus, if one adds this contribution to that resulting from glucagon and epinephrine release one would argue that closer to two thirds of the increase in glucose production seen in response to 5.0 mU·kg^{-1}·min^{-1} insulin infusion in the dog is driven by the CNS.

In summary, an increase in glucose production is a vital part of the normal defense of the body against insulin-induced hypoglycemia. That response involves an early increase in glycogen breakdown and a more slowly (>90 min) developing enhancement of gluconeogenesis. The latter is driven by increases in both glucagon and epinephrine with the latter being predominant. The increases in the level of the counterregulatory hormones are themselves driven primarily by the CNS. In addition, if the hypoglycemia is deep enough there is an additional autoregulatory response of the liver which results in a stimulation of both glycogenolysis and gluconeogenesis. Direct neural input to the liver does not appear to play a significant role in stimulating glucose production under such a circumstance. On the other hand direct neural input to adipose tissue does appear to augment lipolysis and this indirectly influences hepatic glucose production, both by providing a gluconeogenic substrate (glycerol) and a source of energy for gluconeogenesis (FFA).

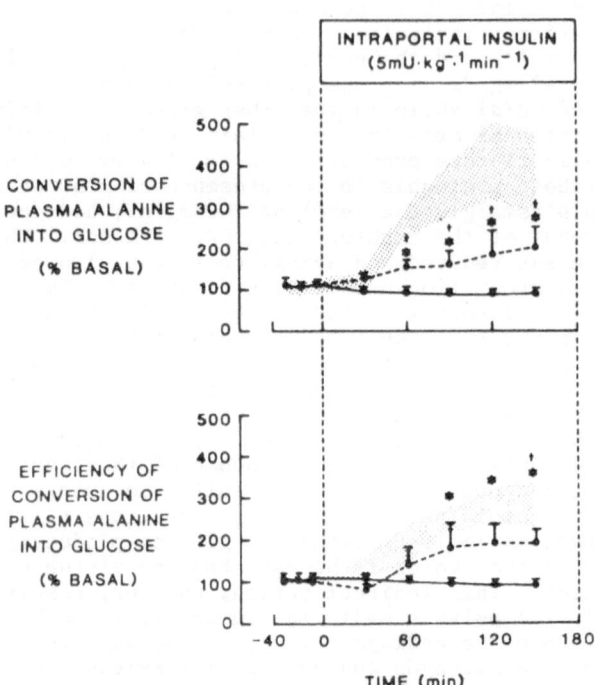

Fig. 5. Gluconeogenic conversion of ^{14}C-alanine to ^{14}C-glucose and the efficiency of that conversion in dogs infused with insulin intraportally at 5.0 mU·kg^{-1}·min^{-1}. The shaded area represents mean ± SEM of the data taken from Figure 2. Data are from animals in which euglycemia was maintained by peripheral glucose infusion and saline (solid line; n=5) or replacement hormone infusions (broken line; n=6) were given. Data are mean ± SEM and are taken from reference 11.

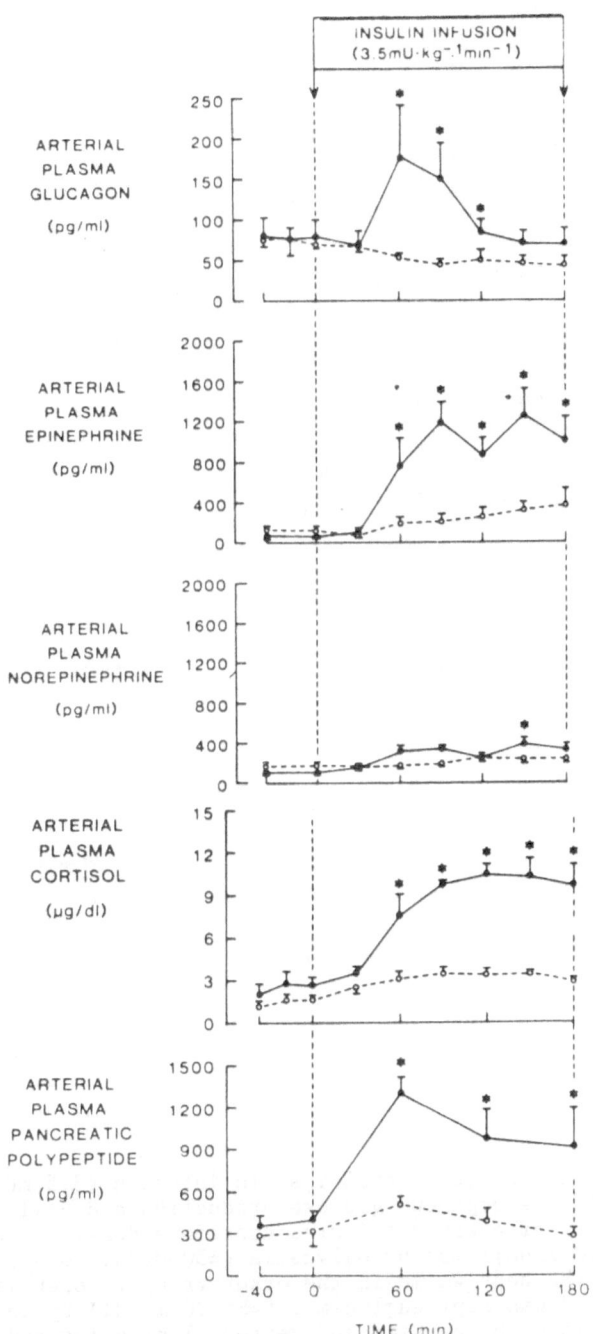

Fig. 6. Effect of peripheral insulin infusion (3.5 mU·kg⁻¹,min⁻¹) on arterial plasma glucagon, epinephrine, norepinephrine, cortisol and pancreatic polypeptide in two groups of overnight-fasted, conscious dogs. In one group (solid line; n=5) hypoglycemia (≃50 mg/dl) developed throughout the body while in the other group (broken line; n=4) the head was kept euglycemic (≃85-90 mg/dl) by infusion of glucose into the carotid and vertebral arteries and the peripheral hypoglycemia was maintained at 50 mg/dl. The data are redrawn from reference 12.

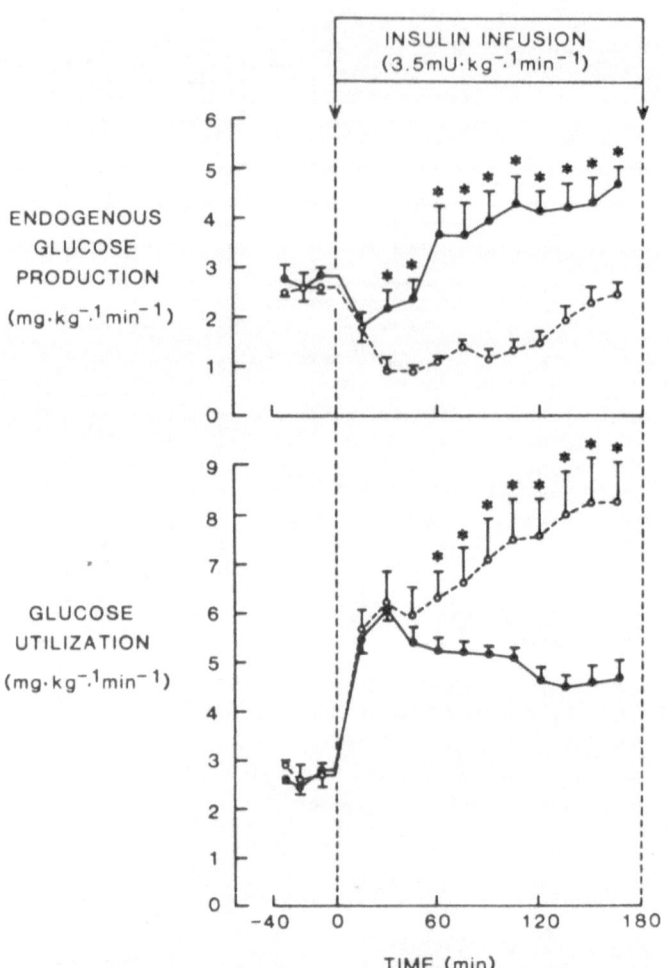

Fig. 7. Effect of peripheral insulin infusion (3.5 mU·
kg⁻¹·min⁻¹) on endogenous glucose production and utilization
in two groups of overnight fasted conscious dogs. In one
group (solid line; n=5) hypoglycemia (≈50 mg/dl) developed
throughout the body while in the other group (broken line;
n=4) the head was kept euglycemic (≈85-90 mg/dl) by infusion
of glucose into the carotid and vertebral arteries and the
peripheral hypoglycemia was maintained at 50 mg/dl. The data
are redrawn from reference 12.

Table I

The effects of insulin-induced ($5.0 \ mU \cdot kg^{-1} \cdot min^{-1}$) hypoglycemia on net hepatic gluconeogenic precursor uptake ($\mu mol \cdot kg^{-1} \cdot min^{-1}$) in the overnight-fasted, conscious dog (n=5). Values are mean ± SEM.

	Control Period	Minutes of Hypoglycemia 30'	60'	120'	180'
Gluconeogenic Amino Acids	4.5±0.6	7.7±1.9	8.8±2.1	13.4±2.1	12.5±3.1
Lactate	-3.9±2.7	2.5±2.8	3.5±2.1	21.5±4.2	19.6±6.0
Glycerol	1.3±0.4	3.5±1.8	8.4±2.1	13.6±2.8	16.5±3.2
Total	1.9	13.7	20.7	48.5	48.6

Table II

Insulin and counterregulatory hormone levels during insulin-induced ($5.0 \ mU \cdot kg^{-1} \cdot min^{-1}$) peripheral hypoglycemia ($\simeq 40$ mg/dl) in adrenalectomized, overnight-fasted, conscious dogs (lacking glucagon, epinephrine and cortisol) with either cerebral euglycemia (n=5), maintained via glucose infusion into the carotid and vertebral arteries, or cerebral hypoglycemia (n=6). Values are mean ±SEM.

	HEAD-EUGLYCEMIC control	peripheral hypoglycemia	HEAD-HYPOGLYCEMIC control	peripheral hypoglycemia
Insulin (µU/ml)	228±49	244±56	263±44	306±51
Glucagon (pg/ml)	<30	<30	<30	<30
3500 M.W. Glucagon	0	0	0	0
Cortisol (µg/dl)	<0.5	<0.5	<0.5	<0.5
Epinephrine (pg/ml)	<30	<30	<30	<30
Norepinephrine (pg/ml)	185±11	324±22	190±59	409±110

Table III

Net hepatic glucose balance ($mg \cdot kg^{-1} \cdot min^{-1}$; positive value indicates output) during insulin-induced ($5.0\ mU \cdot kg^{-1} \cdot min^{-1}$) peripheral hypoglycemia ($\simeq 40\ mg/dl$) in adrenalectomized, overnight-fasted, conscious dogs (lacking glucagon, epinephrine and cortisol) with either cerebral euglycemia (n=5), maintained via glucose infusion into the carotid and vertebral arteries, or cerebral hypoglycemia (n=6). Values are mean $\pm$SEM.

		Minutes of hypoglycemia		
	control	60'	120'	180'
Net Hepatic Glucose Balance ($mg \cdot kg^{-1} \cdot min^{-1}$)				
Head-euglycemic	-1.42 ± 0.64	0.53 ± 0.11	0.47 ± 0.36	0.61 ± 0.28
Head-hypoglycemic	-1.29 ± 0.37	1.18 ± 0.51	1.15 ± 0.63	0.74 ± 0.50

Table IV

Gluconeogenic conversion of alanine to glucose (% basal) and intrahepatic gluconeogenic efficiency during insulin-induced ($5.0\ mU \cdot kg^{-1} \cdot min^{-1}$) peripheral hypoglycemia ($\simeq40\ mg/dl$) in adrenalectomized, overnight-fasted, conscious dogs (lacking glucagon, epinephrine and cortisol) with either cerebral euglycemia (n=5), maintained via glucose infusion into the carotid and vertebral arteries, or cerebral hypoglycemia (n=6).

	control	Minutes of hypoglycemia		
		60'	120'	180'
Gluconeogenic Conversion				
Head-euglycemic	100 ± 2	101 ± 5	98 ± 19	107 ± 16
Head-hypoglycemic	100 ± 2	128 ± 18	189 ± 43	121 ± 14
Gluconeogenic Efficiency				
Head-euglycemic	0.17 ± 0.07	0.24 ± 0.10	0.29 ± 0.12	0.30 ± 0.11
Head-hypoglycemic	0.09 ± 0.01	0.15 ± 0.02	0.21 ± 0.06	0.17 ± 0.04

References

1. Garber, A.J., Cryer, P.E., Santiago, J.V., Haymond, M.W., Pagliara, A.S. and Kipnis, D.M. The role of adrenergic mechanisms in the substrate and hormonal response to insulin-induced hypoglycemia in man. J. Clin. Invest. 8: 7-15, 1976.
2. Clark, W.L., Santiago, J.V., Thomas, L., Ben-Galim, E., Haymond, M.W., and Cryer, P.E.: Adrenergic mechanisms in recovery from hyopoglycemia in man: Adrenergic blockade. Am. J. Physiol. 236(2): E147-E152, 1979.
3. Gauthier C., and Hetenyi, G., Jr.: Origin of glucose released in the regulatory response against hypoglycemia. Metabolism 31: 147-153, 1982.
4. Frizzell, R.T., Hendrick, G.K., Biggers, D.W., Lacy, D.B., Donahue, D.P., Green, D.R., Carr, R.K., Williams, P.E., Stevenson, R.W. and Cherrington, A.D.: The role of gluconeogenesis in sustaining glucose production during hypoglycemia in response to insulin infusion in the conscious dog. Diabetes 37: 749-759, 1988.
5. LeCavalier, L., Bolli, G., and Gerich, J.: Major role for gluconeogenesis during counterregulation in man. Diabetes (Suppl. 1) 1987.
6. Rizza, R., Cryer, P. and Gerich, J.: Role of glucagon, catecholamines, and growth hormone in human glucose counterregulation: effects of somatostatin and combined α- and β-adrenergic blockade on plasma glucose recovery and glucose flux rates after insulin-induced hypoglycemia. J. Clin. Invest. 64: 62-71. 1979.
7. Gerich, J., Davis, J., Lorenz, M., Rizza, R., Karam, J., Lewis, S., Daplan, R., Schultz, T. and Cryer, P.: Hormone mechanisms of recovery from insullin-induced hypoglycemia in man. Am. J. Physiol. 236:380-385, 1979.
8. Cryer,: Glucose counterregulation in man. Diabetes, 30: 261-164, 1981.
9. Ivy, R., Adkins, B., Williams, P., Neal, D., Cherrington, A.D. The role of epinephrine in counterregulation in hypoglycemia induced by insulin infusion. Diabetes 36: 25A, 1987.
10. Dobbins, R.L., Connolly, C.C., Neal, D.W., Palladino, L.S., Parlow, A.F. and Cherrington, A.D.: The role of glucagon in countering the hypoglycemia induced by insulin infusion. Diabetes (submitted).
11. Frizzell, R.T., Hendrick, G.K., Biggers, D.W., Lacy, D.B., Donahue, D.P. Green, D.R., Carr, R.K., Williams, P.W., Stevenson, R.W., and Cherrington, A.D.: Hypoglycemia induced by insulin infusion stimulates glucose production through mechanisms in addition to hormone secretion. Diabetes (in press).
12. Biggers, D.W., Myers, S.R., Neal, D., Stinson, R., Cooper, N.B., Jaspan, J.B., Williams, P.E., Cherrington, A.D., and Frizzell, R.T. Role of brain in counterregulation of insulin-induced hypoglycemia in dogs. Diabetes 38: 7-16, 1989.
13. McGraw, E.F., Peterson, M.J. and Ashmore, J. Autoregulation of glucose metabolism in the isolated perfused rat liver. Proc. Soc. Exp. Biol. Med. 126: 232-236, 1967.
14. Glinsmann, W.H., Hern, E.P. and Lynch, A. Intrinsic regulation of glucose output by rat liver. Am. J. Physiol. 216: 698-703, 1969.
15. Frizzell, R.T., Campbell, P., and Cherrington, A.D. Gluconeogenesis and hypoglycemia. In: Diabetes/Metabolism Reviews. R. DeFronzo (Ed.), John Wiley and Sons, Inc., New York, Vol. 4, pp. 51-70, 1988.

METABOLISM OF GLUCOSE IN THE BRAIN OF IDDM SUBJECTS:

Brain metabolism in diabetes

Valdemar Grill

Dept Endocrinology
Karolinska Hospital
S-104 01 Stockholm, Sweden

INTRODUCTION

Glucose is the predominant fuel used by the brain[1-3]. Dependency on glucose is demonstrated by the deleterious effect of hypoglycemia on brain functioning. It is therefore not surprising that many mechanisms have evolved to ensure an adequate supply of glucose to the brain. Among these the release of so called counterregulatory hormones during hypoglycemia plays an important role for upholding levels of blood glucose[4].

The supply of glucose to the brain is also regulated at the level of the blood-brain-barrier (BBB). It has been shown in animals that uptake of glucose over the BBB occurs by facilitated diffusion[5]. Uptake is saturable and K_m and V_{max} for the process can be defined. K_m values reported are usually close to the normal plasma concentration of glucose. A higher percentage of plasma glucose should thus be taken up by the brain at hypoglycemia than at hyperglycemia.

Evidence exists in animals that glucose uptake and metabolism in the brain can be adapted to long term hypo- or hyperglycemia. Increased uptake of glucose was found during chronic hypoglycemia with attendant hyperinsulinemia[6]. Reciprocally, a decreased uptake of glucose was found in some studies in animal diabetes[7,8] although not in others[9,10]. As to glucose metabolism, studies in hyperglycemic animals have yielded conflicting results, one study showing a decrease[11] and another an increase[12] in glucose metabolism. Notwithstanding the inconsistencies, the studies performed in animals raise the possibility that brain uptake and/or metabolism of glucose is altered in diabetic man as a consequence of the abnormal glucose homeostasis.

The starting point for the present studies were clinical observations in well-controlled subjects with insulin dependent diabetes mellitus (IDDM). During recent years intensified insulin treatment by continuous subcutaneous insulin infusion (CSII) or by multiple injection therapy has offered ways to improve metabolic control. Because good metabolic control is thought to prevent or delay the onset of complications, such therapy has been tried in many patients. Intensively treated

Fuel Homeostasis and the Nervous System, Edited by M. Vranic *et al.*
Plenum Press, New York, 1991

diabetic subjects can achieve mean glucose levels close to
those of non-diabetic subjects as assessed by a long term measure of glycemia i.e. the percent haemoglobin present in glyco-
sylated form in blood (haemoglobin A_1C). A price to be paid is
the occurrence of periods of mild to moderate hypoglycemia
which are usually symptomless[4,13]. It is not known whether the
brain adapts in some or several ways to this type of intermittent hypoglycemia or whether it is, in the long run, detrimen-
tal to brain function. Direct assessment of glucose uptake and
metabolism in the brain of well-controlled diabetic subjects
would be helpful in resolving this question.

In the first part of our study[14] we used positron emission
tomography (PET) with [U-11-C]-D-glucose as a tracer to detect
possible abnormalities in glucose uptake and metabolism. We
chose labelled D-glucose instead of deoxyglucose for this study
because of the change in the so called "lumped constant" (i.e.
the sum of correction factors for differences in uptake and
phosphorylation between glucose and analogue) which occurs
during hypoglycemia[15]. Furthermore, during hypoglycemia the
influx to outflux ratio (k_1/k_2) increased for deoxyglucose[15,16]
but not for glucose[15] indicating that phosphorylated deoxyglucose may compartmentalize in the brain. These observations make
deoxyglucose unsuitable for the study of glucose metabolism
during hypoglycemia.

Measurements were performed during Biostator-controlled
normo- and hypoglycemia in the presence of constant hyperinsulinemia. Results were compared with those obtained from a group
of non-diabetic subjects at the same glucose and insulin
levels. The results indicated i.a. a marked decrease in cerebral metabolism rate for glucose (CMRgl) in the IDDM subjects at
normoglycemia. These findings initiated the second part of the
study[17] which was designed to study the net uptake of glucose
and also the uptake of other potential fuels of the brain by
use of the Fick principle.

Subjects

IDDM subjects were treated by continuous subcutaneous insulin infusion (CSII) or by multiple injection therapy. They
were well controlled as judged by Haemoglobin A1c levels below
127% of the upper limit of normal except in one subject. Duration of disease was 23.7±4.0 years in subjects participating in
the [U-11-C]-glucose and 19.4±2.1 years in the arteriovenous
study. The difference was not significant. Overt signs of neuropathy (lack of Achilleus tendon reflexes) were found in one
subject. All IDDM subjects had undetectable levels of C-pep-
tide, i.e. below 0.05 nmoles/l while fasting. All had experienced hypoglycemic unawareness i.e. symptom-free periods of low
glucose as measured by home monitoring or in the hospital.

Protocols

Healthy volunteers and IDDM subjects alike fasted from 10
p.m. on the night preceding the experiment. IDDM subjects injected their usual dose of short acting insulin at 5 p.m.
before dinner on the day preceding the experiments. They were
then treated by CSII ([U-11-C]-glucose study) or by intravenous
insulin (arteriovenous study) during the night. In the morning
insulin was stopped for 1-3 hours during catheter insertions
and calibration of the Biostator (an "artificial pancreas").

Following positioning of the catheters a constant insulin infusion (Actrapid Human, Novo Co, 2 mU/kg/min) was started together with a Biostator-controlled feed-back infusion of glucose.
After achieving normoglycemia the level of glucose was maintained for a further half hour.

In the first study [U-11-C]-glucose was then injected. The
time course of the total and regional cerebral radioactivity
was measured with PET during 20 min. Approximately 90 min later
blood glucose was lowered to the desired level of hypoglycemia
(2.00-2.5 mmol/l in arterialized venous blood). [U-11-C]-D-glu-
cose was subsequently injected and a second PET scan performed
during constant hypoglycemia.

In the second (arteriovenous) study blood was sampled
from the brachial artery and from the bulbous part of an internal jugular vein. Concomitant with these measurements cerebral
blood flow (CBF) was determined by [11-C]-CH$_3$F, the distribution of which was measured by the PET scanner. Glucose and insulin levels (normoglycemia, hyperinsulinemia) did not differ
from those achieved in first study.

Non-diabetic subjects were kept at hyperinsulinemia during
Biostator directed normoglycemia for 60-90 min preceding either
injection of [U-11-C] glucose (first study) or the combined
measurements of arteriovenous differences and CBF (second
study).

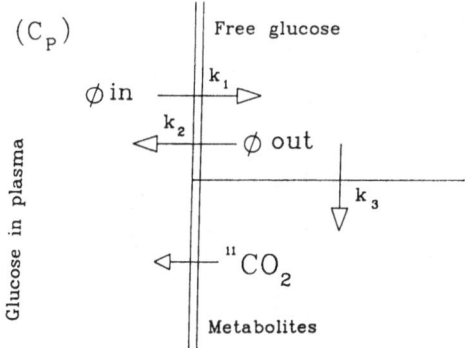

Fig. 1. Compartments of [11-C]-glucose distribution in the
brain.

PET measurements

The time course for the distribution of radioactivity in
the brain was monitored with a 4-ring positron camera (Scanditronix PC-384-7B). When randomly labeled [U-11-C]-glucose was
used as a tracer about 300 MBq was injected before each PET
measurement. The kinetic data obtained were analyzed using a
3-compartment model (Fig. 1). The tracer is considered to be
either in the plasma or in the tissue as [11-C]-D-glucose, or

in the tissue as [11-C]-labeled metabolic products. The mathematical, kinetic model is the same as the one used for 11-C-deoxyglucose[18] except in two regards. First, with labeled natural glucose there is no need for a "lumped constant". Second, loss of [11-C]-CO_2 has to be accounted for by a correction term. This is necessary since use of randomly labeled [11-C]-glucose implies rapid loss of [11-C]-CO_2 from the tissue. The tracer kinetic model requires about 20 minutes of PET measurements whereas an egress of radioactivity in the form of [11-C]CO_2 starts already within a few minutes after tracer injection. This loss was corrected for using data from separate experiments in healthy volunteers in which the loss of [11-C]-CO_2 was directly measured[19].

The inert, freely diffusible gas [11-C]-CH_3F and PET scanning was used to measure CBF in the arteriovenous study. Subjects inhaled the fluoromethane (bolus of 1700- 1900 MBq) through an oral tube in a single breath[20]. To calculate CBF, least square estimate of parameters and the algorithm described by Koeppe et al[21] was applied.

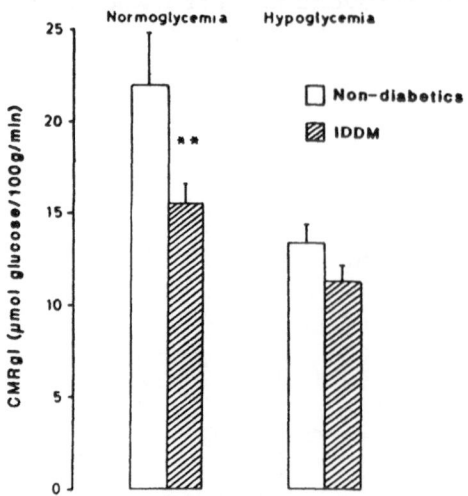

Fig. 2. Calculated CMRgl from [11-C]-glucose data. xx = p<0.02, significance of difference for non-diabetic vs IDDM subjects. Mean ± SE of experiments in 8 non-diabetic and in 6 IDDM subjects. Data obtained from ref 14.

RESULTS

A main finding of the [U-11-C]-glucose study was a difference in calculated CMRgl between non-diabetic and IDDM subjects (Fig. 2). During normoglycemia the calculated CMRgl was 29% lower in the IDDM than in the non-diabetic patients. These findings were similar when analyzed for total brain or for 5 selected regions of the brain[14]. The lower CMRgl in IDDM subjects was not due to decreased uptake of glucose since the blood-brain glucose flux was similar between groups at normoglycemia (Fig. 3).

During moderate hypoglycemia the calculated CMRgl was decreased in non-diabetic and IDDM subjects alike but more in the
non-diabetic subjects (40%) than in IDDM subjects (28%). Accordingly the calculated CMRgl during hypoglycemia did not differ
between groups (Fig. 2).

The blood-brain glucose flux was decreased during hypoglycemia in both groups but to a lesser extent than the fall in
blood glucose. This difference was due to an increase in k_1
during hypoglycemia. This increase was seen both in non-diabetic subjects (from 0.061±0.007 to 0.090±0.006 ml/g/min) and in
IDDM subjects (from 0.061±0.06 to 0.0093±0.013 mg/g/min). Also
k_2 was increased. Such increases are expected if glucose transport over the BBB proceeds by facilitated diffusion[5].

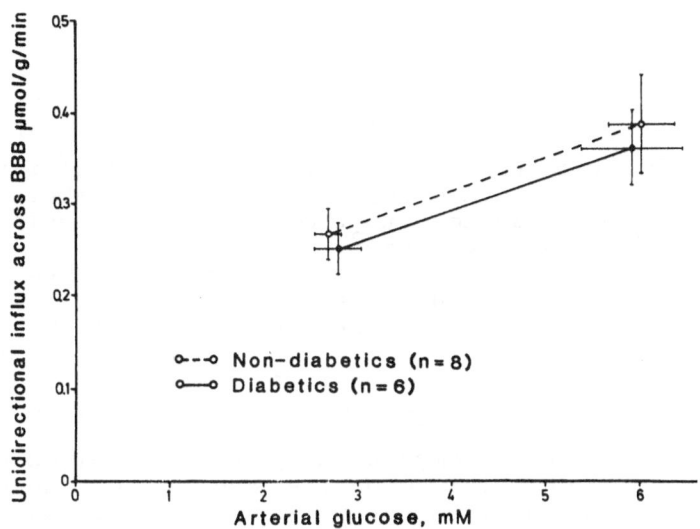

Fig. 3. Unidirectional influx of glucose across BBB. Mean ± SE
        of experiments in 8 non-diabetic and in 6 IDDM sub-
        jects. Data obtained from ref 14.

The arteriovenous study was performed during normoglycemia
only. The experimental conditions were comparable to those of
the tracer study with regard to normoglycemia. As shown in
Table 1, CMRgl did not differ significantly between non-diabetic and IDDM subjects. Neither did brain consumption of oxygen.
However the ratio of glucose to oxygen consumed was lower in
the IDDM subjects: 4.68 in IDDM as compared with 5.50 in non-
diabetic subjects, $p < 0.05$, Wilcoxon test[17]. Other potential
substrates of the brain did not significantly contribute to
brain metabolism in either of the two groups. There was however
small but significant efflux of lactate and pyruvate in the
IDDM but not in the non-diabetic subjects.

Total cerebral blood flow was within normal limits in the
IDDM subjects. It was however significantly higher than in the
non-diabetic subjects (Table 1).

Table 1. Global mean cerebral blood flow, substrate utilization and production during normoglycemia and hyperinsulinemia[*]

Subjects	Cerebral blood flow ml·100 g^{-1}·min^{-1}	O$_2$ utilization µmol·100 g^{-1}·min^{-1}	glucose utilization µmol·100 g^{-1}·min^{-1}	lactate production µmol·100 g^{-1}·min^{-1}	pyruvate production µmol·100 g^{-1}·min^{-1}
IDDM	64.9 ± 5.9[+]	164.2 ± 8.5	34.1 ±2.4	2.2 ± 0.4[+]	0.3 ± 0.0[+]
Non-diabetic	49.3 ± 2.7	154.0 ± 7.5	27.3 ± 1.9	0.5 ± 0.5	0.0 ± 0.0

[*]Mean ± SE for non-diabetic, n—6 and IDDM subjects, n=6. Data obtained from ref 15.
[+]p<0.05 or, less significance of difference non-diabetic subjects vs IDDM subjects.

DISCUSSION

The study demonstrates similar unidirectional flux from blood to brain in the non-diabetic and in the diabetic subjects. These results agree with those of a study using [11-C]-3-0-methyl glucose as a tracer[22].

The present results indicate, however, that the IDDM subjects differ from normal subjects with respect to some aspect of glucose metabolism. The difference is most clearly apparent from the abnormal value of calculated CMRgl during normoglycemia. To my knowledge this is the first time that an abnomality of cerebral glucose metabolism has been demonstrated in IDDM subjects.

It is difficult to compare the present findings with those of animal experiments because of the severity of the experimental diabetes and associated marked ketosis. Since ketones, when available, can substitute for glucose as brain fuel the decrease in CMRgl observed in one animal study[8] could be secondary to a decreased need for glucose. In this context it should be pointed out that the levels of ketones in our IDDM patients during normoglycemia were low. Under these conditions, arterio-jugular vein differences for acetoacetate and beta-hydroxybutyrate were not significant.

An apparent contradiction exists between CMRgl calculated from the tracer data and from arterial-jugular vein differences. Thus, the CMRgl calculated from the tracer model was 29% lower in the IDDM subjects than in the non-diabetic subjects whereas no difference between the two groups was found when CMRgl was estimated according to the Fick principle. Two possible explanations come to mind. The first one is undercorrection for the loss of [11-C]-CO$_2$ in the IDDM subjects. The efflux of [11-C]-CO$_2$ is 30-35 per cent of tissue radioactivity during a 20 min period of measurement[17]. Evidence has been provided that the loss of [11-C]-CO$_2$ is correctly estimated under normal conditions (normoglycemia, non-diabetic subjects). The possibility remains that efflux of [11-C]-CO$_2$ proceeds differently in diabetic subjects during normoglycemia. Although this possibility cannot be excluded it is not supported by a comparison of CMRgl calculated after a short (10 min) as well as after a long time period (25 min) rather than after the standard 20 min period of PET measurement. Such a comparison (which could not be carried out in all subjects) did not reveal differences between CMRgl calculated for the shorter or the longer time period of measurement in the IDDM subjects[14].

The second possibility is preferential release of glucose metabolites other than [11-C]-CO_2 in the diabetic subjects. In experiments on non-diabetic subjects it was observed that the brain released labeled acidic substances during the first few minutes following the injection of labeled glucose[19]. This was observed both with [U-11C]- and [1-11C]-glucose as tracers. With the latter tracer it occurred before any significant egress of [11-C]-CO_2 was noticed. If corrected for this loss of radioactivity the CMRgl would increase by a mean of 12 per cent. This is much more than would be expected from the egress of "cold" lactate and pyruvate, which was found to be minimal in our experiments. However, the [11-C]-lactate produced from [11-C]-glucose should be of high specific activity compared with metabolites of the citric acid cycle and with [11-C]-CO_2[23]. Also the carrier transport of lactate is bidirectional across the blood-brain barrier so that there is both uptake and egress of lactate[3]. When uptake of "cold" arterial lactate occurs simultaneously with release of brain lactate of high specific activity the result will be loss of labeled lactate even in the absence of a net release of lactate.

In the IDDM subjects the net release of "unlabeled" lactate and pyruvate, albeit small, was 5-fold higher than in the non-diabetic subjects. If loss of label increases with increased net release of lactate, then greater loss of label would occur in the diabetic subjects and such loss could account for the lower tracer-calculated CMRgl in the IDDM subjects.

The notion of non-oxidative metabolism of glucose being increased in the IDDM subjects is supported by the finding of a diminished O_2/glucose uptake in the arteriovenous study. The differences in release of lactate and pyruvate however did not seem sufficient to account for the decreased O_2/glucose ratio in the IDDM subjects.

Few conditions are known to be associated with a lower-than-normal O_2/glucose ratio. A recent study reports that the ratio decreases markedly in healthy subjects in areas of the brain with acutely increased neural activity[24]. The change in ratio due to increased neural activity was associated with an increase of the CBF flow and of the regional uptake of glucose but only a slight increase in oxygen uptake. In rats somatosensory stimulation was associated i.a. with an increase in brain lactate suggesting stimulation of glycolysis[25]. Interestingly an association between a low O_2/glucose ratio, increased CBF and evidence for increased lactate production was observed also in our experiments with IDDM subjects. It seems possible that the increased global mean CBF in these subjects (for which no clear explanation can presently be offered) could be coupled to an increase in non-oxidative glucose metabolism.

What do the present findings tell us about adaptation of brain metabolism to intermittent hypoglycemia in IDDM subjects? Table 2 lists conditions that potentially reduce the impact of hypoglycemia on the brain. Our studies fail to show increased utilization of non-glucose nutrients in well-controlled diabetic subjects. Neither did we obtain evidence for increased extraction of glucose in the IDDM subjects. Indeed both K_m and

**Table 2. Mechanisms reducing impact of
hypoglycemia on the brain**

1) Increased utilization of non-glucose
   nutrients in the brain - ketones

2) Increased utilization of glucose
   - increased extraction
   - increased efficiency of glucose
     metabolism

3) Increased levels of energy stores in
   the brain
   - glycogen

V_{max} for facilitated diffusion of glucose were similar between non-diabetic and IDDM subjects.

Theoretically glucose could be more efficiently utilized by increasing the proportion of the sugar that is aerobically metabolized. In this context it is interesting to note that net uptake of glucose falls during moderate hypoglycemia but oxygen uptake is little affected[26]. However, the fact that approximately 90 per cent of glucose is aerobically metabolized in the brain during normal conditions leaves little room for increasing the energy yield of glucose during normal conditions.

Evidence indicates that the relatively unchanged O_2 consumption during hypoglycemia is secondary to the utilization of endogenous energy stores. Of these glycogen in the brain as well as elsewhere should be the most readily exploited. Calculations based on experiments in the rat indicate that the total glycogen stores of the brain would be consumed in less than 10 min in the absence of exogenous glucose[26]. However the stores would last for much longer times during moderate hypoglycemia. The glycogen stores in man are unknown. If the concentration of glycogen is equal or even higher than that of rodents glycogen stores may be very important during short-term hypoglycemia in man since CMRgl is lower than in small mammals.

Some observations indicate that glycogen stores may be increased in intensively insulin-treated diabetic subjects. Hence insulin treatment of diabetic animals leads to glycogen deposition in the brain[27]. Also our finding that non-oxidative metabolism of glucose is likely increased in our IDDM subjects may reflect a state of brain overnutrition. Increased glycogen stores could potentially reduce the impact of hypoglycemia in the IDDM subjects. If so, vulnerability to hypoglycemia would vary depending on levels of glucose and insulin present in the hours before a hypoglycemic event. It remains for future studies to test this hypothesis.

ACKNOWLEDGEMENTS

These studies were supported by the Swedish Medical Research Council (grants no 04540 and 02330), the Juvenile Diabetes Foundation, USA and the Osterman Foundation.

REFERENCES

1. B. K. Siesjö, Utilization of substrates of brain tissues, in: Brain Energy Metabolism. New York: Wiley (1978).
2. L. Sokoloff, Cerebral circulation, energy metabolism and protein synthesis: general characteristics and principles of measurement, in: Positron emission tomography and autoradiography: Principles and applications for the brain and heart, New York, Raven Press pp. 1-71 (1986).
3. W. H. Pardridge, Brain metabolism from the blood-brain barrier, Physiol Rev 63:1481-1535 (1983).
4. P. E. Cryer and J. E. Gerich, Glucose counterregulation, hyperglycemia and intensive insulin therapy in diabetes mellitus, N Eng J Med 313:232-239 (1987).
5. C. Crone, Facilitated transfer of glucose from blood into brain tissue, J Physiol 181:103-113 (1965).
6. A. L. McCall, L. B. Fixman, N. Fleming, K. Tornheim, W. Chick and N. B. Ruderman, Chronic hypoglycemia increases brain glucose transport, Am J Physiol 251:E442-E447 (1986).
7. A. Gjedde and C. Crone, Blood-brain glucose transfer: repression in chronic hyperglycemia, Science 214:456-457 (1981).
8. A. L. McCall, W. R. Millington and R. J. Wurtman, Metabolic fuel and aminoacid transport into the brain in experimental diabetes mellitus, Proc Natl Acad Sci USA 79: 5406-5410 (1982).
9. S. I. Harik, S. A. Gravina and R. N. Kalaria, Glucose transporter of the blood-brain barrier and brain in chronic hyperglycemia, J Neurochem 51:1930-1934 (1988).
10. R. B. Duckrow, Glucose transfer into rat brain during acute and chronic hyperglycemia, Metabolic Brain Disease 3: 201-209 (1988).
11. R. B. Duckrow and R. M. Bryan, Regional cerebral glucose utilizatoin during hyperglycemia, J Neurochem 48: 989-993 (1987).
12. A. M Mans, M. R. De Joseph, D. W. Davis and R. A. Hawkins, Brain energy metabolism in streptozotocin-diabetes, Biochem J 249:57-62 (1988).
13. The Diabetes Control and Complications Trial (DCCT): Results of feasibility study, Diabetes Care 10:1-19 (1987).
14. M. Gutniak, G. Blomqvist, L. Widén, S, Stone-Elander, B. Hamberger and V. Grill, Brain uptake and metabolism of [U-11-C]-D-glucose in insulindependent diabetic subjects, Am J Physiol. In press.
15. K. Mori, N. Cruz, G. Dienel, T. Nelson and L. Sokoloff, Direct chemical measurement of the lumped constant of the [14-C] deoxyglucose method in rat brain: effects of arterial plasma glucose level on the distribution spaces of [14-C] deoxy-glucose and glucose and on lambda, Cereb Blood Flow Metab 9:304-314 (1989).
16. E, Shapiro, M. Cooper, C.-T. Chen, B. D. Given and K. S. Polonsky, Change in hexose distribution volume and fractional utilization of [12F]-2-deoxy-2-fluoro-D-glucose in brain during acute hypoglycemia in humans, Diabetes 39:175-181 (1990).

17. V. Grill, M. Gutniak, O. Björkman et al., Cerebral blood flow and substrate utilization in insulin-treated diabetic subjects, Am J Physiol. In press.

18. L. Sokoloff, M. Reivich, C. Kennedy et al., The 14C-deoxyglucose method for the measurements of local cerebral glucose utilization: Theory, procedure and normal values in the conscoius and anaesthetized albino rat, J Neurochem 28:897-916 (1977).

19. G. Blomqvist, S. Stone-Elander, S. Halldin et al., Measurement of cerebral glucose utilization with PET using [1-11$_C$] D-glucose. J Cereb Blood Flow Metab. In press.

20. P. E. Roland, L. Eriksson, S. Stone-Elander and L. Widén, Does mental activity change the oxidative metabolism of the brain? J Neurosci 7:2373-2389 (1987).

21. R. A. Koeppe, J. E. Holden and W. R. Ip, Performance comparison of parameter estimation techniques for the quantitation of local cerebral blood flow by dynamic positron computed tomography, J Cereb Blood Flow Metab 5: 224-234 (1985).

22. D. J. Brooks, S. R. Gibbs, P. Sharp et al., Regional cerebral glucose transport in insulin-dependent diabetic patients studied using [11-C]3-0-methyl-D-glucose and positron emission tomography, J Cereb Blood Flow Metab 6:240-244 (1986).

23. W. Sacks, Cerebral metabolism of isotopic glucose in normal human subjects, J Appl Physiol 10:37-44 (1957).

24. P. T. Fox, M. E. Raichle, M. A. Mintun and C. Dance, Nonoxidative glucose consumption during focal physiologic neural activity, Science 241:462-464 (1988).

25. K. A. Hossman and F. Linn, Regional energy metabolism during functional activation of the brain, J Cereb Blood Flow Metab 7:S297 (1987).

26. B. K. Siesjö, Hypoglycemia: in: Brain Energy Metabolism. New York, Wiley (1978).

27. S. R. Nelson, D. W. Schultz, J. V. Passonneau and O. H. Lowry, Control of glycogen levels in the brain, J Neuro Chem 15:1271-1279 (1968).

EATING DISORDERS AS ASSESSED BY CRANIAL COMPUTERIZED TOMO-
GRAPHY (CCT, dSPECT, PET)

Jürgen-Christian Krieg

Max Planck Institute of Psychiatry
Kraepelinstraße 10
Munich, FRG

Anorexia nervosa and bulimia nervosa are eating disorders which preferentially occur in female adolescents and young female adults who live in the western civilization. The prominent psychopathological symptoms are being overly concerned with one's shape and the fear of gaining weight or being too fat. To lose weight, the patients reduce their food intake, often in combination with extensive exercising, self-induced vomiting or abuse of diuretics and laxatives. If the patients become underweight, the full clinical picture of an anorexia nervosa emerges. Periods of dieting or fasting may be interrupted by bulimic episodes (i.e. the consumption of large amounts of often high caloric food), which are usually followed by self-induced vomiting or other purging behavior. The bulimic episodes may counterbalance the effect of dieting, resulting in the maintenance of a more or less normal body weight. Thus, patients with the clinical diagnosis of a bulimia nervosa display an anorectic attitude and eating behavior without being emaciated. Table 1 gives the diagnostic criteria of anorexia and bulimia nervosa according to the Diagnostic and Statistical Manual of Mental Disorders, Third Edition, Revised (DSM-III-R, 1987).

Besides the typical psychopathology, patients with anorexia nervosa exhibit a number of somatic abnormalities which can be considered as a direct consequence of malnutrition or starvation (for overview see Ploog and Pirke, 1987). Thus anorexic patients display an increased activity of the hypothalamic-pituitary-adrenocortical (HPA) axis and a decreased activity of the hypothalamic-pituitary-ovarian (HPO) axis resulting in increased plasma levels of cortisol and in decreased levels of the gonadal hormones. As a further mechanism of adaption to starvation, the conversion of the thyroid hormone tetraiodotyronine (T_4) to the more active triiodotyronine (T_3) is reduced resulting in low plasma concentrations of T_3 in the anorexic state. Due to an increased lipolysis, patients with anorexia nervosa show elevated plasma levels of free fatty acids, β-hydroxybutyric acid and acetoacetate. The above-mentioned endocrine and metabolic parameters can be considered as good indicators for the state of starvation.

Fuel Homeostasis and the Nervous System, Edited by M. Vranic *et al.*
Plenum Press, New York, 1991

Table 1. DSM-III-R Diagnostic Criteria for Anorexia Nervosa and Bulimia Nervosa

Anorexia Nervosa

A. Refusal to maintain body weight over a minimal normal weight for age and height, e.g., weight loss leading to maintenance of body weight 15% below that expected; or failure to make expected weight gain during period of growth, leading to body weight 15% below that expected.

B. Intense fear of gaining weight or becoming fat, even though underweight.

C. Disturbance in the way in which one's body weight, size, or shape is experienced, e.g., the person claims to "feel fat" even when emaciated, believes that one area of the body is "too fat" even when obviously underweight.

D. In females, absence of at least three consecutive menstrual cycles when otherwise expected to occur (primary or secondary amenorrhea). (A woman is considered to have amenorrhea if her periods occur only following hormone, e.g., estrogen, administration.)

Bulimia Nervosa

A. Recurrent episodes of binge eating (rapid consumption of a large amount of food in a discrete period of time).

B. A feeling of lack of control over eating behavior during the eating binges.

C. The person regularly engages in either self-induced vomiting, use of laxatives or diuretics, strict dieting or fasting, or vigorous exercise in order to prevent weight gain.

D. A minimum average of two binge eating episodes a week for at least three months.

E. Persistent overconcern with body shape and weight.

Besides a number of other, to some extent rather serious somatic complications, neuropathological alterations have also been described in anorexia nervosa (Gagel, 1953; Martin, 1958). Pneumencephalographic examinations revealed an enlargement of the ventricular system and a widening of the cortical sulci, which was considered as a sign of brain atrophy in anorexia nervosa (Geisler, 1953; Heidrich and Schmidt-Matthias, 1961). These findings were later confirmed by single case studies or studies performed on small groups of anorexic patients by using the technique of cranial computerized tomography (CT)(e.g. Enzmann and Lane, 1977; Kohlmeyer et al. 1983; Datlof et al. 1986).

In our own extensive study we examined the CT scans as well as the hormonal and metabolic states of 50 patients with an anorexia nervosa (Krieg et al. 1988). The mean age of the patients was 21.5 ± 3.4 years and the mean body weight 69 ± 5 percent of the ideal body weight (% IBW). Eighty-six percent of the patients displayed slightly or markedly enlarged external cerebrospinal fluid spaces (i.e. insular cisterns, interhemispheric fissure, cortical and cerebellar sulci), while this was only the case in 16% of the age-matched control subjects. Regarding ventricular size, 70% of the anorexic patients displayed abnormally high ventricle brain ratios (VBR values > 4.7% according to our normative data), indicating an enlarged ventricular system in these patients. There was a significant correlation between the extent of the neuroradiological alterations and the body weight such that patients with the lowest body weight displayed the most enlarged cerebrospinal fluid spaces. It is of special interest, that there was no correlation between the degree of the neuromorphological alterations and the duration of the illness. There was, however, a close relationship between the hormonal and morphological alterations: thus the cortisol plasma levels correlated positively and the T_3 serum concentrations negatively with the size of the sulci and the ventricles, respectively.

In 25 cases CT scans were repeated after the patients had gained weight up to $87 \pm 4\%$ IBW. In nearly one-third of the patients re-examined the previous enlargement of the external cerebrospinal fluid spaces had either disappeared or noticeably diminished; the ventricular brain ratio had significantly decreased from $8 \pm 4\%$ to $6 \pm 2\%$ (mean VBR-value of the controls: $3 \pm 1\%$). This finding can be interpreted to mean that sulcal widening and ventricular dilatation are reversible neuroanatomical alterations in patients with anorexia nervosa; such a reversibility was more frequently observed in patients with a short than with a long history of anorexia nervosa.

We also assessed the CT scans of 50 normal weight bulimic patients with a mean age of 22.5 ± 3.5 years (Krieg et al., 1989). We found enlarged cortical sulci and ventricles in approximately 40% of the patients with bulimia nervosa. As was the case in anorexia nervosa, there was no association between the duration of the illness or frequency of vomiting and the neuroradiological alterations (Lauer et al., 1990). Normal weight bulimic patients with a past history of anorexia nervosa did not more frequently display enlarged cerebro-spinal fluid spaces than did patients without such a history; this finding rejects the explanation, that in normal weight bulimic patients the neuroanatomical alterations are a residue of a previous anorexic state. However, as was the case in anorexic patients, there was a close relationship between the cortisol and T_3 plasma levels and the neuroradiological measurements. Due to these findings we formulated the hypothesis that not weight loss per se but the endocrine and metabolic reactions to starvation are responsible for the morphological brain alterations, regardless of whether starvation leads to emaciation (as is the case in anorexia nervosa) or is counterbalanced by binge attacks (as is the case in bulimia nervosa). This assumption is corroborated by the circumstance that patients with bulimia nervosa - although of normal body weight - show the same metabolic and endocrine indices of starvation, as anorexic patients do (Pirke et al., 1985).

As the structural brain alterations are reversible (this has at least been observed in a number of anorexic patients), it is unlikely that the neuromorphological changes are an expression of a real brain atrophy with loss of neural tissue but rather a sign of brain shrinkage - presumably due to the starvation-induced hormonal and metabolic changes influencing water distribution. It remains to be mentioned that, in the meantime, our X-ray CT findings in patients with anorexia nervosa and in normal weight bulimic patients have been largely confirmed by studies using the magnetic resonance (NMR) imaging technique (Hoffman et al., 1989a; Hoffman et al., 1989b). For illustration of the respective neuromorphological alterations in patients with an eating disorder, figure 1 shows a NMR image of one of our patients with anorexia nervosa, who displays markedly enlarged lateral ventricles.

Due to the above-mentioned findings, the question arises of whether there are functional correlates for the neuroanatomical alterations. We therefore measured the regional cerebral blood flow using the xenon-133 inhalation method in 12 female anorexic patients (mean age: 21.3 $\pm$ 3 years, body weight: 73 $\pm$ 8% IBW) and once again after an average weight gain of 12% IBW (Krieg et al., 1989). Although nine of the patients showed ventricular and/or sulcal widening in their CT scans, all patients except one exhibited normal cerebral blood flow rates. The mean flow rates of the patients in the anorexic and remitted state as well as the flow rates of the control subjects did not differ significantly among each other in all regions assessed. Moreover, no significant differences could be observed between the flow rates of the left and right hemispheres in the subjects studied.

In addition, we measured resting regional cerebral glucose metabolism with ^{18}F-2-fluoro-2-deoxyglucose and positron emission tomography (PET) in seven anorexic patients (mean age: 20 $\pm$ 2 years, body weight: 72 $\pm$ 6% IBW), and once again in five of them after an average weight gain of 16% IBW (Krieg et al., 1986; Herholz et al., 1987). Furthermore, in a recent study the cerebral glucose metabolism of nine normal weight bulimic patients (mean age: 23 $\pm$ 2 years) was assessed (Herholz et al., 1989). The mean global metabolic rates of the anorexic, remitted and bulimic patients did not significantly differ among each other and also not when compared

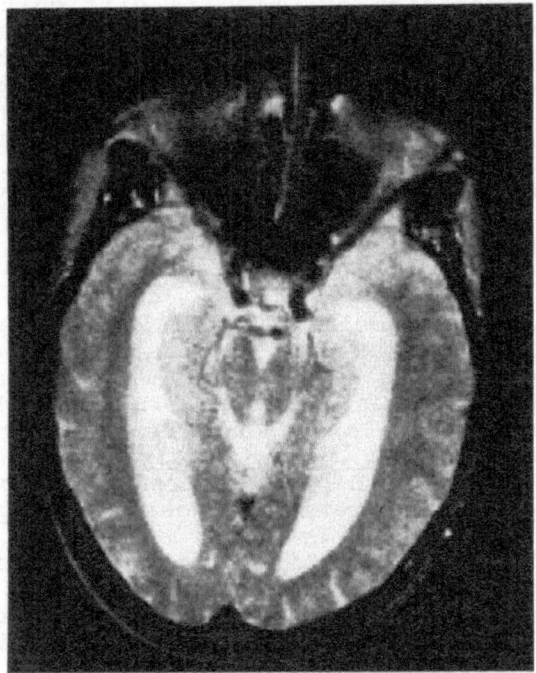

Fig. 1. NMR image of a patient with
anorexia nervosa displaying
dilated ventricles.

with the rates of a control group. Evaluating single brain
regions, the only prominent finding was an absolute and relative (i.e. regional divided by total brain glucose metabolism) glucose hypermetabolism of the caudate nuclei in the
anorexic state. After weight gain the metabolic rates of the
caudate nuclei no longer differed significantly from those
measured in the normal weight bulimic patients and in the
control subjects, respectively.

The functional orientated computerized tomography findings are in good agreement with our very recent neuropsychological study, carried out on 39 female patients with an anorexia and bulimia nervosa (mean age: 25.2 ± 4.9 years): In
the task assigned the patients had to respond to irregularly
presented stimuli. The hit rates (signals detected) were related to the CT scans, resulting in the finding, that patients with ventricular dilatation (VBR > 4.7%) did not perform more poorly (hit rate: 0.56 ± 29) than the patients with
normal sized ventricles (hit rate: 0.54 ± 0.22) (Laessle et
al., 1990).

In summary, until now no functional correlates have been
detected which could be related to the often rather pronounced structural brain alterations observed by X-ray computerized tomography in patients with eating disorders. The
only conspicuous finding was an absolute and relative glucose

hypermetabolism in the caudate nuclei during the anorexic state, the interpretation of which is difficult. Considering that the caudate nuclei are not only involved in the regulation of motor activity but also in the integration of environmental influences (after cortical processing) and limbic afferences, the presented data would suggest that this process is altered in anorexia nervosa. It must first be confirmed in further studies, whether caudate hypermetabolism is indeed a characteristic finding for anorexia nervosa.

References

American Psychiatric Association, 1987, "Diagnostic and Statistical Manual of Mental Disorders", 3. ed., Revised, American Psychiatric Association, Washington, DC.

Datlof, S., Coleman, P.D., Forbes, G.B., and Kreipe, R.E., 1986, Ventricular dilation on CAT scans of patients with anorexia nervosa, Am. J. Psychiatry, 143:96.

Enzmann, D.R., and Lane, B., 1977, Cranial computed tomography findings in anorexia nervosa. J. Comput. Assist. Tomogr., 1:410.

Gagel, O., 1953, Die Erkrankungen des vegetativen Systems, in: "Handbuch der Inneren Medizin", Vol. 5, G. von Bergmann, W. Frey, and H. Schwiegk, eds., Springer, Berlin.

Geisler, E., 1953, Zur Problematik der Pubertätsmagersucht, Psychiatrie, 5:227.

Heidrich, R., and Schmidt-Matthias, H., 1961, Encephalographische Befunde bei Anorexia nervosa, Arch. Psychiatr. Nervenkrankheiten, 202:183.

Herholz, K., Krieg, J.-C., Emrich, H.M., Pawlik, G., Beil, C., Pirke, K.-M., Pahl, J.J., Wagner, R., Wienhard, K., Ploog, D., and Heiss, W.-D., 1987, Regional cerebral glucose metabolism in anorexia nervosa measured by positron emission tomography, Biol. Psychiatry, 22:43.

Herholz, K., Krieg, Ch., Holthoff, V., Emrich, H.M., Pirke, K.M., Heiss, W.-D., and Schreiber, W., 1989, Comparison of anorectic and bulimic patients: Regional cerebral glucose metabolism measured with positron emission tomography, in: "Biological Aspects of Non-Psychotic Disorders (Abstract)", Regional Congress, Jerusalem.

Hoffman, Jr., G.W., Ellinwood, Jr., E.H., Rockwell, W.J.K., Herfkens, R.J., Nishita, J.K., and Guthrie, L.F., 1989a, Cerebral atrophy in anorexia nervosa: A Pilot Study, Biol. Psychiatry, 26:321.

Hoffman, G.W., Ellinwood, Jr., E.H., Rockwell, W.J.K., Herfkens, R.J., Nishita, J.K., and Guthrie, L.F., 1989b, Cerebral atrophy in bulimia, Biol. Psychiatry, 25:894.

Kohlmeyer, K., Lehmkuhl, G., and Poutska, F., 1983, Computed tomography of anorexia nervosa. Am. J. Neuroradiol., 4:437.

Krieg, J.C., Emrich, H.M., Backmund, H., Pirke, K.M., Herholz, K., Pawlik, G., and Heiss, W.D., 1986, Brain morphology (CT) and cerebral metabolism (PET) in anorexia nervosa, in: "Disorders of Eating Behaviour. A Psychoneuroendocrine Approach", Advances in the Biosciences, Vol. 60, E. Ferrari, and F. Brambilla, eds., Pergamon, Oxford.

Krieg, J.-C., Pirke, K.-M., Lauer, C., and Backmund, H., 1988, Endocrine, metabolic and cranial computed tomographic findings in anorexia nervosa. Biol. Psychiatry, 23:377.

Krieg, J.-C., Lauer, C., and Pirke, K.-M., 1989, Structural brain abnormalities in patients with bulimia nervosa, Psychiatry Res., 27:39.

Krieg, J.-C., Lauer, C., Leinsinger, G., Pahl, J., Schreiber, W., Pirke, K.-M., and Moser, E.A., 1989, Biol. Psychiatry, 25:1041.

Laessle, R.G., Krieg, J.C., Fichter, M.M., and Pirke, K.M., 1989, Cerebral atrophy and vigilance performance in patients with anorexia nervosa and bulimia nervosa, Neuropsychobiology, 21:187.

Lauer, C.J., Laessle, R.G., Fichter, M.M., Pirke, K.-M., and Krieg, J.-C., 1990, Structural brain alterations and bingeing and vomiting behavior in eating disorder patients, Int. J. Eating Disorders, 9:161.

Martin, F., 1958, Pathologie des aspects neurologiques et psychiatriques de quelques manifestations carentielles avec troubles digestifs et neuro-endocriniens. Acta Neurol. Belg., 58:816.

Pirke, K.-M., Pahl, J., Schweiger, U., and Warnhoff, M., 1985, Metabolic and endocrine indices of starvation in bulimia: A comparison with anorexia nervosa, Psychiatry Res., 15:33.

Ploog, D.W., and Pirke, K.M., 1987, Psychobiology of anorexia nervosa, Psychol. Med., 17:843.

ABNORMAL BRAIN GLUCOSE METABOLISM IN ALZHEIMER'S DISEASE, AS MEASURED BY POSITRON EMISSION TOMOGRAPHY

Stanley I. Rapoport, Barry Horwitz, Cheryl L. Grady, James V. Haxby, Charles DeCarli, Mark B. Schapiro

Laboratory of Neurosciences
National Institute on Aging
National Institutes of Health
Bethesda, Md. 20892

SUMMARY

Resting glucose metabolism in the association neocortices, measured with positron emission tomography (PET), is disturbed early and throughout the course of Alzheimer's disease (AD), whereas resting metabolism in the primary sensory and motor neocortices is relatively spared. Neocortical metabolic asymmetries precede and predict appropriate deficits in neocortically-mediated cognitive functions in the initial course of disease, indicating that PET can be used for the early diagnosis and characterization of AD. Metabolic abnormalities of the neocortices in late-stage AD correlate with regional densities of neurofibrillary tangles but not of senile plaques post mortem, suggesting that tangle formation is important in disease pathogenesis.

Despite demonstrating reduced resting glucose metabolism, visual association areas demonstrate equivalent (as percent baseline) blood flow responses in mildly-moderately demented AD patients and controls who are performing a face matching task. Thus, viability and integrity of this cortical circuitry is retained into the intermediate stages of the disease, and glucose delivery to the AD brain can be increased.

INTRODUCTION

In the last decade, it has become possible to examine cerebral functional activity in awake men and women by means of positron emission tomography (PET) (Huang et al., 1980). With the appropriate positron emitting isotope, ^{18}F-2-fluoro-2-deoxy-D-glucose (^{18}FDG), we now are able to determine rates of glucose utilization within regions of the human brain as small as 6 mm in diameter, on the cortical surface as well as subcortically. As glucose is the major substrate for brain oxidative metabolism, its rate of consumption is a direct measure of brain functional activity.

In this paper, we review results of studies using PET and ^{18}FDG by the Laboratory of Neurosciences, National Institute on Aging, on patients with Alzheimer's disease (AD). We conclude that AD is a neurodegenerative disorder of brain association regions, and that it may be classified as a phylogenic disease. Reduced quantities of hexose transporter within cerebral capillaries have been reported in AD brain, but we suggest that they are secondary to the neurodegenerative process.

Fuel Homeostasis and the Nervous System, Edited by M. Vranic *et al.*
Plenum Press, New York, 1991

Alzheimer's disease (AD) is a progressive degenerative brain disorder that has no agreed-upon cause. The earliest and most prominent neuropsychological deficit is recent memory impairment. The first cognitive deficits to appear that are related to neocortical dysfunction are impairments of attention, abstract reasoning, language and visuospatial construction (Haxby et al., 1985, 1986; Grady et al., 1988).

Postmortem studies suggest that AD neuropathology is most severe in the association neocortices and connected regions (Rapoport, 1988a, b). This association "system" is organized as follows. Primary sensory areas (visual, somatosensory and auditory) and the primary motor area of the neocortex are connected to adjacent first-order association fields in the parietal, frontal and temporal and occipital lobes. These in turn exchange fibers with modality-specific second-order association areas and with secondary sensory areas. Within each hemisphere, first-order association areas also project reciprocally to the premotor region of the frontal cortex, whereas second-order association regions project to the prefrontal association cortex, polymodal parietal and temporal association areas, and to paralimbic and limbic regions (including the hippocampus, parahippocampal gyrus and entorhinal cortex) (Pandya and Seltzer, 1982; Van Hoesen, 1982).

Association fibers are derived from pyramidal neurons in layers III and V of the association neocortex. Corticocortical fibers within the same hemisphere, and within the corpus callosum connecting left and right hemispheres, arise mainly from pyramidal neurons in layer III, whereas the neurons in layer V give rise mainly to centrifugal fibers to nonneocortical regions (Wise and Jones, 1977; Schwartz and Goldman-Rakic, 1984).

Abundant senile (neuritic) plaques and neurofibrillary tangles with paired helical filaments characterize the AD brain. Association cortical areas of the temporal, parietal and frontal lobes are more severely affected than are primary sensory and motor regions (Brun and Gustafson, 1976; Pearson et al., 1985). Within the association neocortex, senile plaques are found in all cell layers, but most frequently in layers II and III. Neurofibrillary tangles are localized within large pyramidal neurons of layers III and V (Pearson et al., 1985; Rogers and Morrison, 1985; Lewis et al., 1987). Outside of the neocortex, tangles are found in neurons which receive or provide fibers, directly or indirectly, with the association neocortex. These cells are in layers II and IV of the entorhinal cortex, in the subiculum, and in the CA1 region and outer two thirds of the molecular layer of the dentate gyrus of the hippocampus (Ball and Nuttall, 1981; Hyman et al, 1984, 1986; Pearson et al., 1985; Rogers and Morrison, 1985). The subiculum and entorhinal cortex connect the hippocampal formation reciprocally with the association neocortices, as well as with the basal forebrain, thalamus and hypothalamus.

Senile plaques consist of clusters of degenerating neurites surrounded by an amyloid core composed of 5- to 10-nm fibrils which contain a 4.2 kDalton polypeptide, referred to as A4 or β amyloid protein, as well as excess protease-containing amyloid precursor protein. Abnormal quantities of amyloid also are found within vessel walls in the AD brain (Palmert et al., 1988; Abraham et al., 1988).

The paired helical filaments of neurofibrillary tangles contain an abnormally-phosphorylated, microfilament associated protein tau; the protein ubiquitin, and high molecular weight protein aggregates (Grundke-Iqbal et al., 1986). These filaments also are found in the neurites of senile plaques. A precursor, a nonphosphorylated tau-like neurofilament protein, has been demonstrated within neurons of layers III and V of the association neocortices, the subiculum, layers II and IV of the entorhinal cortex, and the pyramidal and hilar regions of the hippocampus, all of which can become pathological in AD (Morrison et al., 1987).

AD also is accompanied by brain atrophy and neuronal loss. Early in the disease course, progressive ventricular dilatation can be demonstrated by serial quantitative CT (Creasey et al., 1986; Luxenberg et al., 1986). Post mortem, atrophy is more severe in association than in primary cortical regions (Najlerahim and Bowen, 1988), where it is evidenced as reduced thickness and reduced length of the cortical ribbon (Duyckaerts et al., 1985; Mann et al., 1985).

Table 1. Cognitive Test Scores in Mildly and Moderate Alzheimer's Disease

Test	N	Controls	Mild AD(10)*	Moderate AD(12)
			Test Scores	
Mattis Dementia Rating Scale	16	141±3	131±6	110±18[2,3]
Wechsler Memory Scale				
Delayed Story Recall	23	17.6±5.7	2.6±3.7[2]	0.8±1.1[2]
Delayed Figure Production	23	6.5±3.3	0.8±1.1[2]	0.3±0.8[2]
WAIS				
Full Scale IQ	25	125±11	117±8	85±15[2,3]
Verbal Comprehension DQ	25	127±11	122±9	96±18[2,3]
Memory and Distractibility DQ	15	118±13	114±9	86±14[2,3]
Perceptual Organization DQ	25	119±14	108±13	76±19[2,3]
Syntax Comprehension (max=26)	23	24.2±2.5	22.7±2.4	15.8-5.4[2,3]
Boston Naming (max=43)	17	37.6±5.7	35.7±6.5	23.8±9.1[2,3]
Controlled Word Association	24	40±14	30±8	23±13[1]
Extended Range Drawing (max=24)	22	20.6±2.6	18.5±4.1	11.7±4.8[2,3]
Benton Facial Recognition	25	44.6±3.8	43.7±3.2	40.3±5.4[1]

Mean±S.D. significantly less than control mean: [1]$p < 0.01$; [2]$p < 0.001$
[3]Mean significantly less than in mild AD, $p < 0.001$
DQ = Factor Deviation Quotient; *No of subjects in parenthesis

The average duration of AD, once diagnosed, is about 7 years (Heston et al., 1981). Thus, neuropathology at end-stage disease may include chronic secondary changes. Consequently, it was important to examine AD in vivo early and throughout its course, using PET to measure rCMRglc and cognitive testing to examine the extent of dementia, and to correlate these measures with regional pathology post mortem in individual AD patients. The following sections summarize the results of these studies.

Brain Metabolism and Cognition in Alzheimer's Disease

To measure rCMRglc in an awake subject, [18]FDG is injected intravenously, and plasma radioactivity and glucose concentration are determined periodically until, after about 45 min, regional brain radioactivity is determined with PET in horizontal cross-sections of the brain. [18]FDG is phosphorylated within the brain, but is not further metabolized nor rapidly dephosphorylated, due to a low activity of brain phosphatase. Its rate of accumulation is used to calculate rCMRglc and to construct metabolic images of the brain (Huang et al., 1980).

PET was performed when the subject was at rest, with eyes covered and ears plugged with cotton. A medium-resolution PET scanner (ECAT II, Life Sciences, Oak Ridge, TN), with a resolution of 17 mm, was employed initially in our studies, followed, when it became available, by a high resolution scanner (PC 1024-7B, Scanditronix, Uppsala, Sweden), with an in-plane resolution of 6 mm and an axial resolution of 10 mm, and with an empirical attenuation correction.

rCMRglc was studied in relation to severity of dementia in AD patients, and in age-matched healthy controls. The AD patients were screened for illnesses other than AD which might contribute to cerebral dysfunction, and the controls were screened for any condition that might influence brain function. AD (possible or probable) was diagnosed according to NINCDS-ADRDA criteria for choosing patients for research purposes (McKhann et al., 1984). Severity of dementia was assessed with the Mini-Mental State Examination (Folstein et al., 1975): mild, score = 21-30; moderate, score = 11-20; severe, score = 0-10. Subjects also were administered an extensive neuropsychological test battery (Haxby et al., 1986).

Mean scores derived from this battery are summarized in Table 1 for mildly and moderately demented AD patients. Whereas moderately demented patients differed from controls on a wide range of neuropsychological tests, mildly demented patients differed only on the tests of memory (p < 0.01). Indeed, a significant and isolated memory impairment, with normal scores on all tests of language and visuospatial function, characterized 5 of the 10 mildly demented patients in Table 1 (Haxby et al., 1986).

Table 2. Left Hemispheric Glucose Metabolism in AD Patients

Parameter	Controls (29)[a]	Dementia Groups		
		Mild (10)	Moderate (7)	Severe (4)
		$rCMR_{glc}$, $mg.100g^{-1}.min^{-1}$		
Frontal Lobe	5.22±1.40	4.54±0.91	5.10±0.81	3.48±0.68[1]
Parietal lobe	5.40±1.54	4.28±1.29	4.65±1.14	2.69±0.76[1]
Temporal lobe	4.48±1.12	3.37±0.81	4.13±0.90	2.46±0.74[1]
Occipital lobe	5.31±1.30	4.43±1.15	5.71±0.97	4.21±0.78
Sensorimotor ctx	5.59±1.55	4.81±1.19	5.59±1.00	3.83±0.91

[a]No of patients in parenthesis; [1]Mean±S.D. differs from control (p<0.05). ECAT II data are from Duara et al. (1986)

Differences between mean rCMRglc values in AD patients and controls were statistically significant for severely but not for moderately or mildly demented patients, due in part to the large 25% standard deviation of the metabolic data (Table 2) (Duara et al., 1986). In the severely demented patients, as illustrated by the example of Figure 1, the association but not primary sensory or motor cortices were most affected. However, when the standard deviation of the data was reduced to about 5% of the mean by calculating ratios of association to primary rCMRglc, the parietal and lateral temporal cortices were found to be metabolically abnormal even in the mildly and moderately demented AD patients (Table 3) (Haxby et al., 1986). Clearly, these association regions are functionally disturbed early and throughout the course of AD.

Table 3. Relative Metabolic Rates in Alzheimer's Disease

Metabolic Ratio	Control(30)[a]	Mild AD(12)	Mod AD(15)	Severe AD(8)
Parietal/Sensorimotor	0.93±0.05	0.83±0.09[1]	0.85±0.09[1]	0.72±0.08[1]
Frontal/Sensorimotor	0.97±0.07	0.94±0.06	0.92±0.11	0.87±0.18[1]
Lat. Temporal/Occipital	0.85±0.08	0.76±0.11[1]	0.77±0.11[1]	0.66±0.09[1]

[a]Number of subjects in parenthesis
[1]Mean±S.D. differs from control by Bonferroni t test (p < 0.05).
ECAT II data from Haxby et al. (1986).

Friedland et al. (1985) also found that within-subject ratios of metabolic rates are more sensitive indicators of cerebral metabolic disorders in AD than are absolute regional rates. On the other hand, Foster et al. (1984) reported significant 24% to 42% reductions in rCMRglc in the frontal, parietal, temporal and occipital lobes in mildly as well as severely demented AD patients as compared with controls. These reductions may have been spurious, because the control rCMRglc values were from a small sample (N = 7) and appear unreasonably elevated (Haxby, 1986).

Some of the discrepancies between reports probably reflect heterogeneity of metabolic and cognitive deficits in AD patients (Friedland et al., 1988). Such heterogeneities make subject selection critical for demonstrating statistically significant group mean differences. For example, individual differences in cognitive performance are a hallmark of AD. Figure 2 illustrates relative cognitive profiles of 10 typical AD patients, divided according to severity with respect to three measures, the difference between predicted and measured WAIS IQ's, performance on the Syntax Comprehension Test (a test of language function), and performance on the Extended Range Drawing Test (a test of visuospatial ability) (Haxby, unpublished observations; Haxby et al., 1986). Patient 10, who had a severely reduced WAIS IQ, showed no apparent deficit in syntax comprehension. Patients 5 and 7, who had relatively mild WAIS deficits, demonstrated worse syntax comprehension than visuospatial relative scores. On the other hand, patients 1 and 8 had language and visuospatial deficits of equal-ranked severity.

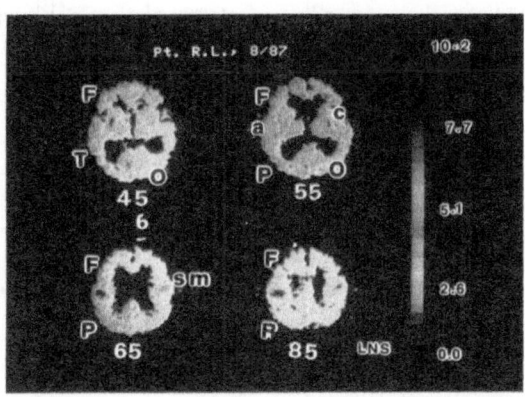

Figure 1. PET scan from severely demented patient with Alzheimer's disease. rCMRglc, derived with a Scanditronix tomograph, in units of mg.$100g^{-1}$.min^{-1}. High metabolic rates are retained in the primary sensorimotor cortex (sm), primary visual cortex (o), primary auditory cortex(a), caudate and lenticular nuclei (c), but are reduced elsewhere in the brain. P, parietal lobe; F, frontal cortex; T, temporal cortex; O, occipital cortex. Laboratory of Neurosciences, National Institute on Aging.

Case	1	2	3	4	5	6	7	8	9	10
Predicted IQ-WAIS IQ	◯	◯	◯	◢	◢	◢	◢	●	●	●
Syntax Comprehension Test	◢	◯	◯	◯	●	◢	●	●	◢	◯
Extended Range Drawing Test	◢	◯	◯	●	◢	●	◢	●	●	●

◯ No deficit   ◢ Mild deficit   ● Moderate to Severe deficit

Figure 2. Heterogeneous cognitive profiles in 10 typical AD patients. Patients are grouped according to the difference between predicted and measured WAIS IQ's, and are ranked with respect to each for this measure, and for reductions in scores on the Syntax Comprehension and Extended Range Drawing tests. From Haxby (unpublished).

Heterogeneous metabolic patterns also are displayed by AD patients. Grady et al. (1989) recently performed a factor analysis of rCMRglc values in 16 cortical and subcortical regions of 36 mildly-severely demented AD patients, and found that the patients could be divided into four orthogonal and independent subgroups ($p < 0.05$), illustrated in Figure 3 according to their normalized metabolic rates (normalized using Z scores with a mean of 0 and SD = 1). Unlike much of the other metabolic data presented in this paper, which were derived with the medium-resolution ECAT II tomograph, these data were obtained with the higher resolution Scanditronix scanner (see above).

The most common metabolic pattern (Group 1, 17 of 36 patients) consisted of reduced rCMRglc in the superior and inferior parietal gyri and both lateral and medial temporal regions. As compared to this pattern, Group 2 patients showed reduced normalized metabolic rates in the orbitofrontal and anterior cingulate gyri of the frontal lobe, whereas parietal regions were relatively spared. Group 3 patients showed reduced left-hemisphere Z scores compared to Group 1 patients, whereas Group 4 patients had marked relative reductions in most frontal lobe regions, as well as in the parietal areas. Thus, although most AD patients have reduced metabolism in parietal and temporal lobes, a smaller but substantial fraction have involvement of different parts of the frontal lobes as well.

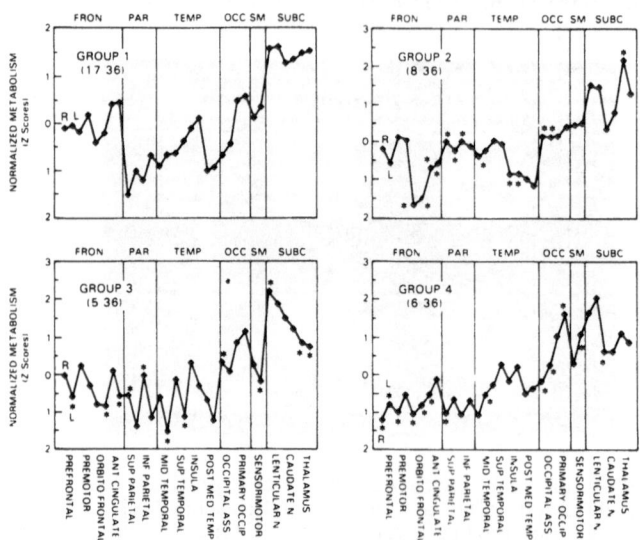

Figure 3. Orthogonal profiles of normalized metabolic rates for glucose in 4 subgroups of AD patients. Scanditronix-derived data. A Z score of ±1 is equivalent to ±1 standard deviation about the mean rCMRglc for the 16 regions examined. R, Right side; L, left side; *, differs significantly (p < 0.05) from normalized rCMRglc in Group 1 subjects.

Clearly, methods in addition to region-by-region comparisons of rCMRglc-derived means are needed to dissect out individual patterns of cognitive and metabolic dysfunction in AD patients. These methods include the analysis of right-left metabolic asymmetries, the construction of matrices of correlations between pairs of normalized rCMRglc values, and the examination of correlations between rCMRglc in brain regions thought to subserve specific cognitive functions, and scores on measures of these functions in individual patients (see below).

Asymmetry of Brain Metabolism and Cognitive Discrepancy in

Alzheimer's Disease

To localize small differences in neocortical metabolism in individual AD patients, Haxby et al. (1985) defined a metabolic asymmetry index (%) for homologous right and left brain regions,

$$\text{Asymmetry Index} = \frac{\text{rCMRglc,right} - \text{rCMRglc,left}}{[\text{rCMRglc,right} + \text{rCMRglc,left}]/2} \times 100 \qquad (1)$$

Asymmetry indices were calculated for mildly and moderately demented

AD patients and for healthy controls, from rCMRglc data obtained with the ECAT II tomograph. Significantly greater variances of asymmetry were demonstrated, as compared with controls, in the frontal, parietal, and temporal association regions, but not in the sensorimotor or occipital cortices (Figure 4) (Haxby et al., 1985), indicating that AD selectively involves the association neocortices, while sparing comparatively primary and motor regions. When the AD patients were divided according to dementia severity, both the mildly and moderately demented patients were shown to have increased variance of asymmetry in the association neocortices, as compared with controls (Table 4) (Haxby et al., 1986).

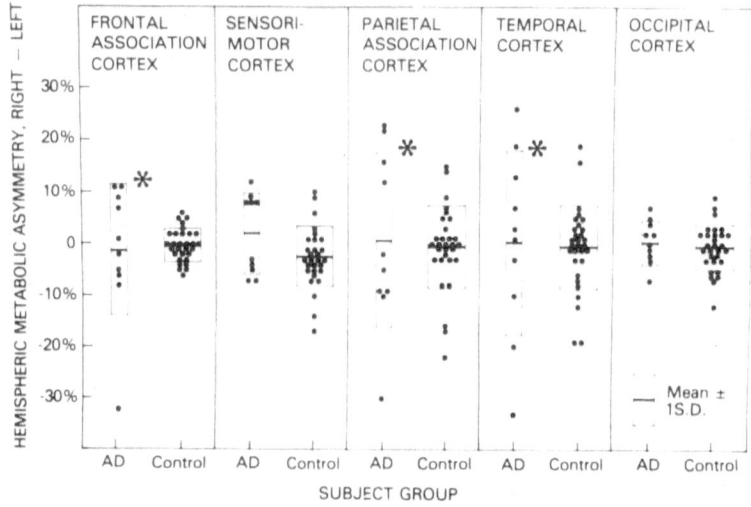

Figure 4. Metabolic asymmetry (Eq. 1, text) in each of 5 cortical regions in mildly-moderately demented AD patients and controls. Values are right minus left rCMRglc values, divided by the mean, in percent. *, coefficient of variation differs from control value. ECAT II data from Haxby et al. (1985).

Table 4. Metabolic Asymmetry Scores in Alzheimer Disease Patients

Glucose Utilization Asymmetry Index	Control (29)*	Mild AD (10)	Moderate (12)
Frontal Association Cortex	0.00±0.03	-0.01±0.08[3]	-0.02±0.12[3]
Parietal Association Cortex	0.00±0.06	0.00±0.12[1]	0.02±0.14[3]
Lateral Temporal Cortex	-0.01±0.08	-0.05±0.18[2]	-0.04±0.19[3]

*Mean ± S.D. is given (number of subjects in parenthesis). Variance is greater than in controls: [1]$p < 0.05$; [2]$p < 0.01$; [3]$p < 0.001$
ECAT II data from Haxby et al. (1986)

Studies of patients with focal brain damage suggest that syntax comprehension, mental arithmetic and immediate verbal memory are related to fuctions of regions in the left parietal and temporal lobes, whereas visuospatial construction is related to right parietal lobe function (Benton, 1985; Haxby et al. 1985, 1986). To see whether the metabolic asymmetries in AD corresponded to appropriate neocortically mediated cognitive deficits, Haxby et al. (1986) used the Syntax Comprehension Test to test left neocortical function, and the Extended Range Drawing Test to examine right neocortical function (cf., Figure 2). AD patients were ranked separately on the test scores; the difference between the ranks was calculated as a "syntax/drawing discrepancy."

Consistent with Table 1, which demonstrates significant deficits in neocortically mediated cognitive function in moderately but not in mildly demented AD patients, the metabolic asymmetry index was correlated significantly and appropriately with the syntax/drawing discrepancy index in moderately demented patients, but not in mildly demented patients or controls (Table 5). Correlations were significant for the frontal and parietal association neocortices, but not for the lateral temporal or sensory or motor regions; a lower left-sided rCMRglc corresponded to a worse language score, and a lower right-sided rCMRglc corresponded to a worse visuoconstruction test score. These cross-sectional results suggest that early metabolic dysfunction in association regions of AD patients precedes and predicts the deficits in language and visuospatial performance that later appear.

Table 5. Correlations between Right-Left Metabolic Asymmetries and Drawing/Syntax Comprehension Discrepancy.

CORTICAL REGION	CONTROLS N=30	MILD AD N = 12	MODERATE AD N = 15
Frontal Association	-0.30	-0.01	0.71*
Parietal Association	-0.11	-0.20	0.73*
Lateral Temporal	-0.08	0.01	0.49

*p < 0.05   From Haxby et al. (1986)

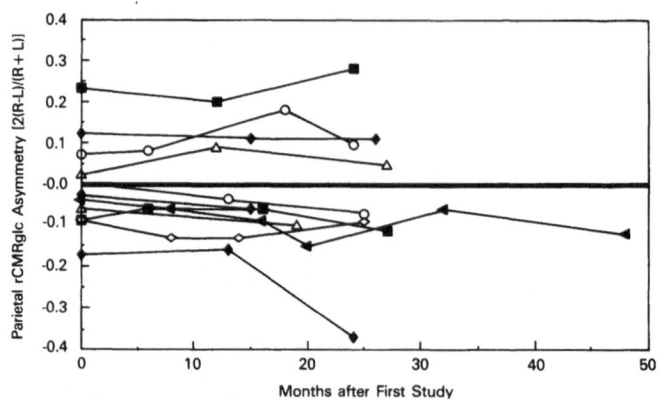

Figure 5. Stability of right-left metabolic asymmetry in parietal association cortices of 11 initially mildly-demented patients. Figure from Haxby et al. (in press).

This conclusion is being examined by longitudinal studies of individual AD patients, the preliminary results of which we present here (Grady et al., 1988; Haxby et al., in press). Figure 5 illustrates right/left metabolic asymmetries in the parietal association cortex of 11 AD patients who first were studied when mildly impaired, and were studied repeatedly for 25 to 47 months thereafter. In no patient did the initial direction of asymmetry change with time (Grady et al., 1986).

This constancy of direction was statistically significant for the prefrontal, parietal and lateral temporal association cortices, but not for the premotor, orbitofrontal, sensorimotor or occipital regions (Table 6) (Haxby et al., in press). Furthermore, the magnitude of asymmetry worsened with time in mildly-demented patients (mean follow-up time equaled 26 months) in prefrontal, parietal and temporal association regions. Finally, whereas mildly demented AD patients on initial evaluation did not display a significant correlation between metabolic asymmetries and discrepancies in language/visuospatial test scores or WAIS Deviation Quotients, significant correlations in the appropriate directions were found on follow-up studies (Haxby et al., 1987).

As far as we know, the above data are the first demonstration, in individual AD patients studied repeatedly over extended periods of time, (1) that the direction of metabolic asymmetry is constant in a given AD

Table 6. Stability over Time of Right-Left Metabolic Asymmetries in
11 AD Patients, with mild Dementia on Initial Evaluation

BRAIN REGION	SPEARMAN CORRELATION INITIAL/FOLLOW UP	RATE OF CHANGE percent/yr
ASSOCIATION CORTICES		
Prefrontal	0.72*	1.0 ± 0.35[1],*
Premotor	0.60	1.3 ± 0.85
Orbitofrontal	0.31	-0.1 ± 1.01
Parietal	0.97***	2.3 ± 0.82*
Lateral temporal	0.82**	1.5 ± 0.60*
PRIMARY CORTICES		
Sensorimotor	0.42	1.5 ± 0.73
Occipital	0.60	0.5 ± 0.70

[1]Mean ± S.E.M.
Differs from 0, *p < 0.05, **p < 0.01, ***p < 0.001
Data from Haxby et al. (in press). Mean follow up time, 26 mo.

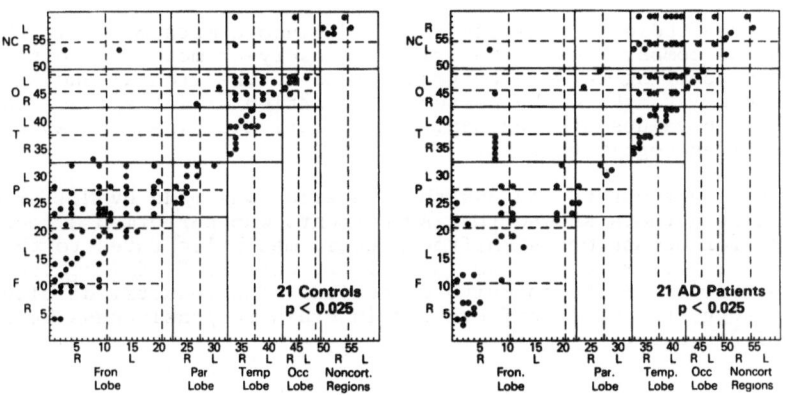

Figure 6. Matrices of positive partial correlation coefficients
between pairs of rCMRglc values for control and Alzheimer subjects.
Significant correlations at p < 0.025 are illustrated. Regions are
arranged according to whether they fall in the left (L) or right (R)
frontal (F), parietal (P), temporal (T), occipital (O) or noncortical (NC)
domains. ECAT II data. See Horwitz et al. (1987) for identities of each
numbered region. ECAT II data.

Table 7. Significant Differences between Numbers of Reliable Partial
Correlations in Mildly-Moderately Demented AD Patients and Controls

	TOTAL POSSIBLE	CONTROL	DAT
Frontal-Parietal	220	32	12*
Homologous (left-right)	28	22	14*

*Significantly different from control by χ^2 (p < 0.05) patient, (2) that the magnitude of metabolic asymmetry increases with time, and (3) that the heterogeneous metabolic asymmetries that appear early in the disease precede and accurately predict the heterogeneous deficits in neocortically-mediated functions that later appear and establish the individual dementia profile.

Involvement of the association neocortices in AD also is evidenced by metabolic uncoupling far beyond that seen in the healthy elderly, as illustrated by the correlation matrix analysis of Figure 6 and Table 7 (Horwitz et al., 1987). The matrix of significant positive correlations (p < 0.025) between pairs of rCMRglc values in AD patients (after partialing out global CMRglc) differs in several ways from the matrix for elderly, healthy controls. It has fewer significant positive partial correlations in the right and left frontal and parietal domains, and fewer significant correlations between corresponding regions in the left and right hemispheres. The latter reduction is displayed by fewer correlations in the diagonals between left and right lobar boxes of the AD matrix in Figure 6. The finding of metabolic uncoupling between ipsilateral and contralateral neocortical association and primary sensory and motor regions (Rapoport et al., 1986, 1987), differs from results of studies by Metter et al. (1984). These authors reported increased numbers of partial correlations in AD patients (43 as compared to 17 in controls), together with a decline in global metabolism, and concluded that functional dependence between brain regions is increased in AD.

PET Activation Studies in Alzheimer's Disease

The above studies demonstrate reduced glucose metabolism in the association as compared with the primary sensory or motor neocortices, and functional uncoupling between cortical areas in the frontal and parietal lobes. We now may ask to what extent association brain regions, whose resting rCMRglc values are reduced, can be activated when the AD patient performs a given cognitive task.

An initial attempt to address this question was made by Miller et al. (1987). These authors showed that AD patients who performed a recognition memory task for common words shifted glucose metabolic rates to their right temporal lobe more than to their left, whereas 5 of 7 controls shifted metabolism in the opposite direction. The results, difficult to interpret, suggested major differences in brain responsiveness between AD and control subjects.

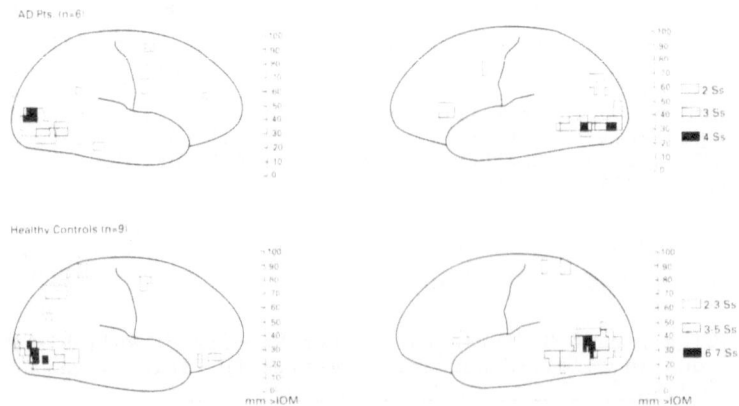

Figure 7. Cerebral blood flow activation patterns in DAT and healthy controls, on lateral hemispheric surfaces. Areas activated are identified if rCBF (normalized to whole brain rCBF) was 30% higher during face-matching task than in control sensorimotor task. Numbers of subjects in which this occurred are indicated. Scale give level (mm) of brain region above inferior orbitomeatal (IOM) line. DAT, dementia of the Alzheimer type. From Grady et al. (1990, in preparation).

On the other hand, Grady et al (1990, in preparation), who examined regional cerebral blood flow (rCBF) with PET in subjects performing a face matching task, found remarkably few differences between mildly-moderately demented AD patients and controls. rCBF was measured with the positron-emitting isotope $H_2{}^{15}O$, using the Scanditronix PC 1024-7B PET scanner (see above). In each subject, rCBF normalized during a control task was subtracted from rCBF during the face-matching task, and flow differences which exceeded 30% of the baseline were identified.

As illustrated in Figure 7, 30% or greater activation of normalized rCBF during face matching occurred bilaterally in occipital and occipito-temporal regions (more medial primary visual cortices [not shown] also were activated) in both control and AD subjects. The areas activated are homologous to visual association areas for object recognition in the rhesus monkey (Haxby et al., 1988; Mishkin et al., 1983). Thus, although resting glucose metabolism is reduced in visual association areas in mildly-moderately demented AD patients, these areas are viable and can respond to the same proportional extent in AD patients as in controls during face matching. Their coherent activation demonstrates that the brain network which subserves face recognition remains intact in mildly-moderately demented AD patients.

Correlation Between rCMRglc Reductions and Local Neuropathology

in Individual Alzheimer Patients

As noted above, controversy exists about which pathological phenotype is most closely related to the pathogenesis of AD. Blessed et al. (1968) first showed that the severity of global dementia in AD patients correlated with the number of senile plaques in the entire brain on postmortem, but did not study the regional distribution of plaques nor analyze neurofibrillary tangles. Later studies have reported significant correlations between global dementia and virtually all postmortem AD neuropathological findings, including the number of neurofibrillary tangles (Wilcock and Esiri, 1982), of tangles and plaques (Delaere et al., 1989; Neary et al., 1986), loss of large cortical neurons in temporal and frontal lobes (Mann et al., 1988; Neary et al., 1986), reduced markers for acetylcholine synthesis (Mann et al., 1988), and cortical atrophy (Duyckaerts et al., 1988; Najlerahim and Bowen, 1988). In one individual AD patient, McGeer et al (1986) reported that most severe reductions in rCMRglc during life, as measured with PET, occurred in those brain regions which demonstrated most neuropathology post mortem (neuronal loss, gliosis, loss of laminar pattern). However, no publication to date has correlated quantitatively PET measures of regional metabolism during life, with regional neuropathology post mortem, in individual AD patients. We report here preliminary results (DeCarli et al., unpublished) which demonstrate that reductions in rCMRglc during life correlate significantly with regional densities of neurofibrillary tangles post mortem, but not with regional densities of senile plaques.

rCMRglc was measured during life in 10 neocortical regions of each of 6 AD patients (De Carli et al., in preparation). Each patient died within 14 to 53 months of his last PET scan, and regional densities of neurofibrillary tangles (number per mm^3) and of senile plaques were determined post mortem, after correction was made for brain shrinkage. In 2 of the 6 patients, regional densities of neurofibrillary tangles were correlated significantly with rCMRglc values (r = -0.69, p < 0.04; r = -0.98, p < 0.001, n = 10). Five of the correlation coefficients were more negative than -0.43. Furthermore, for the 6 patients taken together, the correlation between rCMRglc and neurofibrillary tangle density (55 individual observations) equaled -0.43, and was significant at p < 0.005. Association brain areas (e.g., orbitofrontal gyrus, middle temporal gyrus) had the highest tangle densities and the lowest metabolic rates, whereas primary sensory and motor areas (e.g., calcarine cortex, precentral (motor) cortex and postcentral [somatosensory] cortex) had the lowest tangle densities and highest metabolic rates. On the other hand, in no patient, nor for the 6 patients grouped together, was the correlation coefficient between rCMRglc and senile plaque density statistically significant (p > 0.05). These results confirm selective involvement of association as compared to primary sensory-motor cortices in AD, and indicate that neurofibrillary tangles are more closely related to the primary AD process than are senile plaques. They are consistent with the hypothesis (Schapiro et al., 1989, Schapiro and Rapoport, 1988) that AD-type dementia occurs in older Down syndrome subjects only when large numbers of neurofibrillary tangles appear, which may be some 10-20 years after large numbers of senile plaques appear (Wisniewski et al., 1985).

CONCLUSIONS

Resting glucose metabolism in the association neocortices is disturbed early and throughout the course of AD. Metabolic asymmetries in mildly demented patients precede and predict appropriate deficits in neocortically-mediated cognitive functions as dementia progresses. Metabolic and neocortically-mediated cognitive disturbances in AD are consistently asymmetrical in an individual patient. Both are heterogeneous and differ from patient to patient, but are correlated within individual patients. Metabolic abnormalities of the neocortices in life, furthermore, correlate with densities of neurofibrillary tangles but not of senile plaques post mortem.

The correlation matrix from AD patients differs from that of age-matched elderly controls by having fewer significant correlations between ipsilateral parietal and frontal regions, and between identical regions in the right and left cerebral hemispheres, consistent with disrupted function (Rapoport et al., 1986, 1987; Rapoport and Horwitz, 1989). Nevertheless, brain networks which subserve visual recognition (including primary visual cortex, temporal and occipito-temporal association areas) retain functional integrity and connections late into the course of AD disease (Grady et al., 1990, in preparation).

Significance of Alterations in rCMRglc in Alzheimer's Disease

The above evidence suggests that AD is a primary neurodegenerative disorder in which neurofibrillary tangle formation plays an important role. In contrast, several authors have proposed that abnormalities at the blood-brain barrier are primary in AD (Glenner, 1985; Hardy et al., 1986; Mooradian, 1988), in so far as amyloid frequently accumulates in cerebral blood vessels of the AD brain (Palmert et al., 1988, Abraham et al., 1988), and the hexose transporter at cerebral capillaries, measured by [^{3}H]cytochalasin B binding, is reduced by 50% whereas other capillary markers are not altered (Kalaria and Harik, 1988). We believe that a primary role for reduced glucose transport is unlikely for the following reasons:

1. In vivo studies of blood-brain barrier integrity, using [^{18}Ga]EDTA and PET, do not demonstrate that the permeability of the cerebrovasculature in AD is abnormal (Kessler et al., 1984).
2. Brain glucose metabolism is far from being rate-limited by transport at the blood-brain barrier in normoglycemic individuals (Gjedde, 1983); there is no evidence that AD patients are uniquely hypoglycemic or have rate-limited delivery of glucose to the brain (Friedland et al., 1989; Fukuyama et al., 1989).
3. Cerebrospinal fluid concentrations of glucose in AD patients are not abnormal (Laboratory of Neurosciences, unpublished observations).
4. During a face matching task, brain regions which show reduced resting rCMRglc in AD patients nevertheless have blood flow responses which are equivalent to responses in healthy controls (see above, Grady et al., 1990, in preparation). As rCBF is coupled to cerebral glucose consumption (Reivich, 1974; Raichle, in press), these equivalent responses suggest that glucose delivery is not limited in the AD brain.
5. ^{31}P NMR spectroscopy demonstrates normal ratios of PCr, ATP, ADP and Pi in the AD brain (Smith et al., 1986), suggesting that glucose delivery is not limited in the AD brain.

242

Reductions of the capillary hexose transporter in AD may be due to down-regulation secondary to reduced metabolic demand from local neurodegeneration. Animal studies suggest that the hexose transporter can be up-regulated during chronic hypoglycemia (McCall et al., 1986), or down-regulated during chronic hyperglycemia (Gjedde and Crone, 1981; Pardridge, in press). If the signal for regulating the glucose transporter were the capillary endothelial glucose concentration, reduced metabolic demand might be equivalent to hyperglycemia, as either hyperglycemia or reduced demand would tend to increase endothelial cell glucose concentration.

A Phylogenic Hypothesis for Alzheimer's Disease

The PET-derived measures of regional cerebral metabolic rates for glucose (rCMRglc), when correlated with cognitive test scores and postmortem neuropathology, suggest that AD is a primary neurodegenerative disease which affects selectively regions of the brain association system -- the association neocortices and subcortical regions which form direct and important reciprocal connections with the association neocortices. As the association system is most extensive in higher primates and particularly in humans, its preferential involvement has suggested that AD is a phylogenic neurodegenerative disease, made likely by changes in the primate genome during evolution (Rapoport, 1988a, b; 1989). No animal model for AD has yet been demonstrated.

These evolutionary changes in the primate genome might possibly be related to genetic factors which contribute to AD. AD can be familial with autosomal dominant transmission (Heston et al., 1981). Chromosome 21 has been implicated in some cases of early-onset familial AD (although not in other familial cases) (St George-Hyslop et al., 1987; Roses et al., 1988; Schellenberg et al., 1988). Alzheimer pathology is observed invariably in older subjects with trisomy 21 (Wisniewski et al., 1985), suggesting a genetic basis involving chromosome 21 for at least some types of the disease. Brains of Down subjects older than 40 years exhibit neurochemical abnormalities, and neuropathology with the same density, chemical and antigenic properties, and regional distributions as do brains of Alzheimer patients (Ball et al., 1981; Mann et al., 1984; Casanova et al., 1985; Wisniewski et al., 1985). In life, glucose metabolism is abnormal in the association but not sensory or motor neocortices of older Down subjects (Schapiro et al., 1986).

Rapid evolution of the brain and particularly of its association regions in higher primates probably did not arise by point mutations coding for structural proteins, but rather by genomic changes leading to increased or otherwise altered expression of genes coding for products in the association neocortices and their connections (King and Wilson, 1975; Rapoport, 1988a, b; 1989). These events might have included regulatory mutations, leading to altered expression of a regulatory enzyme, growth or recognition factor within association regions; gene duplication, causing a discontinuous increase in expression of regulatory proteins; or chromosomal rearrangement, changing gene expression by altering gene position within a chromosome. Indeed, although nuclear DNA differs by less than 1% between humans and the chimpanzee, multiple gene rearrangements distinguish the genomes of these species (King and Wilson, 1975).

In view of the correlation between rCMRglc and neurofibrillary tangles in AD, and the suggested role of neurofibrillary tangles in the dementia of Down syndrome (see above) (De Carli et al., in preparation; Schapiro et al., 1989; Schapiro and Rapoport, 1988), one might speculate further that a genomic event during primate evolution enhanced the tendency for the brain to accumulate abnormally phosphorylated neurofilaments with neurofibrillary tangles (Grundke-Iqbal et al., 1986; Rapoport, 1989). Excess expression of genes on chromosome 21 in Down syndrome may augment this tendency (Rapoport, 1988a).

REFERENCES

Abraham, C. R., Selkoe, D. J., and Potter, H., 1988, Immunochemical identification of the serine protease inhibitor alpha 1-antichymotrypsin in the brain amyloid deposits of Alzheimer's disease, Cell, 52:487.

Ball, M. J., and Nuttall, K., 1981, Topography of neurofibrillary tangles and granulovacuoles in hippocampi of patients with Down's syndrome: quantitative comparison with normal ageing and Alzheimer's disease, Neuropathol. Appl. Neurobiol., 7:13.

Ball, M. J., Fisman, M., Hachinski, V., Blume, W., Fox, A., Kral, V. A., Kirshen, A. J., Fox, H., and Merskey, H., 1985, A new definition of Alzheimer's disease: a hippocampal dementia, Lancet, Jan., 14.

Benton, A., 1985, Visuoperceptual, visuospatial and visuoconstructive disorders, in: "Clinical Neuropsychology," 2nd Ed., K. M. Heilman, E. Valenstein, eds., Oxford University Press, Oxford.

Blessed, B., Tomlinson, B. E., and Roth, M., 1968, The association between quantitative measures of dementia and of senile change in the cerebral gray matter of elderly subjects, Brit. J. Psychiatry 114:797.

Brun, A., and Gustafson, L., 1976, Distribution of cerebral degeneration in Alzheimer's disease. A clinico-pathological study, Arch. Psychiatr. Nervenkr. 223:15.

Casanova, M. F., Walker, L. C., Whitehouse, P. J., and Price, D. L., 1985, Abnormalities of the nucleus basalis in Down's syndrome, Ann. Neurol. 18:310.

Creasey, H., Schwartz, M., Frederickson, H., Haxby, J. V., and Rapoport, S. I., 1986, Quantitative computed tomography in dementia of the Alzheimer type, Neurology 36:1563.

DeCarli, C., Atack, J. R., Ball, M. J., Kaye, J. A., Grady, C. L., Fewster, P., Katz, D., Schapiro, M. B., and Rapoport, S. I., in preparation, Regional neurofibrillary tangle densities, but not regional senile plaque densities, correlate with regional reductions in cerebral glucose utilization during life in Alzheimer disease patients.

Delaere, P., Duyckaerts, C., Brion, J. B., Poulain, V., Hauw, J. J., 1989, Tau, paired helical filaments and amyloid in the neocortex: a morphometric study of 15 cases with graded intellectual status in aging and senile dementia of the Alzheimer type, Acta Neuropathol. (Berlin) 77:645.

Duara, R., Grady, C. L., Haxby, J. V., Sundaram, M., Cutler, N. R., Heston, L., Moore, A., Schlageter, N. L., Larson, S., and Rapoport, S. I., 1986, Positron emission tomography in Alzheimer's disease, Neurology 36:879.

Duyckaerts, C., Hauw, J. J., Piette, F., Rainsard, C., Poulain, V., Berthaux, P., and Escourolle, R., 1985, Cortical atrophy in senile dementia of the Alzheimer type is mainly due to a decrease in cortical length, Acta Neuropathol. (Berl.) 66:72.

Folstein, M. F., Folstein, S. E., and McHugh, P. R., 1975, "Mini-Mental State." A practical method for grading the cognitive state of patients for the clinician, J. Psychiat. Res. 12:189.

Foster, N. L., Chase, T. N., Mansi, L., Brooks, R., Fedio, P., Patronas, N. J., and Di Chiro, G., 1984, Cortical abnormalities in Alzheimer's disease, Ann. Neurol. 16:649.

Friedland, R. P., Budinger, T. F., Koss, E., and Ober, B. A., 1985, Alzheimer's disease: anterior-posterior and lateral hemispheric alterations in cortical glucose utilization, Neurosci. Lett. 53:235.

Friedland, R. P., Koss, E., Haxby, J. V., Grady, C. L., Luxenberg, J., Schapiro, M. B., and Kaye, J., 1988, Alzheimer disease: clinical and biological heterogeneity, Ann. Int. Med. 109:298.

Friedland, R. P., Jagust, W. J., Huesman, R. H., Koss, E., Knittel, B., Mathis, C. A., Ober, B. A., Mazoyer, B. M., and Budinger, T. F., 1989, Regional cerebral glucose transport and utilization in Alzheimer's disease, Neurology 39:1427.

Fukuyama, H., Kameyama, M., Harada, K., Nishizawa, S., Senda, M., Mukai, T., Yonekura, Y., and Torizuka, K., 1989, Glucose metabolism and rate constants in Alzheimer's disease examined with dynamic positron emission tomography scan, Acta Neurol. Scand. 80:307.

Gjedde, A., 1983, Modulation of substrate transport to the brain, Acta Neurol. Scand. 67:3.

Gjedde, A., and Crone, C., 1981, Blood-brain glucose transfer: repression in chronic hyperglycemia, Science 214:456.

Glenner, G. G., 1985, On causative theories in Alzheimer's disease, Hum. Pathol. 16:433.

Grady, C., Haxby, J., Horwitz, B., Schapiro, M., Carson, R., Herscovitch, P., and Rapoport, S. I., 1990, Activation of regional cerebral blood flow (rCBF) in extrastriate cortex during a face matching task in patients with dementia of the Alzheimer type (DAT), Soc. Neurosci. Abstr.

Grady, C. L., Haxby J. V., Horwitz, B., Sundaram, M., Berg, G., Schapiro, M., Friedland, R. P., and Rapoport, S.I., 1988, A longitudinal study of the early neuropsychological and cerebral metabolic changes in dementia of the Alzheimer type, J. Clin. Exp. Neuropsychol. 10:576.

Grady, C. L., Haxby, J., Schapiro, M. B., Kumar, A., Friedland, R. P. and Rapoport, S. I., 1989, Heterogeneity in dementia of the Alzheimer type (DAT): subgroups identified from cerebral metabolic patterns using positron emission tomography (PET), Neurology 39(Suppl. 1):167.

Grady, C. L., Haxby, J. V., Schlageter, N. L., Berg, G., and Rapoport, S. I., 1986, Stability of metabolic and neuropsychological asymmetries in dementia of the Alzheimer type, Neurology 36:1390.

Grundke-Iqbal, I., Iqbal, K., Tung, Y.-C., Quinlan, M., Wisniewski, H.M., and Binder, L. I., 1986, Abnormal phosphorylation of the microtubule-associated protein τ (tau) in Alzheimer cytoskeletal pathology, Proc. Natl. Acad. Sci. (USA). 83:4913.

Hardy, J. A., Mann, D. M. A., Wester, P., and Winbland, B., 1986, An integrative hypothesis concerning the pathogenesis and progression of Alzheimer's disease, Neurobiol. Aging 7:489.

Haxby, J. V., 1986, Cerebral metabolic rate of glucose and Alzheimer's disease: Reply, J. Cereb. Blood Flow Metab. 6:125.

Haxby, J. V., Duara, R., Grady, C. L., Cutler, N. R., and Rapoport, S. I., 1985, Relations between neuropsychological and cerebral metabolic asymmetries in early Alzheimer's disease, J. Cereb. Blood Flow Metab. 5:193.

Haxby, J. V., Grady, C. L., Duara, R., Schlageter, N., Berg, G., and Rapoport, S. I., 1986, Neocortical metabolic abnormalities precede non-memory cognitive deficits in early Alzheimer's-type dementia, Arch. Neurol. 43:882.

Haxby, J. V., Grady, C. L., Friedland, R. P., and Rapoport, S. I., 1987, Neocortical metabolic abnormalities precede nonmemory cognitive impairments in early dementia of the Alzheimer type, J. Neural Transmission, 24 (Suppl):49.

Haxby, J. V., Grady, C. L., Horwitz, B., Schapiro, M. B., Carson, R. E., Ungerleider, L. G., Mishkin, M., Herscovitch, P., Friedland, R. P. and Rapoport, S. I., 1988, Mapping two visual pathways in man with regional cerebral blood flow (rCBF) as measured by positron emission tomography and $H_2{}^{15}O$, Soc. Neurosci. Abstr., 14:750.

Haxby, J. V., Grady, C. L., Koss, E., Horwitz, B., Schapiro, M. B., Katz, D., Friedland, R. P., and Rapoport, S. I., in press, Longitudinal study of cerebral metabolic asymmetries and associated neuropsychological deficits in early dementia of the Alzheimer type.

Heston, L. L., Mastri, A. R., Anderson, V. E., and White, J., 1981, Dementia of the Alzheimer type. Clinical genetics, natural history, and associated conditions, Arch. Gen. Psychiatry 38:1085.

Horwitz, B., Grady C. L., Schlageter, N. L., Duara, R., and Rapoport, S. I., 1987, Intercorrelations of regional cerebral glucose metabolic rates in Alzheimer's disease, Brain Res. 407:294.

Huang, S.-C, Phelps, M. E., Hoffman, E. J., Sideris, K., Selin, C. J., and Kuhl, D. E., 1980, Non-invasive determination of local cerebral metabolic rate of glucose in man, Am. J. Physiol. 238:E69.

Hyman, B. T., Van Hoesen, G. W., Damasio A. R., and Barnes, C. L., 1984, Alzheimer's disease: cell-specific pathology isolates the hippocampal formation, Science (Wash.) 225:1168.

Hyman, B. T., Van Hoesen, G. W., Kromer, L. J., and Damasio, A. R., 1986, Perforant pathway changes and the memory impairment of Alzheimer's disease, Ann. Neurol. 20:472.

Kalaria, R. N., and Harik, S. I., 1988, Reduced glucose transporter at the blood-brain barrier and in cerebral cortex in Alzheimer disease, J. Neurochem. 53:1083.

Kessler, R. M., Goble, J. C., Bird, J. H., Girton, M. E., Doppman, J. L., Rapoport, S. I., and Barranger, J. A., 1984, Measurement of blood-brain barrier permeability with positron emission tomography and ^{68}Ga EDTA., J. Cerebral Blood Flow Metab., 4:323.

King, M.-C., Wilson, A. C., 1975, Evolution at two levels in humans and chimpanzees, Science (Wash.) 188:107.

Lewis, D. A., Campbell, M. J., Terry, R. D., and Morrison, J. H., 1987, Laminar and regional distributions of neurofibrillary tangles and neuritic plaques in Alzheimer's disease: a quantitative study of visual and auditory cortices, J. Neurosci. 7:1799.

Luxenberg, J. S., Haxby, J. V., Creasey, H., Sundaram, M., and Rapoport, S. I., 1987, Rate of ventricular enlargement in dementia of the Alzheimer type correlates with rate of neuropsychological deterioration, Neurology 37:1135.

Mann, D. M., Marcyniuk, B., Yates, P. D., Neary, D., and Snowden, J. S., 1988, The progression of the pathological changes of Alzheimer's disease in frontal and temporal neocortex examined both at biopsy and at autopsy, Neuropathol. Appl. Neurobiol. 14:177.

Mann, D. M. A., Yates, P. O., and Marcyniuk, B., 1984, Alzheimer's presenile dementia, senile dementia of Alzheimer type and Down's syndrome in middle age form an age related continuum of pathological changes, Neuropathol. Appl. Neurobiol. 10:185.

Mann, D. M. A., Yates P. O., and Marcyniuk, B., 1985, Correlation between senile plaque and neurofibrillary tangle counts in cerebral cortex and neuronal counts in cortex and subcortical structures. Neurosci. Lett. 56:51.

McCall, A. L., Fixman, L. B., Tornheim, K., Chick, W., and Ruderman, N. B., 1986, Chronic hypoglycemia increases brain glucose transport, Am. J. Physiol. 251:E442.

McGeer, P. L., Kamo, H., Harrop, R., McGeer, E. G., Martin, W. R. W., Pate, B. D., and Li, D. K. B., 1986, Comparison of PET, MRI, and CT with pathology in a proven case of Alzheimer's disease. Neurology 36:1569.

McKhann, G., Drachman, D., Folstein, M., Katzman, R., Price, D., and Stadlan, E. M., 1984, Clinical diagnosis of Alzheimer's disease: Report of the NINCDS-ADRDA Work Group under the auspices of Department of Health and Human Services Task Force on Alzheimer's disease, Neurology 34:939.

Metter, E. J., Riege, W. H., Kameyama, M., Kuhl, D. E., and Phelps, M. E., 1984, Cerebral metabolic relationships for selected brain regions in Alzheimer's, Huntington's, and Parkinson's diseases, J. Cereb. Blood Flow Metab. 4:500.

Miller, J. D., De Leon, M. J., Ferris, S. H., Kluger, A., George, A. E., Reisberg, B., Sachs, H. J., and Wolf, A. P., 1987, Abnormal temporal lobe response in Alzheimer's disease during cognitive processing as measured by ^{11}C-2-deoxy-D-glucose and PET. J. Cerebral Blood Flow Metab., 7:248.

Mishkin, M., Ungerleider, L. G., and Macko, K. A., 1983, Object vision and spatial vision: two cortical pathways. Trends Neurosci., 6:414.

Morrison, J. H., Lewis, D. A., Campbell, M. J., Huntley, G. W., Benson, D. L., and Bouras, C., 1987, A monoclonal antibody to non-phosphorylated neurofilament protein marks the vulnerable cortical neurons in Alzheimer's disease, Brain Res. 416:331.

Najlerahim, A., and Bowen, D. M., 1988, Regional weight loss of the cerebral cortex and some subcortical nuclei in senile dementia of the Alzheimer type, Acta Neuropathol. (Berl.) 75:509.

Neary, D., Snowden, J. S., Mann, D. M. A., Bowen, D. M., Sims, N. R., Northern, B., Yates, P. O., and Davison, A. N., 1986, Alzheimer's disease: a correlative study. J. Neurol. Neurosurg. Psychiatry 49:229.

Palmert, M. R., Golde, T. E., Cohen, M. L., Kovacs, D. M., Tanzi, R. E., Gusella, J. F., Usiak, M. F., Younkin, L. H., and Younkin, S. G., 1988, Amyloid protein precursor messenger RNAs: differential expression in Alzheimer's disease. Science 241:1080.

Pandya, D. N., and Seltzer, B., 1982, Association areas of the cerebral cortex, Trends Neurosci. 5:386.

Pardridge, W. M., In Press, Blood-brain barrier transport of glucose, ketone bodies and free fatty acids, In "1st Toronto-Stockholm Symposium on Perspectives in Diabetes Research, The Nervous System and Fuel Homeostasis," Plenum Press, New York.

Pearson, R. C., Esiri, M. M., Hiorns, R. W., Wilcock, G. K., and Powell, T. P. S., 1985, Anatomical correlates of the distribution of the pathological changes in the neocortex in Alzheimer's disease, Proc. Natl. Acad. Sci. (USA) 82:4531.

Raichle, M. E., In Press, Nonoxidative glucose consumption and normal brain function - positron emission tomography studies in normal humans, In "1st Toronto-Stockholm Symposium on Perspectives in Diabetes Research, The Nervous System and Fuel Homeostasis," Plenum Press, New York.

Rapoport, S. I., 1988a, Brain evolution and Alzheimer's disease, Rev. Neurol. 144:79.

Rapoport, S. I., 1988b, A phylogenetic hypothesis for Alzheimer's disease, in: "Genetics and Alzheimer's Disease. Research and Perspectives in Alzheimer's Disease," P. M. Sinet, Y. Lamour and Y. Christen, eds., Fondation Ipsen, Springer-Verlag, Berlin.

Rapoport, S. I., 1989, Hypothesis: Alzheimer's disease is a phylogenetic disease, Med. Hypotheses 29:147.

Rapoport, S. I., and Horwitz, B., 1989, Use of positron emission tomography to study patterns of brain metabolism in relation to age and disease: a correlation matrix approach, in: "Regulatory Mechanisms of Neuron to Vessel Communication in the Brain," F. Battaini, S. Govoni, M. S. Magnoni and M. Trabucci, eds., NATO Advanced Science Institute Series, Cell Biology, Vol 33, Springer-Verlag, Berlin.

Rapoport, S. I., Horwitz, B., Grady, C. L., and Haxby, J. V., 1987, Alzheimer's disease causes metabolic uncoupling of associative brain regions beyond that seen in the healthy elderly, in: "Modifications of Cell to Cell Signals During Normal and Pathological Aging," S. Gouani and F. Battaini, eds., NATO ASI Series, Vol. H9, Springer-Verlag, Berlin.

Rapoport, S. I., Horwitz, B., Haxby, J. V., and Grady, C. L., 1986, Alzheimer's disease: metabolic uncoupling of associative brain regions, Can. J. Neurol Sci. 13:540.

Reivich, M., 1974, Blood flow metabolism coupling in brain, Res. Publ. Assoc. Res. Nerv. Ment. Dis. 53:125.

Rogers, J., and Morrison, J. H., 1985, Quantitative morphology and regional and laminar distributions of senile plaques in Alzheimer's disease, J. Neurosci. 5:2801.

Roses, A. D., Pericak-Vance, M. A., Haynes, C. S., Haines, J. L., Gaskell, P. A., Yamaoka, L. H., Hung, W-Y., Clark, C. M., Alberts, M. J., Lee, J. E., Siddique, T., and Heyman, A. L., 1988, Genetic linkage studies in Alzheimer's disease (AD). Neurology 38(Suppl 1):173.

Schapiro, M. B., Haxby, J. V., Grady, C. L., Rapoport, S. I., 1986, Cerebral glucose utilization, quantitative tomography, and cognitive function in adult Down syndrome, in: " The Neurobiology of Down Syndrome," C. J. Epstein, ed., Raven Press, New York.

Schapiro, M. B., Luxenberg, J. S., Kaye, J. A., Haxby, J. V., Friedland, R. P., and Rapoport, S. I., 1989, Serial quantitative computed tomography analysis of brain morphometrics in adult Down syndrome at different ages, Neurology 39:1349.

Schapiro, M. B., and Rapoport, S. I., 1988, Alzheimer's disease in premorbidly normal and Down's syndrome individuals: selective involvement of hippocampus and neocortical associative brain regions, Brain Dysfunction 1:2.

Schellenberg, G. D., Bird, T. D., Wijsman, E. M., Moore, D. K., Boehnke, M., Bryant, E. M., Lampe, T. H., Nochlin, D., Sumi, S. M., Deeb, S. S., Beyreuther, K., and Martin, G. M., 1988, Absence of linkage of chromosome 21q21 markers to familial Alzheimer's disease, Science (Wash.) 241:1507.

Schwartz, M. L., Goldman-Rakic, P. S., 1984, Callosal and intrahemispheric connectivity of the prefrontal association cortex in Rhesus monkey: relation between intraparietal and principal sulcal cortex, J. Comp. Neurol. 226:403.

Smith, L. S., Bottomley, P. A., Drayer, B. P. and Herfkens, R. J., 1986, Localized clinical ^{31}P NMR sprectroscopy in Huntington's, Parkinson's, Alzheimer's and Binswanger's diseases, Abstr. Fifth Annual Meeting Soc. Magnetic Resonance in Medicine, August 19-22, 1986, Montreal, 4:1386.

Sokoloff, L., Reivich, M., Kennedy, C., Des Rosiers, M. H., Patlak, C. S., Pettigrew, K. D., Sakurada, O., and Shinohara, M., 1977, The [^{14}C]deoxyglucose methods for the measurement of local cerebral glucose utilization: theory, procedure, and normal values in the conscious and anesthetized rat, J. Neurochem., 28:897.

St George-Hyslop, P. H., Tanzi, R. E., Polinsky, R. J., Haines, J. L., Nee, L., Watkins, P. C., Myers, R. H., Feldman, R. G., Pollen, D., Drachman, D., Growdon, J., Bruni, A., Foncin, J.-F., Salmon, D., Frommelt, P., Amaducci, L., Sorbi, S., Piacentini, S., Stewart, G. D., Hobbs, W. J., Conneally, P. M., and Gusella, J. F., 1987, The genetic defect causing familial Alzheimer's disease maps on chromosome 21, Science (Wash.) 235:885.

Van Hoesen, G. W., 1982, The primate parahippocampal gyrus: new insights regarding its cortical connections, Trends Neurosci. 5:345.

Wilcock, G. K., and Esiri, M. M., 1982, Plaques, tangles and dementia: a quantitative study, J. Neurol. Sci. 56:343.

Wise, S. P., and Jones, E. P., 1977, Cells of origin and terminal distribution of descending projections of the rat somatic sensory cortex, J. Comp. Neurol. 175:129.

Wisniewski, K. E., Wisniewski, H. M., and Wen, G. Y., 1985, Occurrence of neuropathological changes and dementia of Alzheimer's disease in Down's syndrome, Ann. Neurol. 17:278.

A CORRELATION BETWEEN GENE TRANSCRIPTIONAL ACTIVITY AND CEREBRAL GLUCOSE

METABOLISM IN ALZHEIMER'S DISEASE-AFFECTED NEOCORTEX: CAUSE OR EFFECT?

W.J. Lukiw, P. Handley, M.K. Sutherland, L. Wong,
and D.R. McLachlan

Center for Research in Neurodegenerative Diseases
University of Toronto, Toronto, Canada, M5S 1A8

ABSTRACT

Our laboratory has measured mRNA pool sizes in neocortex afflicted with Alzheimer's disease (AD). We have observed a repression of gene expression in the temporal and parietal regions compared to age-matched control neocortex. These changes in messenger RNA pool size closely parallel the observed alterations in local cerebral metabolic rates for glucose (LCMR-g), as detected by positron emission tomography (PET). For example, deficits in both gene transcription and glucose metabolism appear to be the greatest in AD-affected superior temporal neocortex (Brodmann area 22) but are less apparent in the primary visual cortex (Brodmann area 17) or in the cerebellum. The unresolved question is whether changes in gene expression are the cause or effect of altered glucose metabolism. However, the non-random reductions in the pool size for certain neocortical mRNAs argue in favour of altered gene expression as the primary event.

INTRODUCTION

Local cerebral metabolic rates for glucose (LCMR-g), as detected by 11-C-2 deoxyglucose and 18-F-2 fluorodeoxyglucose positron emission tomography (PET), indicate that the largest reductions in glucose metabolism, above those which can be explained by tissue atrophy alone, occur in a neocortical strip extending from the posterior parietal lobe to the temporal lobe in AD (Friedland et al., 1985; Montaldi et al., 1990). Other neocortical association areas show only small metabolic changes whereas the primary sensory areas, the motor neocortices, and the cerebellum show changes which are only marginal or statistically insignificant. These focal deficits in energy metabolism become larger with the progression of AD (Hoyer, 1990). In addition, these metabolic deficits correspond well to the distribution of altered biochemical and morphological changes in the AD-affected neocortex (Terry, 1983; Mazziotta and Phelps, 1990).

Comprehensive post-mortem examination of AD-affected brains in several laboratories using _in situ_ and molecular probe hybridization techniques has revealed a reduced yield of RNA message in the AD-afflicted cerebral cortex when compared to controls (Taylor et al., 1986;

Guillemette et al., 1986; Crapper-McLachlan et al., 1988; Clark et al., 1989; Kittur et al., 1990). For 20 RNA messages isolated from temporal and parietal neocortex, we have measured a reduction in message expression in AD-affected brains to an average of 41.8 per cent of control, based on wet weight. The reductions in primary transcription products, however, are not uniform in a given region of neocortex.

We postulate that insufficiencies in the genetic output of neocortical nuclei in AD may restrict the ability of neurons to effectively maintain normal homeostasis and may be a contributing factor in the alteration of normal brain energy metabolism during the disease process.

MATERIALS AND METHODS

Pathology of Brain Tissue

At autopsy, 45 human brains were bisected in the sagittal plane; one half was frozen at -80 degrees C. and the other half was fixed in buffered formalin. On the basis of extensive histopathological examination, a control group was allocated to brains without neuropathological changes or brains from neurological disorders in which neither senile plaques nor neurofibrillary degeneration occurred in the cerebral cortex. Brains exhibiting extensive AD neuropathological changes such as neuritic plaques and neurons with neurofibrillary degeneration (the AD group) were selected for RNA analysis.

RNA Purification and Northern Analysis

Reagents and chemicals of the highest grades commercially available were used throughout these experiments. Neocortices were dissected in the frozen state cooled to -45 degrees C. to minimize RNA degradation (Taylor et al., 1986). Total cellular RNA was isolated by a method modified from Chirgwin et al. (1979) and Guillemette et al., (1986). Northern transfers, Northern and dot blot hybridizations (performed on duplicate sets of membranes) and quantitative autoradiography onto Kodak XAR-5 film were carried out as previously described (Crapper McLachlan et al., 1988). Hybridization signals were quantitated using a Hoefer GS300 transmittance scanning densitometer and an IBM-XT supported Hoefer GS350 data integration system.

Correlative Data

Our gene transcription and RNA message level data were correlated with the brain protein synthesis data of Naber and Dahnke (1979) and Heiss and Phelps (1983), the LCMR-g PET data of Friedland et al., (1985), McGeer et al., (1986), Rapoport et al. (1986) and Montaldi et al., (1990) and the abnormal neuropathological inclusion (presence of senile plaques or neurofibrillary degeneration) data of Terry (1983) and McGeer et al., (1986).

RESULTS

Table 1 shows that upon statistical analyses, the 20 RNA messages from AD-afflicted neocortex fall into four defined groups. ANOVA analysis of Group I hybridization signals indicated that the RNA messages for GFAP, the most abundant intermediate filament of glial cells, the estrogen receptor, the cytoskeletal components beta-actin and tau protein, the calcium binding protein calbindin D28K, the human prion

TABLE 1 - POOL SIZE OF RNA MESSAGES IN HUMAN NEOCORTEX - PERCENT ALZHEIMER/CONTROL. TOTAL RNA ISOLATED FROM HUMAN CEREBRAL CORTEX (NS = Not Significant; ND = Not Detected; ANOVA = analysis of variance).

Group	DNA Probe	Based on Total RNA	Statistical Significance	Based on Wet Weight
I	GFAP	125	NS	66
	ER	117	NS	62
	beta-actin	106	NS	56
	Prion PrP	106	NS	56
	CaBP	100	NS	53
	somatostatin	96	NS	51
	C-erbA-T	88	NS	47
	tau	84	NS	45
	Alu element (pBLUR8)	73	NS	39
II	rDNA (18S)	68	0.03	36
	calmodulin	68	0.04	36
	C-erb-A	63	0.03	33
	a-tubulin	58	0.01	31
	superoxide dismutase	51	0.03	27
III	HNF-L	33	0.001	17
	BC200	27	0.001	14
IV	PSTV (pAV401)	ND		
	pUC8	ND		
	pUC9	ND		
	pBR322	ND		

Total RNA isolated from control and Alzheimer-affected neocortex, was analyzed by Northern and dot blot molecular probe hybridization and quantified by computer-assisted densitometry as described by Crapper Mclachlan et al., 1988. Sources of probes are acknowledged later and in Crapper McLachlan et al., 1988. GFAP = glial fibrillary acidic protein cDNA (glial specific); ER = estrogen receptor (human genomic); beta-actin = beta actin cDNA probe; Prion PrP = prion protein cDNA coding for the 27Kd-30Kd infectious particle; CaBP = 28Kd human calcium binding protein cDNA; C-erbA-T = nervous system expressed thyroid hormone receptor cDNA probe; somatostatin = human somatostatin cDNA (neuromodulator); Alu element = Alu repetitive DNA element found in hnRNA and mRNA leaders and trailers cloned into plasmid BLUR8; rDNA(18S) = DNA probe coding for the 18S (1900 nucleotide) ribosomal RNA; calmodulin = human calmodulin cDNA; c-erb-A = general thyroid hormone receptor cDNA probe; tau = neurofilament associated linker protein; a-tubulin = alpha-tubulin cDNA probe; superoxide dismutase = superoxide dismutase (free radical scavenger) human cDNA probe; HNF-L = human neuron-specific genomic light chain neurofilament probe; BC200 = brain (neuron-specific) cytoplasmic 200 base pair genomic DNA; PSTV = potato spindle tuber viroid.

proteinaceous component, somatostatin, the thyroid hormone receptor message for the c-erbA-T receptor which is highly expressed in brain, and the Alu repetitive element are not altered in statistical significance although there was a trend towards reduction in relative abundance. Group II, which consisted of the 18S structural ribosomal RNA, and the RNA messages for the calcium binding protein calmodulin, the general thyroid hormone receptor c-erb-A, the microtubule component alpha-tubulin and the free radical scavenger superoxide dismutase, all showed a statistically significant trend toward reduction in the abundance of RNA message (ANOVA 0.01-0.04; Table 1). These Group II reductions might be explained by neuronal cell loss in AD pathogenesis or may reflect sub-lethal changes in surviving neurons. Group III, consisting of HNF-L, the 10 nm intermediate filament component unique to neurons which regulates axonal caliber (Chin and Liem, 1987), and BC200, a 200 base pair brain cytoplasmic RNA which may be important in regulating neuron-specific gene transcription (Watson and Sutcliffe, 1987), consistently showed the most significant reductions. Finally, Group IV, which included pAV401, a plant viroid probe (hybridized at both relaxed and high stringencies), and the empty cloning vehicles pUC8, pUC9 and pBR322, all gave non detectable hybridization signals in these experiments.

TABLE 2 - CORRELATION OF HNF-L MESSENGER, TOTAL PROTEIN SYNTHESIS, LCMR-g DEFICITS AND PREVALENCE OF ABNORMAL NEUROPATHOLOGICAL INCLUSIONS IN ALZHEIMER-AFFECTED NEOCORTEX COMPARED TO AGE-MATCHED CONTROL NEOCORTEX. N.D.= not determined.

BRAIN AREA (Brodmann)	HNF-L(1)	Total Protein Synthesis(2)	LCMR-g(3)	Abnormal Neuropathological Inclusions(4)
temporal A22	0.160	0.600(F)	0.610	++++
cerebellum	0.822	N.D.	0.810	+
occipital A17	1.000	0.938	0.911	+

(1) Based on the relative strength of hybridization signal of mononucleosomes (generated by brief micrococcal nuclease digestion) to HNF-L 5' promoter segments. Data of Lukiw and Mclachlan (1990).
(2) Total protein synthesis values expressed are the incorporation of nanomoles of methionine per minute per gram of brain. (F) = data for frontal lobe. Data of Naber and Dahnke (1979) and of Heiss and Phelps (1983).
(3) LCMR-g = Local cerebral metabolic rate for glucose; ratios expressed from micromoles of glucose metabolized per 100 grams of tissue per minute as determined by 18-F-2 fluorodeoxyglucose PET. Data of McGeer et al., (1986).
(4) Based on quantitative electron photomicrography and quantitative cytometrics (presence of neuritic plaques and neurofibrillary degeneration). Data of Terry (1983) and McGeer et al., (1986).

HNF-L was found to be the most reduced RNA message coding for protein (to 17 per cent of the control value) while the RNA abundance of Group 1 and Group 11 messages ranged between 27 and 66 percent of controls, based on wet weight (Table 1). Table 2 shows the correlation between Brodmann area, HNF-L message reduction, protein synthetic capability, LCMR-g and the presence of abnormal neuropathological inclusions.

DISCUSSION

Cerebral glycolytic breakdown of glucose yields 20 per cent of total ATP whereas 80 per cent of total ATP is formed by glucose oxidation. Rates of ATP hydrolysis and glucose utilization are relatively higher in brain cells when compared to other cell types (Thompson, 1973; Milner and Sutcliffe, 1983; Hoyer, 1990). Nuclear processes such as transcription are highly dependent upon the abundance of high energy nucleotide triphosphates.

The nuclei of normally functioning neocortical cells generate a larger quantity and a more complex population of messenger RNA (mRNA) when compared to glia, liver or kidney cells (Thompson, 1973; Naber and Dahnke, 1979). The extending branches of neuron dendrites and axons require that neuronal nuclei must supply RNA transcripts for a cell that is 100 to 100,000 times larger in volume than a non-neuronal cell. Up to 150,000 genes are expressed in normal brain cells and about 40 per cent of the RNA messages generated represent brain specific transcripts (Morrison and Griffin, 1981; Milner and Sutcliffe, 1983; Johnson et al. 1986). This substantial transcriptional output of genetic information suggests that brain cells are highly dependent upon high rates of DNA transcription yielding unusually high levels of RNA messages to maintain normal neocortical homeostasis and a high level of ATP.

For most eukaryotic genes, gene expression is proportional to RNA message concentrations which in turn are proportional to the abundance of cellular protein. Restrictions in the availability of RNA messages may therefore limit the cytoplasmic abundance of the protein for which they code. For example, calmodulin RNA message is reduced to 68 per cent of control brain (based on total RNA; Table 1) while the levels of this calcium binding protein are reduced in a lobe dependent manner, to 66 per cent of control neocortex (Crapper McLachlan et al., 1987). Calmodulin is thought to be a universal cellular receptor and transducer of the calcium signal. Calcium-calmodulin complexes subsequently bind to, and regulate, the function of a very large number of other proteins. Calmodulin, when bound to cytosolic calcium, acts as an important intracellular messenger/regulator of neuronal pinocytosis, glycolytic enzyme phosphorylation, glycogen metabolism, protein kinase C activation, cyclic nucleotide metabolism and the energy intensive processes of gene repair and regulation, neurotransmitter synthesis, release and reuptake and cytoskeletal assembly and disassembly (reviewed by Goldman et al., 1986). Thus any decrement in the amount of this calcium binding protein would be deleterious to both neuronal homeostasis and survival. Similarly, the deficit in the abundance of the thyroid hormone receptor RNA message c-erb-A to 63 per cent of control (based on total RNA) suggests that this would compromise the full effects of thyroxine hormone as an enhancer of gene transcription and hence, the generation of gene products involved in calorigenesis and protein, lipid and carbohydrate metabolism. The reduction of 18S ribosomal RNA to 68 per cent of control (based on total RNA) would similarly restrict the availability of ribosomes for RNA message translation and may be responsible for the observed decrement in neocortical protein synthesis in AD-affected brain (Heiss and Phelps, 1983;

Table 2). Lastly, both glia and neurons are a rich source of filamentous proteins that perform critical roles in the maintenance of neocortical form and function and the integrity of the neural network. Significant reductions in alpha-tubulin and the HNF-L gene product would result in the attenuation of dendritic and axonal caliber and cause neurite shrinkage, both of which are observed in AD-afflicted neurons. Considering that the BC200 RNA transcript is postulated to function in the regulation of neuron-specific gene expression (Watson and Sutcliffe, 1987), a reduction in abundance of the BC200 RNA transcript to 14 per cent of control (based on wet weight; Table 1) may be influential in compromising critical functions that are metabolically unique to neurons.

It is noteworthy that the general pattern of metabolic deficits in clinically diagnosed AD closely approximates the pattern of neuronal loss and plaque and tangle formation as seen in autopsied AD brains (Friedland et al., 1985; McGeer et al., 1986). It has been suggested that abnormalities of cytoskeletal and in particular, neurofilament gene expression, may be responsible for some of the pathological features seen in the AD-affected neuron, such as the appearance of neurofibrillary tangles (McLachlan et al., 1988; Kittur et al., 1990). If the primary deficit in AD resulted in a deficiency in glucose utilization, it would be predicted that the regulation of all genes important to a class of neuronal structural proteins, like the cytoskeletal system, would be equally affected. However, the RNA signal for the HNF-L component of the neurofilament is profoundly depressed, whereas the RNA message coding for alpha-tubulin is only moderately depressed and the RNA transcript coding for tau, a microtubule associated protein known to bind to and crosslink neurofilaments (via HNF-L) and microtubules, is not significantly reduced (Chin and Liem, 1987; Table 1). Indeed, tau accumulates in the neuronal cytoplasm as an integral part of the Alzheimer type neurofibrillary tangle. The non-random pattern of changes in RNA pool sizes within neurons argues against a general down regulation of gene transcription due to reduced energy metabolism and favors a causal, rather than a secondary role, for altered gene expression in AD.

ACKNOWLEDGMENTS

This work was supported in part by the National Science and Engineering Research Council, The Ontario Mental Health Foundation, Medical Research Council of Canada, and the Scottish Rite Charitable Foundation of Canada. Normal and pathological tissues were kindly supplied by the Canadian Tissue Brain Bank, Toronto, Canada. Thanks are extended to C. Bergeron and J. Deck, Toronto General Hospital, Toronto, Canada for critical neuropathological evaluations.

We gratefully acknowledge the gifts of the following probes used in this study: Estrogen receptor (lambdaGHER1): P. Chambon, CNRS, Strasbourg, France; C-erbA-T (erbA-T B1): M. Pfahl, La Jolla Cancer Research Foundation, La Jolla, California; CaBP(HP1.3): M. Parmentier, University of Brussels, Belgium; c-erb A (pheA12): R. Evans, Howard Hughes Medical Institute, University of California, San Diego; BC200 RNA: J. Watson, Research Institute of Scripps Clinic, La Jolla, California; PSTV (pAV401), potato spindle tuber viroid, P. Vos, University of Wageningen, Netherlands.

REFERENCES

Chin S. S. M. and Liem R. K., 1987, Neurofilaments: A Review and an Update. in: "Alterations in the Neuronal Cytoskeleton in Alzheimer's Disease, Advances in Behavioral Biology 34", G. Perry, ed., Plenum Press, New York.

Chirgwin, J. M., Przybyla, A. E., MacDonald, R. J., and Rutter, W. J., 1979, Isolation of biologically active ribonucleic acid from sources enriched in ribonuclease, Biochemistry, 18:5294.

Clark, A. W., Krekoski C. A. and Parhad I. M., 1989, Altered expression of genes for amyloid and cytoskeletal proteins in Alzheimer cortex. Ann. Neurol., 25:331.

Crapper McLachlan, D. R., Wong, L., Bergeron, C., and Baimbridge, K. D., 1987, Calmodulin and calbindin D28K in Alzheimer disease, Alzheimer's Disease and Associated Disorders, 1:171.

Crapper McLachlan D. R., Lukiw W. J., Wong, L., Bergeron, C., and Bech-Hansen N. T., 1988, Selective messenger RNA reduction in Alzheimer's disease, Mol. Brain Res., 3:255.

Friedland, R. P., Brun, A., and Budinger, T. F., 1985, Pathological and positron emission tomography correlations in Alzheimer's disease, Lancet, i:228.

Goldman, J. E., 1986, Cytoskeletal protein abnormalities in neurodegenerative diseases, Ann. of Neurol., 19:209.

Guillemette J. G., Wong, L., Crapper McLachlan, D. R., and Lewis, P. N., 1986, Characterization of messenger RNA from the cerebral cortex of control and Alzheimer brain, J. of Neurochem., 47:987.

Heiss W. D., and Phelps M. E., 1983, Brain protein metabolism in dementia and schizophrenia, in: "Positron Emission Tomography of the Brain," Springer Verlag, Berlin.

Hoyer, S., 1990, Brain energy metabolism and its significance for Alzheimer's disease. Abstract 33. Second International Conference on Alzheimer's Disease, Neurobiol. Aging, 11:260.

Johnson, S A., Morgan, D. G., and Finch, C. E., 1986, Extensive post-mortem stability of RNA from rat and human brain, J. Neurosci. Res., 16:267.

Kittur, S., Hoh, J., Kawas, C., Tourtellotte, W., Marksberry, W., and Adler, W., 1990, Neurofilament gene expression in Alzheimer's disease. Abstract 135. Second International Conference on Alzheimer's Disease, Neurobiol. Aging, 11:285.

Lukiw, W. J. and Crapper McLachlan D. R., 1990, Chromatin structure and gene expression in Alzheimer's disease, Mol. Brain Res., 7:227.

Mazziotta, J.C. and Phelps, M. E., 1990, The pathophysiology of Alzheimer's disease as identified with PET. Abstract 36. Second International Conference on Alzheimer's Disease, Neurobiol. Aging, 11:261.

McGeer, P. L., Kamo, H., Harrop, R., McGeer, E. G., Martin, W. R. W., Pate D. and Li, D. K. B., 1986, Comparison of PET, MRI and CT with pathology in a proven case of Alzheimer's disease, <u>Neurology</u>, 36:1569.

Milner, R. J., and Sutcliffe, J. G., 1983, Gene expression in rat brain, <u>Nucleic Acids Res</u>., 11:5497.

Montaldi, D., Brooks, D. N., McColl., J. H., Wyper, D., Patterson, J., Barron, E. and McCulloch J., 1990, Measurements of regional cerebral blood flow and cognitive performance in Alzheimer's disease, <u>J. Neurol</u>., 53:33.

Morrison, M. R., and Griffin, W. S. T., 1981, The isolation and in vitro translation of undegraded messenger RNAs from human post-mortem brain, <u>Analyt. Biochem</u>., 113:318.

Naber, D., and Dahnke, H. G., 1979, Protein and nucleic acid content in the aging brain. <u>Neuropathol. App. Neurobiol</u>., 5:17.

Rapoport, S. I., Horwitz, B., Haxby, J. V., and Grady, C.L., 1986, Alzheimer's disease: metabolic uncoupling of associative brain regions, <u>Can. J. Neurol. Sci</u>., 13:540.

Taylor G. R., Carte, G. I., Grow, T. J., Johnson, J. A., Fairbairn, A. F., Perry, E. K. and Perry, R. H., 1986, Recovery and measurement of specific RNA species from post-mortem brain tissue: A general reduction in Alzheimer's disease detected by molecular hybridization, <u>Exp. Mol. Pathol</u>., 44: 111.

Terry, R. D., 1983, Cortical morphometry in Alzheimer's disease, <u>in</u>: "Biological Aspects of Alzheimer's Disease," R. Katzman, ed., Cold Spring Harbor Laboratory, New York.

Thompson R. J., 1973, Studies on RNA synthesis in two populations of nuclei from the mammalian cerebral cortex, <u>J. Neurochem</u>., 21:19.

Watson, J. B., and Sutcliffe, J. G., 1987, Primate brain specific cytoplasmic transcript of the Alu Repeat Family, <u>Mol. Cell. Biol</u>., 7: 3324.

OPTIC NEUROPATHY IN THE DIABETIC BB-RAT

Subrata Chakrabarti, Wei-Xian Zhang, and Anders A.F. Sima

Neuropathology Research Laboratories, Dept. of Pathology
University of Manitoba, Winnipeg, Manitoba, Canada

Investigations of diabetic neuropathy have primarily been concerned with the distal symmetric mainly sensory and autonomic polyneuropathies. However in both clinical and experimental diabetes there are evidence to suggest both structural and functional involvements of the central nervous system[1,2]. In a recent report it was demonstrated that spinal cord conduction velocity declines in parallel with peripheral nerve conduction slowing suggesting a common mechanism responsible for both central and peripheral neuropathies in diabetes[3].

The optic nerves and the retina form part of the central nervous system and are affected by diabetes. Diabetic retinopathy and ischemic optic neuropathy are believed to be primarily the result of microvascular insults[4,5]. Although a vascular genesis is important, several neurosensory impairments point towards an early possibly metabolic involvement of the neuronal component of the optic system in diabetes[6].

A deficit in color vision, reduced contrast sensitivity at high frequencies, and abnormal visual evoked cortical potentials (VEP) have been demonstrated in diabetic patients with minimal or no retinopathy[7-9]. It has been suggested that prolonged latencies of the VEP in diabetes, similar to nerve conduction slowing in peripheral nerve, may be due to a demyelinating process[10]. These neurosensory abnormalities may suggest that the underlying pathophysiology may be induced by metabolic abnormalities secondary to hyperglycemia as in peripheral nerve[11]. Biochemical abnormalities, like activation of the polyol pathway with sorbitol accumulation, myo-inositol depletion, and Na-K ATPase inactivation have been demonstrated in the retina[12-13]. Furthermore aldose reductase, the key enzyme in the polyol pathway, has been localized to both neuronal and glial components and the microvasculature of the retina and to the glia of the optic nerve[14,15].

In the peripheral nervous system, diabetic polyneuropathies affect both ganglion cells and there axons[2,16,17,18]. In the present study we investigated the functional and structural integrities of retinal ganglion cells and their axons contained in the optic nerve in the diabetic BB/W rat.

Fuel Homeostasis and the Nervous System, Edited by M. Vranic *et al.*
Plenum Press, New York, 1991

Male prediabetic and age-matched non-diabetes prone male BB/W-rats, obtained from the NIH colony (University of Massachussets, Worchester, MA, USA), were individually housed in air filtered metabolic cages and given free access to water and rat chow (Wayne Lab. Blox F-6, Wayne Feed Division, Chicago, IL, USA). Diabetic BB/W-rats were maintained at fasting hyperglyemic levels of 15-25 m.mol/L by small daily doses of protamine zinc insulin (Cannaught-Novo Inc., Toronto, ON). The animals were monitored regularly as to blood glucose (Glucometer, Miles Laboratories, Ames Division, Elkhart, IN) & glycated hemoglobin (Glycotest, Pierce Chemical, Co., Rockford, Ill). Diabetic animals showed a significantly lower body weight, hyperglycemia and elevated glycated hemoglobin levels when compared to age-matched controls (Table 1).

ELECTROPHYSIOLOGICAL STUDIES

On the day before sacrifice, animals were dark-adapted overnight for VEP recordings. They were sedated by intraperitoneal injections with a 10:1 mixture of ketamine hydrochloride (100mg/ml, Rogar) and acepromazine maleate (25mg/ml, Ayerst), 1ml/kg body weight. Following pupillary dilatation with tropicamide (1%) and coneal anesthetia with proparacaine hydrochloride (0.5%), VEPs were recorded from the midline of the occipital crest (Oz), referenced to the midline of the forehead (Fz) using platinum subdermal electrodes (type E2, Grass Instrument Company, Quincy MA 02169). Responses were averaged 16 times with the aid of a VER averager and recorded with the ERG/EOG recorder (Life-Tech Instrument, Inc., Houston, Texas) on chart paper with pass filters set at 0.1 and 500 Hz and the storage time at 250 ms. Light stimuli with an intensity of 4.82×10^{-3} lumen-sec/cm and duration of 10 μsec were generated automatically at a frequency of 2Hz by the ERG/EOG stimulator. The light source was fitted with a white diffuser and placed 0.5 m from the animal. The latencies of P_2 and N_2 peaks were significantly prolonged in diabetic rats compared to those of controls (48% and 20% respectively). The amplitudes of P_2 and N_2 in diabetic animals did not differ from those of controls (Table 1). Prolonged VEP latencies in diabetic animals is in keeping with similar findings in diabetic patients and may reflect damage to nerve fibers beyond the level of retina[10,19].

At the time of sacrifice the animals were anesthetized intraperitoneally with sodium pentobarbitol (50 mg/kg body weight) and perfused through the left cardiac ventricle with 0.1M cacodylate buffered (pH 7.4) glutaraldehyde (2.5%). The intracranial portion of the right optic nerve and posterior eyecups of right eye within 3 mm of the optic nerve head were dissected and post fixed in the same fixative at $4^{o}C$ overnight. The tissues were then osmified and processed for electronmicroscopy.

MORPHOLOGICAL STUDIES OF THE OPTIC NERVE

Morphometric studies were all performed with the aid of a 9872A HP digitizer interfaced with a 9825A HP desk computer and plotter (Hewlett-Packard Co., Fort Collins, CO). Electronmicrographs with a final magnificantion of 7,000 times were used to measure the areas of 1,000 systematic randomly selected fibers per optic nerve for the study of myelinated fiber size frequency distributions, fiber densities and occupancies for each animal. The myelinated fibers of the optic nerve showed a unimodal fiber size distribution. Mean fiber size in diabetic animals was significantly (p<0.05) smaller than in control rats (Table 2) indicating fiber atrophy of the optic nerve in diabetic animals. Myelinated fiber density and occupancy were unchanged in diabetic rats (Table 2). Electronmicrographs of 100 randomly chosen fibers with a final magnification of 23,200 times were used to calculate the axon-myelin ratios. Analysis of axon-myelin ratio plots showed a significant (p<0.001)

Table 1. Clinical and Electrophysiological Data
(Mean ± SE)

				VISUAL EVOKED POTENTIAL			
	Body Weight (G)	Blood Glucose (m.mol/L)	Glycated Hb (%)	P2 Latency (ms)	Amplitude (uv)	N2 Latency (ms)	Amplitude (uV)
Diabetic (n=5)	418.8 ± 21.1	21.3 ± 2.3	8.7 ± 1.4	45.5 ± 3.2	25.6 ± 4.1	92.5 ± 4.9	20.4 ± 1.0
	$p<0.005$	$p<0.001$	$p<0.05$	$p<0.001$		$p<0.02$	
Control (n=5)	516.2 ± 7.6	3.8 ± 0.6	4.7 ± 0.3	25.0 ± 1.6	17.8 ± 0.9	76.2 ± 1.9	21.4 ± 2.0

Significance judge by students t-test

Table 2.   Morphometric Analysis of Optic Nerve
(Mean ± SE)

	Fiber Size (um)	Occupancy (%)	Density (x 10^3/mm^2)	Axon-Myelin Ratio a	b
Diabetic (n=5)	1.9 ± 0.06	59.0 ± 1.3	311.3 ± 13.4	-1.97 ± 0.14	0.11 ± 0.003
	p<0.05			p<0.01	p<0.002
Control (n=5)	2.32 ± 0.2	63.0 ± 3.7	276.0 ± 10.0	-1.49 ± 0.05	0.09 ± 0.003

Significance judged by students t-test

260

difference in the y-intercept (a) and slope (b), with the intercept in diabetic animals being smaller and slope being higher than in controls (Table 2) suggesting axonal atrophy predominantly of smaller sized myelinated fibers. Nerve fiber atrophy demonstrated in the size-frequency distribution assessment is therefore probably secondary to axonal atrophy as indicated by decreased axon-myelin ratio in diabetic rats compared to their age- and sex-matched controls. These findings are similar to data previously reported by this laboratory on the somatosensory and autonomic neuropathies in the BB/W-rat[2,16-18,20].

MORPHOLOGICAL STUDIES OF THE RETINA

Twenty retinal ganglion cells and 400 axons of the nerve fiber layer were randomly selected and photographed from each animal. These micrographs were examined for the presence of dystrophic changes and their frequencies were determined on the basis of the predominant substructure and expressed as percentages of examined ganglion cells or axons. In diabetic animals 9% of retinal ganglion cells showed the presence of dystrophic changes. No dystrophic changes were detected in ganglion cells of control animals. The most frequent dystrophic change in ganglion cells consisted of the accumulation tubulovesicular profiles and was followed by accumulation of neurofilaments and layered membranes. Proximal axons of nerve fiber layer demonstrated similar dystrophic changes. The most frequent occurring abnormality consisted of electrondens bodies, followed by accumulation of neurofilaments and tubulovesicular profiles. Layered membranous dystrophic changes were not observed in axons. Occasionally electrondense bodies were observed in the axons of control animals (Table 3, Fig. 1).

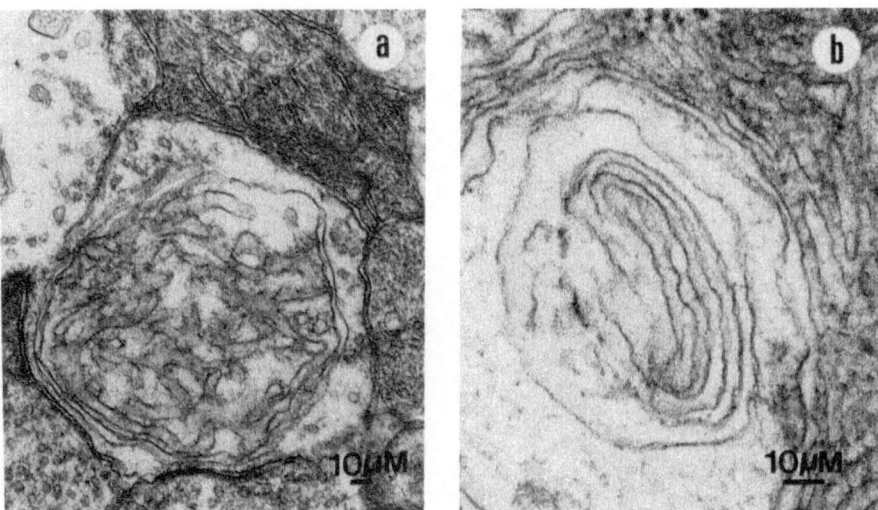

Fig. 1. Electronmicrographs from the retina of diabetic BB/W rats showing a) Tubulovesicular change in the nerve fiber layer axon and b) lamellar membraneous change in the ganglion cell.

Table 3. Dystrophic Changes (%) in Ganglion Cells and Proximal Axons
(Mean $\pm$ SE)

Ganglion Cells:	Tubulo-vesicular	Lamellar Membrane	Neurofilament Accumulation	Electron Dense Bodies	Swelling	Total
Diabetic (n=5)	5.2 ± 2.2	0.8 + 0.6	2.5 ± 1.0	0 ± 0	0 ± 0	8.8 ± 2.0
	p<0.05		p<0.05			p<0.05
Control (n=5)	0 ± 0	0 ± 0	0 ± 0	0 ± 0	0 ± 0	0 ± 0
Proximal Axons:						
Diabetic (n=5)	0.3 ± 0.2	0 ± 0	0.6 ± 0.8	0.8 ± 0.2	4.8 ± 0.8	6.2 ± 1.1
	p<0.05		p<0.01	p<0.02		p<0.05
Control (n=5)	0 ± 0	0 ± 0	0 ± 0	0.2 ± 0.3	2.8 ± 1.0	2.9 ± 0.9

Significance judged by students t-test

These dystrophic changes in the ganglion cells and in the proximal axons of optic nerve, were qualitatively identical to those previously described in autonomic ganglia and axons, and in the fasiculus gracialis of the spinal cord in BB/W-rats[2,16-18,20]. We have previously speculated whether dystrophic changes may reflect abnormalities in axonal transport mechanisms[2,16-18,20], which is affected by diabetes in the optic nerve[21].

The present findings suggest the presence of optic neuropathy in the BB/W-rat, which both functionally and structurally appears to be similar to somatic and autonomic polyneuropathies occurring in the same animal model[16-18]. Whether a common sorbitol -myoinositol related mechanism, thought to be responsible for peripheral neuropathy and other diabetic complications[11] is the culprit for the presently demonstrated abnormalities of a central sensory system needs further careful biochemical evaluation. It is interesting to note that similar biochemical alterations have been demonstrated in the retina in experimental animals with diabetes[12,13]. A significant increase in the optic nerve sorbitol content, although lower than that in peripheral nerves has been demonstrated in diabetes[22], and may suggest that indeed the polyol pathway may be of pathogenetic relevance for optic neuropathy.

In summary the present study suggest that optic nerve fibers and their ganglion cells are affected by diabetes in the BB/W-rat with similar functional and structural changes as those characterizing peripheral diabetic neuropathy.

ACKNOWLEDGEMENTS

The present study was supported by grants from Medical Research Council of Canada and a post-doctoral fellowship (S.C.) from Diabetes Canada.

REFERENCES

1. D. Greenbaum, P.C. Richardson, M.V. Salmon, and H. Urich, Pathological observations on six cases of diabetic neuropathy, Brain, 87:201 (1964).
2. A. A. F. Sima, and S. Yagihashi, Distal central axonpathy in the spontaneously diabetic BB-Wistar rat. A sequential ultrastructural and morphometric study, Diab. Res. Clin. Prac. 1:289 (1986).
3. E. Ronald, L. Carsten, R. Whalen, and D. N. Ishii, Impairment of spinal cord conduction velocity in diabetic rats, Diabetes 38:730 (1989).
4. R. N. Frank, On the pathogenesis of diabetic retinopathy, Ophthalmology 91:933 (1984).
5. I. A. D. O'Brien, I. G. Lewin, J. P. O'Hare, R. J. M. Corrall, Papilloedema in diabetes: An ischemic optic mononeuropathy, Lancet 1:267 (1984).
6. G. H. Bresnick, Diabetic neuropathy viewed as a neurosensory disorder, Arch. Ophthalmol. 104:989 (1986).
7. M. S. Roy, R. D. Gunkel, and M. J. Podgor, Color vision defects in early diabetic retinopathy, Arch. Ophthalmol. 104:225 (1986).
8. S. Sokol, A. Moskowitz, B. Skarf, R. Evand, M. Molitch, and B. Senior, Contrast sensitivity in diabetes with or without background retinopathy, Arch. Ophthalmol. 103:51 (1985).
9. H. Yamazaki, E. Adachi-Usami, and J. Chiba, Contrast thresolds of diabetic patients determined by VECP and psycophysical methods, Acta Ophthalmol. 60:386 (1982).
10. K. Puvanendran, G. Devathasan, and P. K. Wong, Visual evoked responses in diabetes, J. Neurol. Neurosurg. Psychiatry, 46:643 (1983).
11. A. L. Winegrad, Banting lecture 1986: Does a common mechanism induce the diverse complications of diabetes, Diabetes 36:396 (1987).

12. L. C. MacGregor, L. R. Rosecan, A. M. Laties, and F. M. Matschinsky, Altered retinal metabolism in diabetes: I. Microanalysis of lipid, glucose, sorbitol, and myo-inositol, in the choroid and in the individual layers of the rabbit retina, J. Biol. Chem. 261:4045 (1986).
13. L. C. MacGregor, and F. M. Matschinsky, Altered retinal metabolism in diabetes: II. Measurement of sodium-potassium ATPase and total sodium and potassium in the individual layers of the rabbit retina, J. Biol. Chem. 261:4052 (1986).
14. S. Chakrabarti, A. A. F. Sima, T. Nakajima, S. Yagihashi, and D. A. Greene, Aldose reductase in the BB-rat: isolation, immunological identification and localization in the retina and in the peripheral nerve, Diabetologia 30:244 (1987).
15. M. A. Ludvigson, and R. L. Sorenson, Immunohistochemical localization of aldose reductase. II. Rat eye and kidney, Diabetes 29:450 (1980).
16. A. A. F. Sima, Annotation: can the BB-rat help to unravel diabetic neuropathy, Neuropath. App. Neurobiol. 11:253 (1985).
17. S. Yagihashi, and A. A. F. Sima, Diabetic autonomic neuropathy in the BB rat: Ultrastructural and morphometric changes in the sympathetic nerves, Diabetes 34:558 (1985).
18. S. Yagihashi, and A. A. F. Sima, Diabetic autonomic neuropathy in the BB rat: Ultrastructural and morphometric changes in the parasympathetic nerves, Diabetes 35:733 (1985).
19. M. Algan, O. Ziegler, P. Gehin, I. Got, A. Raspiller, M. Weber, P. Genton, E. Saudax, and P. Drouin, Visual evoked potentials in diabetic patients, Diabetes Care 12:227 (1989).
20. S. Yagihashi, and A. A. F. Sima, Neuroaxonal and dendritic dystrophy in diabetic autonomic neuropathy. Classification and tophographic distribution in the BB rat, J. Neuropath. Exp. Neurol. 45:545 (1986).
21. T. Tsukada, and E. Chihara, Changes in the components of fast axonally transported proteins in the optic nerves of diabetic rabbits, Invest. Ophthalmol. Vis. Sci. 27:1115 (1986).
22. P. Naeser, S. E. Brolin, and U. J. Erikson, Sorbitol metabolism in the retina, optic nerve, and sural nerve of diabetic rats treated with an aldose reductase inhibitor, Metabolism 12:1143 (1988).

PERIPHERAL NERVE REPAIR FOLLOWING ARI TREATMENT

Anders A.F. Sima, Virgil Nathaniel,
and *Douglas A. Greene

Neuropathology Research Laboratories,
University of Manitoba, Winnipeg, Canada, and
*Michigan Diabetes Research and Training Center
University of Michigan, Ann Arbor, MI, USA

INTRODUCTION

Distal symmetric polyneuropathy accompanying diabetes mellitus is a major contributor to the morbidity associated with diabetes. Its pathogenesis remains controversial, although hyperglycemia induced metabolic abnormalities as well as microangiopathic changes are believed to be important [1,2].

In diabetic rodents a series of interrelated metabolic abnormalities associated with hyperglycemia accompanies the acute nerve conduction slowing and later occurring structural defects affecting the node of Ranvier and myelinated axons [3].

An early step in this cascade of metabolic abnormalities involves the conversion of excess glucose to sorbitol by the enzyme aldose reductase [4]. The consequent increase in nerve sorbitol levels is associated with depletion of nerve myo-inositol and impaired (Na,K)-ATPase activity [1]. These early interrelated and reversible biochemical and functional abnormalities in the diabetic rat precede a series of sequential less readily reversible neuroanatomical changes. The abnormalities affecting the functionally relevant node of Ranvier include; nodal swelling, axo-glial dysjunction, paranodal demyelination and subsequent remyelination of the nodal and paranodal areas [5]. The axonal changes involve progressive axonal atrophy, Wallerian degeneration resulting in loss of myelinated fibers [3,6]. Similar sequences of structural changes characterize human diabetic polyneuropathy and constitute the structural substrate for overt clinical neuropathy [7].

Investigational drugs that inhibit the rate limiting enzyme in the polyol pathway, aldose reductase inhibitors (ARI's), prevent or reverse nerve sorbitol accumulation and subsequent biochemical abnormalities as well as the nerve conduction slowing in acutely diabetic rats [8,9]. In long term prevention studies, these drugs delay in a significant way the development of the characteristic structural abnormalities [10].

In a recent randomized, placebo-controlled, double blind clinical trial, twelve months treatment with the ARI sorbinil decreased nerve sorbitol levels and promoted repair of the characteristic neuroanatomical lesions and facilitated nerve fiber regeneration. These effects on peripheral nerve structure were accompanied by small but significant improvements in nerve function and patients' symptom scores [11].

These observations suggest that sustained activity of the polyol pathway has a continued effect on the progression of diabetic nerve damage.

STUDY DESIGN

The entrance criteria, patient demographics and study design of the 12 months clinical trial in patients with clinical overt diabetic neuropathy have been described in detail[7,11]. After having given written consent to participate in the study, patients underwent a 2 months single-blind placebo run-in period, during which baseline neurological testing and function were assessed. Patients were then randomly assigned to either placebo or active treatment with sorbinil (250 mg/d). Of 28 patients who completed the trial, 16 patients consented to both a baseline and a follow-up sural nerve biopsy at the termination of the 12 months trial. Ten patients were randomized to receive drug treatment and 6 patients to placebo treatment. Each fascicular sural nerve biopsy was prepared for light microscopic and ultrastructural morphometric analysis, and for single teased nerve fiber assessments as previously described in detail[7,11]. Evaluation on the effect of sorbinil on nodal abnormalities included determinations of the frequencies of axo-glial dysjunction, paranodal demyelination and remyelination. The effect on axonal pathology was assessed by the frequency and severity of axonal atrophy, and the frequency of Wallerian degeneration. Furthermore, peripheral nerve repair was also evaluated by the frequency of regenerating fibers and changes in myelinated fiber density.

RESULTS

The patients who underwent repeat biopsies were representative of the larger group of 28 patients who participated in the study. Patients assigned to active drug treatment (n=10) did not differ in their baseline data from the patients randomized to placebo treatment (n=6)[11].

Clinical and Functional Responses

The overall assessment of neuropathy by the clinical investigator showed a significantly greater improvement in treated patients versus placebo patients (+8.1 ± 0.6 vs +3.3 ± 2.0; p<0.01). Sural nerve conduction velocity showed a significant (p<0.05) 2.2 m/sec improvement in sorbinil-treated patients, while the nerve conduction in placebo-treated patients showed a non-significant 0.8 m/sec decrease[11]. Sural nerve compound action potential increased from 2.6 ± 0.5 μV to 3.6 ± 0.8 μV at follow-up (p<0.025) in sorbinil-treated patients compared to 2.5 ± 1.1 vs 2.7 ± 1.5 μV (p=N.S.) in placebo-treated patients.

Nerve sorbitol content decreased by 42% (p<0.01) in sorbinil-treated patients, whereas the decrease in placebo-treated patients reached a non-significant 9%, suggesting that the active drug indeed had an inhibitory effect on aldose reductase activity.

Structural Responses at the Node of Ranvier

Preservation of the structural integrity of the node of Ranvier is fundamental to the normal propagation of the nerve impulse. Axo-glial dysjunction devotes the detachment of paranodal terminal myelin loops from the axolemma mediated by the loss of intercellular adhesion complexes[12]. It is believed that axo-glial junctions are responsible for the necessarily high concentration of voltage sensitive sodium channels of the nodal axolemma[13], and that their loss initiates the characteristic paranodal demyelination in diabetic nerve[10].

At baseline both sorbinil- and placebo-treated patients showed a doubling in the frequency of axo-glial dysjunction compared to age-matched non-diabetic patients (Figure 1). Following 12 months of sorbinil treatment the frequency of axo-glial dysjunction was significantly decreased (p<0.02) and was indistinguishable from normal values. Placebo-treated patients showed no significant change in the frequency of axo-glial dysjunction over the 12 months period (p=N.S.) (Figure 1).

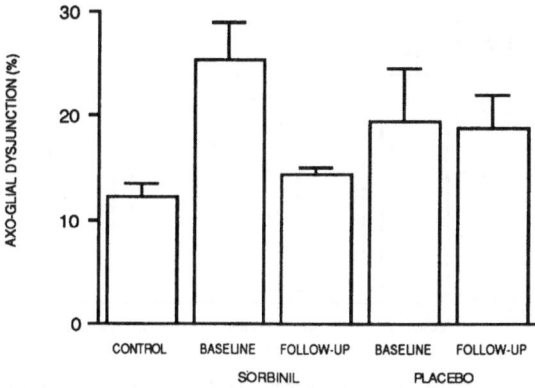

Figure 1. Axo-glial dysjunction was increased approximately 2-fold in baseline biopsies. Following 12 months of sorbinil treatment axo-glial dysjunction was returned to normal values, while placebo treatment had no effect on axo-glial dysjunction.

In both experimental and human diabetic neuropathy, paranodal demyelination is believed to be a consequence of axo-glial dysjunction. At baseline both sorbinil-treated and placebo patients showed a 7-fold increase in paranodal demyelination compared to non-diabetic controls. Sorbinil-treatment was associated with a 20% reduction ($p<0.002$) in the frequency of paranodal demyelination, whereas in placebo patients no significant change in the frequency of paranodal demyelination was obtained.

Remyelinated nodes represent repaired previously demyelinated nodes. Their frequency was approximately double that of age-matched non-diabetic controls at baseline. Neither sorbinil-treated nor placebo patients showed any change in the frequency of remyelinated nodes following the 12 months trial, which may not be surprising since they represent the static sequela of the preceding dynamic nodal changes.

Effect of Sorbinil-Treatment on Axonal Abnormalities

Axonal atrophy, characterized by excessive myelin wrinkling of single teased fibers, was 10 times as frequent in both sorbinil- and placebo-treated patients at baseline compared to non-diabetic control subjects. In sorbinil-treated patients the frequency of excessive myelin wrinkling decreased significantly ($p<0.0001$) to about half the baseline values, whereas no change was demonstrable in placebo patients at the termination of the trial. Excessive myelin wrinkling reflects the frequency of axonal atrophy, whereas the internodal length-fiber diameter ratio of myelinated fibers signifies the severity of axonal atrophy. The internodal length/diameter ratio increases with the severity of axonal atrophy. Both sorbinil and placebo assigned patients showed a 32% increase in the internodal length/diameter ratio at baseline compared to non-diabetic control subjects (Figure 2). Sorbinil-treatment was followed by a significant ($p<0.0000$) 32% reduction in internodal length/diameter ratio, whereas placebo-treatment resulted in a non-significant 4% decrease (Figure 2). Hence sorbinil-treatment was accompanied by marked improvements in both the extent and the severity of axonal atrophy. The ultimate consequence of axonal abnormalities in diabetic neuropathy is Wallerian degeneration, which is likely to be responsible for the characteristic nerve fiber loss, since degenerated fibers are not replaced by regenerating fibers (see below). The frequency of myelinated fibers undergoing Wallerian degeneration was not altered by 12 months of sorbinil-treatment ($p=$N.S.), whereas a significant increase in placebo patients ($p<0.05$) was observed resulting in a significant treatment-effect ($p<0.025$).

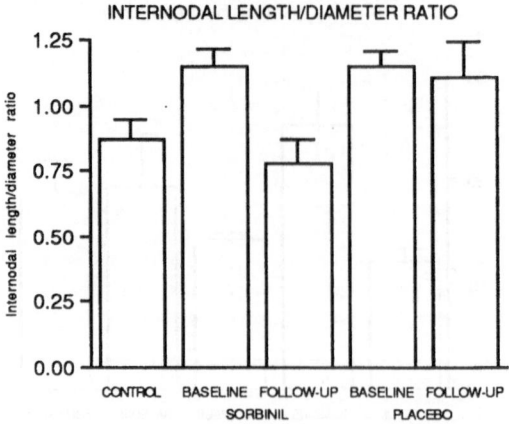

Figure 2. The internodal length/diameter ratio was similarly increased in sorbinil and placebo treated patients at baseline, indicating a similar severity of axonal atrophy. Twelve months of sorbinil-treatment was accompanied by a normalization of internnodal length/diameter ratio. Placebo treated patients exhibited no change in internodal length/diameter ratio over the 12 month trial period.

Effect of Sorbinil-Treatment on Myelinated Fiber Regeneration and Density

In contrast to the central nervous system, peripheral nerves possess the ability to regenerate following nerve fiber damage. Such regenerated fibers are characterized by proportionately short and thinly myelinated internodes. Although they may restore peripheral nerve function, their functional capacity will not restore fully that of the fibers they replace. In diabetic neuropathy the regenerative capacity appears to be impaired, since diabetic nerve shows a progressive net loss of myelinated fibers, although the frequency of regenerating fibers in the present material at baseline was 4 times that seen in non-diabetic control nerves. Twelve months of sorbinil treatment was accompanied by a further 4-fold increase (p<0.0001) in the frequency of regenerating fibers. Placebo patients showed no increase in regenerating fibers (Figure 3). This burst of myelinated fiber regeneration resulted in 33% (p<0.02) increase in myelinated fiber density. As would be expected, patients treated with placebo demonstrated no increase in fiber density.

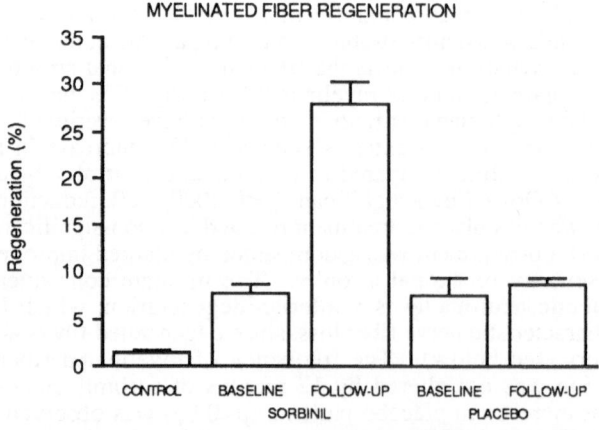

Figure 3. The frequency of myelinated fiber regeneration is increased in diabetic patients. Sorbinil treatment was associated with a four-fold increase in the frequency of regenerated fibers, whereas placebo treatment did not alter the frequency of regenerating myelinated fibers.

SUMMARY AND CONCLUSIONS

In this placebo-controlled double-blind clinical trial, ARI-treatment was accompanied by small but statistically significant improvements in clinical and electrophysiological indices. These improvements were associated with a lowering of nerve sorbitol levels and significant improvements in quantitative structural parameters. These data suggest that an activated polyol-pathway plays a continuous pathogenetic role even in advanced clinically overt diabetic neuropathy, and that the metabolic abnormalities mediated by the polyol pathway are intimately associated with structural and functional changes and ultimately clinical symptoms[14]. The regenerative and reparative responses to a potent ARI like sorbinil suggest that advanced neuroanatomical changes believed to underlie overt clinic polyneuropathy are at least partly reversible, and that extrapolated over longer treatment periods ARI's may substantially ameliorate the structural changes and the clinical symptoms associated with diabetic neuropathy. As indicated by animal studies[10,15] ARI's may become valuable adjuncts to the therapeutical arsenal not only in the treatment of diabetic neuropathy and other chronic diabetic complications, but also as a prophylactic regimen, provided that safe compounds can be developed.

REFERENCES

1.   Greene, D.A., Lattimer, S.A., Sima, A.A.F., 1987, Sorbitol, phosphoinositides, and sodium-potassium-ATPase in the pathogenesis of diabetic complications. N. Engl. J. Med., 316:599.

2.   Dyck, P.J., Karnes, J.L, O'Brien P., Okazaki, H., Lais, A., Engelstad, J., 1986, The spatial distribution of fiber loss in diabetic polyneuropathy suggests ischemia. Ann. Neurol., 19:440.

3.   Sima, A.A.F., 1985, Can the BB-rat help to unravel diabetic neuropathy? Neuropath. Appl. Neurobiol., 11:253.

4.   Gabbay, K.H., Merola, L.O., Field, R.A., 1966, Sorbitol pathway: presence in nerve and cord with substrate accumulation in diabetes. Science, 151:209.

5.   Sima, A.A.F., Lattimer, S.A., Yagihashi, S., Greene, D.A., 1986, Axo-glial dysjunction: A novel structural lesion that accounts for poorly reversible slowing of nerve conduction in the spontaneously diabetic Bio-Breeding rat. J. Clin. Invest., 77:474.

6.   Sima, A.A.F., Bouchier, M, Christensen, H., 1983, Axonal atrophy in sensory nerves of the diabetic BB-Wistar rat: a possible early correlate of human diabetic neuropathy. Ann. Neurol., 13:262.

7.   Sima, A.A.F., Nathaniel, V., Bril, V., McEwen, T.A.J., Greene, D.A., 1988, Histopathological heterogeneity of neuropathy in insulin-dependent and non-insulin-dependent diabetes, and demonstration of axo-glial dysjunction in human diabetic neuropathy. J. Clin. Invest., 81:349.

8.   Greene, D.A., Chakrabarti, S., Lattimer, S.A., Sima, A.A.F., 1987, Role of sorbitol accumulation and myo-inositol depletion in paranodal swelling of large myelinated nerve fibers in the insulin-deficient spontaneously diabetic Bio-Breeding rat. Reversal by insulin replacement, an aldose reductase inhibitor and myo-inositol. J. Clin. Invest., 79:1479.

9.   Mayer, J.H., Tomlinson, D.R., 1983, Prevention of defects of axonal transport and nerve conduction velocity by oral administration of myo-inositol or an aldose reductase inhibitor in streptozotocin-diabetic rats. Diabetologia, 25:433.

10.   Sima, A.A.F., Prashar, A., Zhang, W-X, Chakrabarti, S., Greene, D.A., 1990, The preventive effect of long term aldose reductase inhibition (Ponalrestat) on nerve conduction and sural nerve structure in the spontaneously diabetic BB-rat. J. Clin. Invest., 85:1410.

11.   Sima, A.A.F., Bril, V., Nathaniel, V., McEwen, T.A.J., Brown, M.B., Lattimer, S.A., Greene, D.A., 1988, Regeneration and repair of myelinated fibers in sural nerve specimens from patients with diabetic neuropathy treated with sorbinil. N. Engl. J. Med., 319:548.

12.   Sima, A.A.F., Brismar, T., 1985, Reversible diabetic nerve dysfunction: structural correlates to electrophysiological abnormalities. Ann. Neurol., 18:21.

13.   Brismar, T., Sima, A.A.F., Greene, D.A., 1987, Reversible and irreversible nodal dysfunction in diabetic neuropathy. Ann. Neurol., 21:504.

14.   Greene, D.A., Sima, A.A.F., Albers, J., Pfeifer, M., Diabetic neuropathy in: "Ellenberg and Rifkin Diabetes Mellitus", H. Rifkin and D. Porte, eds., Elseviers Publishing Company, New York, (1989).

15.   Yagihashi, S., Kamijo, M., Ido, Y., Mirrlees, D.J., 1990, Effects of long-term aldose reductase inhibition on development of experimental diabetic neuropathy: ultrastructural and morphometric studies of sural nerve in streptozocin-induced diabetic rats. Diabetes, 39:690.

INDEX

Acetoacetate, 51, 218, 223
Acetylcholine, 10, 11, 21,
    91, 109, 130, 161,
    162, 174
Acidic fibroblast growth
    factor, 100
ACTH, 10, 11, 13-15, 17,
    92, 163, 164
ACTH-cortisol axis, 163,
    176, 177
Actin, 47, 49, 148, 250,
    251
Adenohypophysis, 101, 102
Adenylate cyclase, 112,
    133, 135
Adipocytes, 99, 162, 179,
    205
    effect of neuropeptides
        on differentiation
        of, 100-103
ADP, 242
Adrenal cortex, 204
Adrenal gland, 90, 91, 92,
    108, 163
Adrenal medulla, 10, 12-15,
    107, 108, 193, 204
Adrenaline, 164
Adrenergic blockade, 118,
    130, 171, 173
Adrenergic receptors, 111,
    112, 114-117
Adrenocorticotropic hormone
    see ACTH
Aldose reductase, 257
    inhibitors(ARI), 265-269
Alloxan, 165, 167, 168,
    169, 171-173, 175,
    177
Alzheimer's disease

abnormal brain
    metabolism, 233-
    243
neuropathology, 231-233
repression of gene
    expression in, 249
Amino acid
    transport across blood-
        brain barrier, 55-
        63, 74
γ-aminobutyric acid(GABA),
    3, 21, 25, 78, 91,
    100
Ammonia
    brain, 55, 66, 78
    blood, 66
Amygdala, 26, 34, 38, 78
Anorexia nervosa, 75
    diagnostic criteria, 223,
        224
    functional
        considerations, 10-
        12, 62
    influence of
        neuropeptides on,
        100, 129-136
    neuropathology, 225-228
Anterior cortex, 189, 190
Association areas, 231-241
ATP, 242, 253
ATP-dependent potassium
    channels, 91, 112,
    131, 132
Atropine, 109, 129, 130,
    133
Auditory cortex, 26, 34

Basic fibroblast growth
    factor, 100, 101

2-deoxyglucose, 9, 23-33, 35-37, 39, 185, 190, 214, 216, 249
Dexfenfluramine, 81, 82
Diabetes, 22, 80, 157, 162
  effect of neuropeptide release on, 92, 167-170
  effect of stress on, 161-179
  imbalance in brain amino acid concentrations in, 56,65
  juvenile onset, 173
  neuropathies in, 265-269
  neurosensory abnormalities, 257-263
  regulation of blood-brain barrier permeability in, 47-49
Diabetes mellitus, 46
  insulin-dependent, 154, cerebral glucose metabolism in, 213-220
  non insulin-dependent, 93, 103
Diacylglycerol, 134, 135
Diagnostic and Statistical Manual of Mental Disorders, 223
Diazepam, 47
Dibutyryl cyclic AMP(dbcAMP), 148, 149, 152, 153
5,7-dihydroxytryptamine, 81
Dopamine, 55, 78, 81, 91, 100, 168
Down's syndrome, 242, 243
Dynorphin, 14, 15

β-endorphins, 10, 11, 78, 161, 164-166
Epinephrine(EPI), 11, 14-17
  effect of stress on levels, 107-121, 161-164
  hypothalamic stimulation, 185-191
  in hypoglycemia, 197, 200, 205, 207, 209,210

role in glucoregulation, 15, 152, 172-176
Estrogen receptor(ER), 251

Feeding behavior
  effect of neuropeptides on, 78-81, 91, 92, 100-103
  glucostatic theory, 75-76
  hypothalamic control, 99-103
  regulation, 74
Forskolin, 133, 148, 150, 152, 153
Free fatty acids(FFA), 77, 163, 205, 223
  effect of neuropeptides on, 90
  reesterification, 162, 177, 178, 179
  transport, 49-52

G-protein, 90, 91, 112, 132, 133
Galanin, 155
  effect of stress on induction, 107-108
  effect on feeding behavior, 91
  effect on islet hormone secretion, 109-121, 131-133, 136
  inhibition of insulin secretion, 89-91, 93, 99, 100
  receptors, 115, 131
  structure, 90-91
Ganglion
  celiac, 108, 109, 111, 129, 130, 131
  parasympathetic, 108, 109
Gastric inhibitory peptide(GIP), 91, 155
Gastrin releasing peptide(GRP), 14, 133, 136, 155, 156
Glial fibrillary acidic protein(GFAP), 251
Glial membrane, 44, 46, 47, 55, 64
Glibenclamide, 91
Glicentin, 143-147, 150-152